T0192337

Sustainability of Biomass through Bio-based Chemistry

SUSTAINABILITY: CONTRIBUTIONS THROUGH SCIENCE AND TECHNOLOGY

Series Editor: Michael C. Cann, PhD
Professor of Chemistry and Co-Director of Environmental Science
University of Scranton, Pennsylvania

Preface to the Series

Sustainability is rapidly moving from the wings to centre stage. Overconsumption of non-renewable and renewable resources, as well as the concomitant production of waste has brought the world to a crossroad. Green chemistry, along with other green science technologies, must play a leading role in bringing about a sustainable society. The Sustainability: Contributions through Science and Technology series focuses on the role science can play in developing technologies that lessen our environmental impact. This highly interdisciplinary series discusses significant and timely topics ranging from energy research to the implementation of sustainable technologies. Our intention is for scientists from a variety of disciplines to provide contributions that recognize how the development of green technologies affects the triple bottom line (society, economic, and environment). The series will be of interest to academics, researchers, professionals, business leaders, policy makers, and students, as well as individuals who want to know the basics of the science and technology of sustainability.

Michael C. Cann

Sustainability of Biomass through Bio-based Chemistry

Edited by

Valentin I. Popa

CRC Press
Taylor & Francis Group
Boca Raton London New York

CRC Press is an imprint of the
Taylor & Francis Group, an **informa** business

First edition published 2021
by CRC Press
6000 Broken Sound Parkway NW, Suite 300, Boca Raton, FL 33487-2742

and by CRC Press
2 Park Square, Milton Park, Abingdon, Oxon, OX14 4RN

CRC Press is an imprint of Taylor & Francis Group, LLC

Library of Congress Cataloging-in-Publication Data

Names: Popa, Valentin I., editor.
Title: Sustainability of biomass through bio-based chemistry / Valentin I. Popa.
Description: First edition. I Boca Raton : CRC Press, 2021. I Series:
Sustainability : contributions through science and technology I Includes bibliographical references and index.
Identifiers: LCCN 2020047296 I ISBN 9780367365950 (hardback) I ISBN 9780429347993 (ebook)
Subjects: LCSH: Biomass chemicals. I Biomass energy. I Sustainable engineering. I Sustainable development.
Classification: LCC TP248.B55 S87 2021 I DDC 662/.88--dc23
LC record available at https://lccn.loc.gov/2020047296

ISBN: 978-0-367-36595-0 (hbk)
ISBN: 978-0-367-71012-5 (pbk)
ISBN: 978-0-429-34799-3 (ebk)

Typeset in Times
by SPi Global, India

Contents

Preface

The United Nations elaborated the 2030 Agenda for sustainable development for transforming our world (sustainabledevelopment.un.org). The 17 Sustainable Development Goals and 169 targets are proposed to demonstrate the scale and ambition of this new universal Agenda. They hope to build on the Millennium Development Goals and complete what they did not achieve. They seek to realize the human rights of all and to achieve gender equality and the empowerment of all women and girls. They are integrated and indivisible and balance the three dimensions of sustainable development: the economic, social, and environmental.

One of the main objectives is to protect the planet from degradation, including through sustainable consumption and production, sustainably managing its natural resources, and taking urgent action on climate change so that it can support the needs of the present and future generations.

The development of economic and environmentally benign processes for the scale-up production of materials, chemicals, and fuels is one of the challenges for the 21st century, and social and economic development depends on the sustainable management of our planet's natural resources.

Consideration of the sustainability of biomass to bioenergy and bioproducts programmes based on utilizing lignocellulosic feedstocks is both timely and important in terms of the current plans for commercial valorization of this sector. First-generation biofuels face both social and environmental challenges, largely because they use food crops that could lead to food price increases and possibly indirect land use change (ILUC). Non-food biomass, for example, lignocellulosic feedstocks such as organic wastes, forestry residues, high-yielding woody or grass energy, crops, and algae have the potential to provide possible solution to this problem, if developed and managed in a sustainable manner. The use of these feedstocks for second-generation biofuel production would significantly decrease the potential pressure on land use, improve GHG emission reductions when compared to some first-generation biofuels, and result in lower environmental and social risks. The use of conventional crops for energy and bioproducts use can also be expanded, with careful consideration of land availability and food demand. For sustainable bioenergy development, lignocellulosic crops (both herbaceous and woody) could be produced on marginal, degraded, and surplus agricultural lands and, in theory, could provide the bulk of the biomass resource in the medium term along with aquatic biomass (algae) as a significant contribution in the longer term. However, significant progress needs to be made to scale up algal production and processing in an economic manner to make algal biomass to bioenergy a commercially viable option.

In addition there has been a growing interest in biorefinery processes of lignocellulosic biomass in light of the advantages of efficient fractionation into secondary components, cellulose, hemicelluloses, and lignin which could be converted to fuels, power, functional materials, and value added chemicals. Several nanomaterials such as nanocellulose and lignin nanoparticles have been prepared as intermediate products during the biorefinery process, and these have shown great potential to be applied in the biomedical and pharmaceutical fields. Taking nanocellulose as an example it is reported that nanocellulose with excellent mechanical stability, low toxicity, biocompatibility, and biodegradability have been widely demonstrated for use in a broad range of biomedical applications such as drug delivery, tissue engineering, wound dressing, cancer treatment, biosensors, to name just a few. The research in this field is expected to provide a possibility for the prevention and treatment of infectious diseases in the near future.

Lignin is one of the most important components of biomass resources. With the increasing consumption of petroleum resource, lignin transformation is of strategic significance and has attracted wide interest. As lignin is a random construction of aromatic monomers, the degradation products are usually very complex, which limits the scaling application of lignin as feedstock for valuable chemicals. But the methods for selective transformation of functional groups of lignin into phenols, aldehydes, carboxylic acids, alkanes, and arenes have been elaborated, which could be developed to substitute some products obtained by petrochemistry. At present the lignin is accessible in huge quantities by extending the biorefining applications that have helped not only in obtaining pulp but also in developing the process of hydrolysis with the aim of ethanol manufacture? The platform to fabricate products based on lignin was proposed.

Nanocellulose has emerged as a sustainable and promising nanomaterial owing to its unique structures, superb properties, and natural abundance. The high mechanical strength and modulus, low coefficient of thermal expansion, superior gas-barrier, and high electrical resistivity properties make nanocellulose attractive in the application of many fields, such as wearable flexible electronic equipment, paper-based generator, OLED, supercapacitor, and battery. For instance three different nanocellulose-based materials (nanopaper, hydrogel, and carbon aerogel), which can be applied in different fields such as the solar cell, wearable sensors, and supercapacitors were obtained. A superhydrophobic, highly transparent, and hazy nanopaper made of cellulose nanofibrils and polysiloxanes was produced. This novel nanopaper that simultaneously achieved light-management and self-cleaning capabilities not only led to an enhancement in overall energy conversion efficiency (10.43%) of solar cell by simply coating but also recovered most of the photovoltaic performance losses due to dust accumulation by a self-cleaning process, showing great application in solar cells. In order to explore the application of nanocellulose in energy storage, a novel and facial method for the fabrication of a cellulose carbon aerogel by using the melamine foam as the skeleton was designed. This method provides an effective and low-cost strategy to fabricate carbon aerogel with the compressible and conductive property, showing great potential for applications in compressible electrodes for supercapacitors. As an important subfield of wearable devices, wearable and flexible strain nanocellulose-based sensors were prepared for electronic skins. A tough, self-healing, and self-adhesive ionic cellulose gel was designed with superior strain-sensitive features. The ionic gels display biocompatible and repeatable adhesiveness to various substrates including human skin tissue, revealing a promising candidate for wearable sensors.

At present many studies have been performed concerning re-use of various wastes, such as agro- and paper wastes, enzymes, pigments, etc. to prepare helpful products for food, chemical processing, and medicine. The environmental impact of paper is very significant all over the world, which has led to the development of modern technologies. Due to this, for instance, disposable paper became a relatively cheap commodity and this provided a high level of consumption and waste. The use of paper wastes as a renewable resource of lignocellulosic constituents has provided the opportunity to promote a cleaner environment and to prepare valuable materials. Therefore, the problem of recycling of trash paper becomes one of the most important ecological problems to be solved for the sustainable development of the world community. The design and assessment of bio-based chemical processes to re-use waste paper and cardboard for making valuable materials could be developed. The products including microcrystalline and nanocrystalline cellulose have attracted considerable attention in the last decades as a profitable source for chemical, medical, and industrial implementation. This is why the isolation of powder fibrous cellulose from newspaper waste and cardboard wrapper is possible and the impact of pretreatment of this waste on the processing of powder celluloses and on their physico-chemical properties has been investigated. The pretreatment has been performed in several steps including maceration, alkali treatment, bleaching, and subsequent mild acid hydrolysis in acid media. An effect of these steps on physicochemical properties and morphology of the prepared powder samples has been evaluated with FTIR and CP/MAS 13C-NMR spectroscopy, WAXS and optical and scanning electron microscopy. These methods demonstrated that the powder cellulose fibres can be obtained and the products have exhibited the properties of purified celluloses. The sorption capacity of the powder celluloses for dyes has been estimated and it was shown that their specific surface area is rather high. As an example of the possibility to functionalize the powder celluloses, they have been subjected to chemical processing to prepare bio-nanocomposites with noble metals. The treatment included diffusion–adsorption interaction between cellulose and metal ions and subsequent reduction of the ions to silver(0) and gold(0) directly in the cellulose scaffold. The bio-nanocomposites contained nanometals in dispersed form.

Bio-based resources are of high scientific importance these days due to their abundance, renewability, environment compatibility, recyclability, and many more aspects of sustainable material production and its consumption. Cellulose, the naturally occurring plant polymer has been a widely worked out substrate material for value addition as their supply from forest products industry and agricultural wastes is huge and inexpensive. Cellulose is a fascinating polymer, with the repeating units of monosaccharide molecule containing secondary and tertiary – OH functional groups. These – OH groups are the primary target locations to modify cellulose chemically or physically. Hybridization of cellulose with synthetic polymers or inorganic particles allows the creation of multitude of functional materials. Topochemical engineering is a method of directed assembly or directed disassembly of functional materials. Directed assembly and disassembly are controlled via the design of molecular and intermolecular interactions in a topological space. The directed assembly uses electrostatic interactions, hydrogen bonding, hydrophobic interactions, or solid-state cross-linking reactions of components to create functional shapes and interfaces. The directed disassembly of a component from a multicomponent system was designed on the basis of the controlled cleavage of molecular and intermolecular bonds and the formation of new molecular and intermolecular interactions that enhance separation and selected fractionation.

The continued increase in demand for products made from sustainable and non-oil resources is a major concern and an important driving force for the development of new materials. In this respect, research has been oriented towards the development of innovative materials from renewable resources, which are available, cheap, and have little impact on the environment. Lignocellulosic biomass is the most common renewable resource, its main polymers being cellulose, lignin, and hemicelluloses, all of which have great potential of application. Although there has been noticeable progress in cellulose applicability, in case of lignin and hemicelluloses additional investigations are needed to unlock all applicative potential and to increase attractiveness for industry. The widespread use of lignin and hemicelluloses is still limited due to the existence of a sub-developed marketplace, besides technical and economic aspects related to product development, as well as the existence of competition from petroleum products. In order to achieve high economic efficiency and to minimize market risks it is necessary to transform them into high-quality products with an increased economic value. In this regard, intense efforts have been made to develop advanced materials and hydrogels that have found numerous applications in various research fields, starting from biomedical fields such as health products, drug delivery, wound healing, and tissue engineering, as well as in agriculture, textiles, and industrial applications as intelligent materials. All the main compounds of biomass (cellulose, lignin, and hemicelluloses) could be used to obtain hydrogels with interesting physicochemical properties that recommend them for biomedical applications.

Rice husks, mainly composed of cellulose, lignin, and hemicelluloses represent a significant agricultural by-product and are commonly used to generate heat in industries, owing to their high calorific potential. However, through purification and acetylation, cellulose may be converted into cellulose acetate to produce membranes for several applications, including reverse osmosis. The rice husks may be converted into flat sheet membranes. These membranes are put in an equilibrium cell, separating a saline solution from distilled water. The evolution of salt concentration in the distilled water reservoir is measured by using a conductometer. In most cases, there are no significant variations in conductivity of the distilled water. To improve these results, cross-linking of the membranes is performed by reaction of the cellulose acetate with oxalic acid. The equilibrium experiments are repeated for several oxalic acid concentrations and optimum value is found.

Hydrogels are cross-linked three-dimensional (3D) networks made from hydrophilic polymeric chains which are capable to absorb large amounts of water or biological fluids. Since water is the greatest component of the human body, hydrogels display an undeniable potential as prime candidates for biomedical purposes, in fields like drug delivery, self-healing materials, and carriers or matrices for tissue engineering. These remarkable materials include several benefits, such as increased biocompatibility, tunable biodegradability, tailorable porous architectures, proper mechanical strength, and a degree of flexibility close to natural tissues. Although covering a wide range of applications, hydrogels have basic characteristics with common physico-chemical origins. The construction of the three-dimensional porous network is one of the essential factors that influence the final properties, from swelling and mechanical properties to transport kinetics and release of active principles. While much of the research on hydrogels is centred on identifying increasingly advanced pathways to create novel materials of superior quality, there will always be a primary need to obtain detailed and complex information about the characteristics of the porous architecture, supramolecular organization, morphology, and structure of these peculiar materials. Pore arrangement, size, and interconnectivity play a major role in cell survival, proliferation, and migration as to fabricate functional 3D hydrogel network scaffolds and SEM is able to provide solid and realistic information on the appearance, integrity, organization, quality, and degree of uniformity of these materials. The obtained data are multiple and allow the construction on several hierarchical levels of a true image of the porous assembly, in a wide dimensional range.

Due to their biocompatibility, high water content, soft and rubbery consistence, hydrogels have the possibilities to resemble natural living tissue more than any other type of synthetic biomaterials. Stimuli-responsive hydrogels, also referred to as 'environmentally-sensitive', 'smart', or 'intelligent' hydrogels, represent a broad class of hydrogels undergoing switchable gel-to-solution or gel-to-solid transitions, as a response to various external stimuli. These can include physical stimuli, such as thermal, magnetic, ultrasonic, electrochemical, and light, or chemical stimuli such as, pH, redox reactions, supramolecular complexes, and biocatalytically driven reactions. Polysaccharide-based hydrogels are versatile materials with properties which are not found in other class of materials. Biocompatibility, biodegradability, low or nontoxicity, hydrophilicity, similarity to biological environments, biological functionality at molecular level, and immune recognition are good examples of the desirable properties attributed to the stimuli-responsive hydrogels. Furthermore, the use of polysaccharides is also encouraged in respect to environmental concerns, these being the cheapest and most abundant, available, and renewable organic materials on earth. Cellulose, starch, alginate, chitosan, xanthan, collagen, and other natural polymers have been frequently used to prepare hydrogels matrices that provide some response to internal/external stimuli that recommend them for biomedical applications.

The use of microorganism for biosynthesis represents an interesting way to obtain the new biomaterials (i.e., carriers, vectors, scaffolds) with medical and pharmaceutical applications. From this point of view, in the last time the attention of researchers was focused on curdlan, a polysaccharide with native antitumour proprieties. Curdlan is a neutral bacterial exopolysaccharide produced by *Agrobacterium* sp. under nitrogen-limited condition, composed primarily of linear β-(1,3)-glycosidic linkages. The insolubility of curdlan, generally attributed to its structure, drastically limits its applications. The ability to control the curdlan structure has led to synthesis of new derivatives and has therefore encouraged their use in development of new biomaterials with applications in medical and pharmaceutical fields.

These examples demonstrate that biomaterials and sustainable resources are two complementary terms supporting the development of new sustainable emerging processes. In this context, many interdisciplinary approaches including biomass waste valorization and proper usage of green technologies, etc. were brought forward to tackle future challenges pertaining to declining fossil resources, energy conservation, and related environmental issues. The implementation of these approaches impels its potential effect on the economy of particular countries and also reduces unnecessary overburden on the environment. On the other hand, new specializations of engineering could be created that could handle incremental transformation of product manufacturing towards more environmentally appropriate practice.

In the future, the Agenda of sustainability will be influenced by COVID-19. In the context of deep integration and globalization of the world economy, COVID-19 has brought severe challenges to global industrial supply chains. Thus, an example could be the pulp and paper industry that involves many fields such as forestry, agriculture, chemicals, biology, distribution, and transportation, thus occupying an important position in the global economy. For example, copier paper and printing paper have seen a notable decrease in demand recently due to the shutdown of colleges and universities. In these conditions the structure of the industry was oriented to personal hygiene paper products, food packaging products (paper boxes, straws, paper bags, food packaging papers), corrugated packaging materials, paper-based medicinal materials/devices/parts, medical specialty papers. As a special type of carrier, paper can be endowed with many functions such as filtration, antibacterial absorption, and detection by chemical/physical/biological design. Some specialty paper can be used for the production of various medical products (e.g., paper electrodes, paper-based microfluidic chips, paper-based biosensors, and biological test paper).

The COVID-19 pandemic is posing serious challenges to the pulp and paper industry due to its disruption to global industrial supply chains. However, it has potential to create positive demand for a variety of paper products such as personal hygiene paper products, food packaging products, corrugated packaging materials, and medical specialty papers. Thus, traditional pulp and paper manufacturing operations are expected to be transformed, upgraded, and integrated to reduce the risk associated with this pandemic. In addition, the biorefinery process may open an opportunity for the traditional pulp and paper industry.

[Kun Liu, Hui Wang, Huayu Liu, Shuanxi Nie, Haishun Du, and Chuanling Si, COVID-19: Challenges and Perspectives for the Pulp and Paper Industry Worldwide, BioResources 15(3), 4638-4641 (2020)]

The project of editing this book would not have been possible without the precious contributions of Hilary Lafoe and Jessica Poile who helped me to overcome difficulty caused in communicating with some of the authors of some chapters during this pandemic situation. That is why special thanks are extended to them and their colleagues involved in this project, who took the risk of bringing out such a book to ensure the excellent quality of publication.

Finally, I appreciate the efforts of recognized specialists in this field who have agreed to collaborate on this project and whom I thank. Last but not least I would like to thank my family for the time I spent on this project, which kept me away from them.

I hope that this book will be very useful for many scientists, students, and post-graduates working in the field of biomass and aimed at opening a new era of renewable resources that contribute to sustainability.

Valentin I. Popa

Editor

Valentin I. Popa is professor emeritus of Wood Chemistry and Biotechnology at Gheorghe Asachi Technical University of Iasi, Romania. Professor Popa earned a BSc and an MSc in chemical engineering (1969) and a PhD in polysaccharide chemistry (1976) at Polytechnic Institute of Iasi, Romania. He was awarded the Romanian Academy Prize for his contributions in the field of seaweed chemistry (1976). He has published more than 600 papers in the fields of wood chemistry and biotechnology, biomass complex processing, biosynthesis, and biodegradation of natural compounds, allelochemicals, bioadhesives, and bioremediation. Professor Popa is the author and co-author of 38 books and book chapters.

Contributors

Sheng Chen
Beijing Key Laboratory of
  Lignocellulosic Chemistry
Beijing Forestry University
Beijing, China

Diana Ciolacu
"Petru Poni" Institute of
  Macromolecular Chemistry
Iași, Romania

Ana Carolina de Oliveira
Chemical Engineering Faculty
Universidade do Vale do Itajaí (UNIVALI)
Itajaí, Brazil

Pedro Fardim
Department of Chemical Engineering
KU Leuven
Leuven, Belgium

Franciélle Girardi-Alves
Chemical Engineering Faculty
Universidade do Vale do Itajaí (UNIVALI)
Itajaí, Brazil

Lokesh Kesavana
Turku University Centre for Materials
  and Surfaces (MatSurf)
University of Turku
Turku, Finland

Nina E. Kotelnikova
Institute of Macromolecular
  Compounds
Russian Academy of Sciences
St. Petersburg, Russia

Aleksandra M. Mikhailidi
St. Petersburg State University of
  Industrial Technologies and Design
St. Petersburg, Russia

Raluca Nicu
"Petru Poni" Institute of
  Macromolecular Chemistry
Iași, Romania

Luizildo Pitol-Filho
Chemical Engineering Faculty
Universidade do Vale do Itajaí (UNIVALI)
Itajaí, Brazil

Valentin I. Popa
'Gheorghe Asachi' Technical University
Iași, Romania

Daniela Rusu
"Petru Poni" Institute of
  Macromolecular Chemistry
Iași, Romania

Changyou Shao
Beijing Key Laboratory of
  Lignocellulosic Chemistry
Beijing Forestry University
Beijing, China

Xiaojun Shen
State Key Laboratory of Catalysis
  (SKLC)
Dalian National Laboratory for Clean
  Energy (DNL)
Dalian, China

Liji Sobhana
Laboratory of Fibre and Cellulose
  Technology
Åbo Akademi University
Åbo, Finland

Dana M. Suflet
"Petru Poni" Institute of
    Macromolecular Chemistry
Iasi, Romania

Runcang Sun
Center for Lignocellulose Science and
    Engineering
Dalian Polytechnic University
Dalian, China

Roxana Vlase
S.C. Vrancart S.A.
Adjud, Romania

Meng Wang
Beijing Key Laboratory of
    Lignocellulosic Chemistry
Beijing Forestry University
Beijing, China

Feng Xu
Beijing Key Laboratory of
    Lignocellulosic Chemistry
Beijing Forestry University
Beijing, China

1 Biomass and Sustainability

Valentin I. Popa

CONTENTS

1.1 INTRODUCTION

1.1.1 WHAT IS SUSTAINABILITY?

There have been many approaches to sustainability over the years and many different private and public organizations have demonstrated leadership. Most recently, there have been actions to broaden the approach to sustainability and to increasingly recognize how social, economic, and environmental systems are interrelated. An example of both identifying and working to address these interconnections are the Sustainable Development Goals (SDGs). The United Nations SDGs are 17 goals with 169 targets that all 191 UN Member States have agreed to try to achieve by the year 2030.

Thus, United Nations defined the sustainable development goals as the following: (1) no poverty; (2) zero hunger; (3) good health and well-being; (4) quality education; (5) gender equality; (6) clean water and sanitation; (7) affordable and clean energy; (8) decent work and economic growth; (9) industry, innovation, and infrastructure; (10) reduced inequalities; (11) sustainable cities and communities; (12) responsible consumption and production, (13) climate action; (14) life below water; (15) life on land; (16) place, justice, and string institutions; (17) partnership for the goals (1).

1.2 HOW THE BIOMASS COULD CONTRIBUTE TO THE SUSTAINABILITY?

The term biomass is defined as any organic matter that is available on a renewable basis, including dedicated energy crops and trees, agricultural food and feed crop residues, aquatic plants, wood and wood residues, animal wastes, and other waste materials. The annual production of biomass is about $1.7–2.0×10^{11}$ tons; however, only $6×10^9$ tons are currently used for food and non-food applications. Food applications are by far most important (96.5–97%). The remainder is used in non-food applications, for example, as a feedstock for the chemical industry.

The chemical composition of biomass depends strongly on its source. Generally biomass consists of 38–50% cellulose, 23–32% hemicelluloses, and 15–25% lignin. Cellulose is a non-branched water-insoluble polysaccharide consisting of several hundred up to tens of thousands of glucose units. Cellulose is the most abundant biopolymer synthesized by nature; its amount is estimated at approximately 2×10^9 tons year^{-1}. Hemicelluloses are polymeric materials although lower in molecular weight than cellulose, consisting of C6-sugars (glucose, mannose, and galactose) and C5-sugars (mainly xylose and arabinose) The third component (lignin) is a highly cross-linked polymer made from substituted phenylpropene units. It acts as glue, holding together the cellulose and hemicelluloses fibres (2).

A wide variety of biomass sources is available for further conversion and utilization. Selection of the biomass feedstock is of paramount importance from both techno- and socio-economical points of view. For ethical reasons, the biomass feedstock should not compete with the food chain. Waste streams with low or even negative value, such as agricultural waste are preferred. Furthermore, it is also advantageous to select sources that are not prone to diseases, only require a limited amount of fertilizer, have a high growth rate per hectare per year and are preferably available throughout the year.

We are entering a new age, the age of science, high technology, and science-based industry, agroforetsry, and services but we are entering the age of environmentalism as well.

1.3 BIOREFINING AS A POSSIBILITY TO OBTAIN ENERGY AND BIOPRODUCTS

The concept of biorefinery was originated in late 1990s as a result of scarcity of fossil fuels and increasing trends of use of biomass as a renewable feedstock for production of non-food products. The term of 'Green Bioefinery' was first introduced in 1997 as: 'Green biorefineries represent complex (to fully integrated) systems of sustainable, environmentally and resource-friendly technologies for the comprehensive (holistic) material and energetic utilization as well as exploitation of biological raw materials in form of green and residue biomass from a targeted sustainable regional land utilization'.

According to the US Department of Energy (DOE), 'A biorefinery is an overall concept of a processing plant where biomass feedstocks are converted and extracted into a spectrum of valuable products'. The American National Renewable Energy Laboratory (NREL) defined biorefinery as follows: 'A biorefinery is a facility that integrates biomass conversion process and equipment to produce fuels, power and chemicals from biomass'. These definitions of biorefinery are analogous to today's integrated petroleum refinery and petrochemicals industry to produce multitude of fuels and organic chemicals from petroleum (3).

However, we think that we have a priority because we have introduced this concept in the paper (4):

> In our days, the idea that vegetable biomass represents a source of liquid fuel and of different new materials has led to the development of various research programs in this field. Our investigations in this direction are based on the following premises: (1) all

kinds of vegetable biomass include almost the same components; (2) the macromolecular compounds existing in the vegetable biomass incorporate biosynthesis energy, and their conversion to useful products seems to be considered; (3) the complex and total processing technology may be modulated depending on the chemical composition of the vegetable source, as well as on the utilization of the obtained chemical compounds. The possibilities of complex processing of soft- and hardwood bark, agricultural wastes, and some energetic cultures of *Helianthus tuberosus* and *Asclepias syriaca* are exemplified.

In order to face the present state of affairs, the manifested tendency is that of adopting the existing classical technologies of carbo- and petrochemical fields in processes of converting biomass into products possessing energetic and/or chemical value. The technology of integral and complex valorization of biomass has been proposed is to be performed on several stages and modules, depending on the chemical composition of the available vegetal resources and on the corresponding field of application for the obtained products as well.

A plant for the fractionation and refining of biomass and the use of its entire components is a 'biorefinery' plant that will have to display a high level of process integration and optimization to be competitive in the near future. Forest products companies may increase revenue by producing biofuels and chemicals in addition to wood, pulp, and paper products in a so-called Integrated Forest Biorefinery (IFBR). The concept of an IFBR is being advanced by a number of investigators who envision converting cellulose, hemicelluloses, and lignin from woody biomass, dedicated annual crops, industrial and municipal waste in bioenergy, and basic chemicals (5, 6).

1.4 RESOURCES

All sorts of biomass are suitable for sustainable energy: industrial and public wastes, organic household, agricultural residues; garden and roadside chippings, clean residues of all sorts of wood and from forests; energy crops (energy plantations – e.g., poplar and willow, oilseed crops, latex bearing plants-*Asclepias syriaca*, a.s.o.). At present on a worldwide basis: 55% of all wood consumed is for fuel, 30% for paper, and 15% for solid wood products. From an environmental perspective wood is preferable to fossil fuels for energy, to agricultural fibres for paper, and to steel or plastics for material applications. We must take into account there is a huge potential to increase world wood growth through use biotechnology and modern silvicultural practices.

1.4.1 EVALUATION OF RESOURCES

In the evaluation of resources, the following aspects have to be considered: evaluation of the removable biomass (green weight dried basis); energy content, energy potential; cost of biomass production at the forest roadsides and biomass user gate (including harvesting, skidding, comminution of biomass, transportation); 'zero-yield transport distances' – in which the total energy content of the crops equals total use of energy – the following maximum distances are mentioned: 600 km by truck, 2000 km by train and 10,000 km by ship; policies must be coherent, integrated and coordinated; innovation in plant and industrial biotechnology should be supported;

policies should support development of the whole supply chain; a communication strategy is essential; pilot projects have a role to play; measurable sustainability indicators should be developed.

Biofuels are more sustainable and environmentally friendly because of the reiterative cycles of burning, followed by carbon fixation by plants, and then by burning of biofuels. Bio-renewables are sustainable means of providing the essential products needed for society.

Biorefineries and production of bioproducts in developing countries could readily deliver social and economic benefits through the production of biofuels and energy for local use, integrated with bioproducts for export. These productive activities, based on market-led innovations, developing technology, and innovation would provide access to new and growing markets. Poverty reduction through the revitalization of the agro-industrial sector would the tangible outcome of the production of feedstocks and the development of bioproducts in developing countries.

The main topics approached at present in the field of biomass are connected with the following: sustainability and land use, biomass recalcitrance, development of new or improved biomass sources, better enzymatic and microbial catalysts, advances in the development of hydrocarbons and algae-based biofuels, progress in biorefinery deployment and infrastructure, and recent improvement in pretreatment, fractionation, and related separation technologies; an industry for the production of clean, renewable biofuels from agricultural and forest feedstocks has begun to emerge; the biotechnology can be used to convert the plant biomass, residues, and wastes in biofuels and biochemicals rather than food and feed.

Our studies carried out, starting several years ago, lead us to the conclusion that phytomass could represent a convenient resource of chemical compounds and energy, if the processing of raw materials is done keeping in view their different sources and different chemical composition. This technology allows us to separate each compound as a function of accessible resource, being similar to the petrochemistry, like refining. Thus using different kinds of phytomass, both in laboratory and pilot plant conditions, we have investigated various possibilities of biomass processing.

In this context, we have carried out the researches approaching some directions such as the following: (i) direct use of the individual chemical compounds isolated from biosystems; (ii) chemical processing of biomass and its components by destruction, thus assuring raw materials for the synthesis of polymers and chemical or energy resources; (iii) chemical or biochemical transformation of both integral biomass and its components (functionalization or functionality) for specific uses; (iv) elucidation of structures and functions of the natural compounds in biosystem aiming at using them in structures with advanced properties and at simulating their behaviours against physical, chemical, and biological agents; (v) 'in vitro' simulation the synthesis of natural chemical compounds (4).

1.4.1.1 Forestry and Wood Processing Wastes

In order to be able to evaluate the sustainability of present consumption and feasibility of introducing modern biomass fuels and bioproducts-based applications, an assessment of the resources and their availability is necessary. Among them different types of residues have to be considered: wood residues from logging and wood processing (saw-milling and manufacturing plywood, particle board, and pulp); wood residues generated by management of forestry (thinning of young stands and removal of dead and dying trees), and perennial crop plantation and replanting of tree; residues resulted in clearing forest lands for agricultural purposes, cutting or lopping trees purely for fuel wood, collecting deadwood, trees growing on agricultural land, communal lands, on waste lands, on private land such as gardens, trees growing along roads, etc. It will be mentioned only gross amounts that may be recommended as raw material for energy and bioproducts (7).

1.4.1.1.1 Logging Residues

Logging residues consist of branches, leaves, lops, tops, damaged, or unwanted stem wood. Such residues are often left in the forests for various reasons of which the low demand for fuel (with high moisture content) in some areas is probably an important one as well as logistics. This is not to suggest that forest-residues recovery is not undertaken. There are countries where there is considerable recovery for use in industries as well as domestic purposes (e.g., as wood chips with bulk density of about 300 kg/m^3). Recovery rates vary considerably depending on local conditions. A 50/50 ratio is found in the literature, for example, for every cubic metre of log removed, a cubic metre of waste remains in the forest (including the less commercial species).

1.4.1.1.2 Saw-milling

Recovery rates vary with local practices as well as species. After receiving the logs, about 12% is waste in the form of bark. Slabs, edgings, and trimmings amount to about 34% while sawdust constituents another 12% of the log input. After kiln-drying the wood, further processing may take place resulting in another 8% waste (of log input) in the form of sawdust and trim end (2%) and planer shavings (6%). For calculation purposes a yield factor of 50% has been used (38% solid wood waste and 12% sawdust). Sawmill residues are used for various purposes but much depends on local conditions such as demand centres nearby. Part of the residues is used by sawmills themselves, basically for stem generation for timber drying purposes. However, the bulk remains unused. Where a local demand exists, wood residues are used for various purposes, mainly as a source of energy for brick and lime burning, other small industrial applications as well as a source of raw material such as for parquet making and blackboards, or composite materials. Sawdust sometimes is briquetted and carbonized and solid as a high-grade charcoal, which commands a higher price than normal charcoal. Considerable quantities are also used to cover charcoal mound kilns.

1.4.1.1.3 Plywood Production

Plywood making is a large-scale operation and involves the cutting of the logs to the length required and debarking the logs. After the preparatory operations, sizing, debarking, and cleaning, the logs are sliced, i.e., the logs are rotated in a machine. While rotating, a knife slices or peels off the veneer. Then the sliced veneer is cut into size required and it is dried after which it is ready for further processing. The dry veneer slices are sorted, with sheets having holes or other irregularities being rejected.

The sheets are glued and hot-pressed into plywood sheets. Finally, the plywood sheets are trimmed (cutting into standard sizes), sanded, and graded. Recovery rates vary from 45 to 50% with the main variable being the diameter and quality of the log. Of the log input, the main forms of waste are log ends and trims (7%), bark (5%), log cores (10%), green veneer waste (12%), dry veneer waste (8%), trimmings (4%), and rejected plywood (1%). These form the largest amount of waste while sanding the plywood sheets results in another loss of 5% in the form of sander dust. For calculation purposes a yield factor of 50% has been used, with 45% solid residues and 5% in the form of dust.

1.4.1.1.4 Particle Board Production

Particle board production basically involves size reduction of the wood, drying, screening, mixing with resins and additives, forming of the so-called mat, pressing, and finishing. All types of wood are used for the production of particle board such as solid wood, solid wood residues (off cuts, trimmings), low-grade waste such as hogged sawmill waste, sawdust, planet shavings, etc. During the production process about 17% residues are generated in the form of trimmings. However, this amount is recycled. In addition about 5% screening fines and about 5% sanding dust are generated as residues, which are mainly used as boiler fuel for process steam generation. For calculation purposes a residue factor of 10% has been taken, consisting of screening fines and dust while 17% of the residues are assumed to be recycled.

At present, these residues are used to produce energy for the industry of wood processing, or to supply energy for the needs of the surrounding community, but in function of their accessibility they could be taken into account to be used as a raw material to obtain bioproducts (8).

1.4.1.1.5 Pulp Industry

The same kinds of residues (bark, coarse residues, sawdust) are characteristic also for the pulp industry and they are used as fuel to obtain steam and energy. In the cases when we discuss the challenges to change the industry according to the principles of biorefining, it is possible to reconsider the use of residues as raw materials to recover chemical compounds or to manufacture other bioproducts. These include industrial chemicals, pharmaceuticals, textiles, renewable materials, personal care products, and other manufactured goods. Using biomass in these ways represents a potential to generate higher value returns than when using it primarily to produce energy. The waste product generated during wood pulping, called black liquor, is another example of industrial waste (2, 9).

1.4.1.2 Agricultural and Food Processing Residues

Agricultural residues primarily comprise of stalks and leaves that are generally not harvested from fields for commercial use. Sugar cane bagasse, corn stover (stalks, leaves, husks, and cobs), wheat straw, rice straw, rice hulls, nut hulls, barely straw, sweet straw bagasse, olive stones, etc. are some of the examples of agricultural residues. The use of agricultural residues for biorefinery is beneficial as it eliminates the need of sacrificing arable lands. The wastes such as animal manure (from cattle, chicken, and pigs) are also included within the agricultural residues. The refuge-derived waste generated from either domestic or industrial sources is another source of biomass.

Large amounts of lignocellulosic wastes are generated through *agro-industrial* activities each year and these materials are underused and disposed off in the environment without any treatment, leading to serious environmental pollution problems. These agricultural wastes can potentially be bioconverted into value-added products such as pulp, animal feed, and biofertilizer through the action lignin-degrading enzyme secreting fungi, such as mushrooms. Currently, biofuels such as bioethanol, biodiesel, biohydrogen, and methane from lignocellulosic biomass are produced from agro-wastes rather than from energy crops, as they compete with food crops.

Oil prices volatility and limited fossil resources are pushing the chemical industry's giants – but not exclusively – towards the use of alternative raw materials: agricultural and food processing residues. Agricultural residues constitute a major part of the total annual production of biomass residues and are an important source of energy for domestic as well as industrial purposes. Sometimes residues are used as fuel, but a large amount is burnt in the field.

1.4.1.2.1 Rice

Rice straw: In many countries rice straw is burnt in the field with the ash used as inorganic fertilizer. Relatively small quantities are used as animal fodder, animal bedding, raw material for pulp, paper and board making, or building material.

Rice husk: Husks are often burnt at the rice mill to get rid of the husk but in some countries it used extensively for power generation in large rice mills and also for the brick industry as a source of energy.

1.4.1.2.2 Cotton

Cotton stalks are often burnt in the field as leaving them in the field may result in damage to future crops due to diseases, infestation, etc. Part of them is possibly to be used as domestic fuel and raw materials for energy (10, 11).

1.4.1.2.3 Sugar Cane

Bagasse and sugar cane tops and leaves are the main residues of which the former is normally used as an energy source for steam generation, while the latter is normally used as cattle feed or is burnt in the field. Most sugar factories burn all the bagasse they generate even at very low efficiency. This is done to ensure that all bagasse is burnt, as dry bagasse is known to be a fire hazard. In some countries bagasse is also used as raw material for the furfural or pulp, paper, and board industries. Increasing the combustion efficiency in the sugar industry could result in the saving of considerable quantities of bagasse which either could be sold to pulp and paper factories or used to generate power products as well (12).

1.4.1.3 Food Processing

Until the end of the 20th century, disposal of food wastes was not considered as a matter of concern. Particularly, increase of food production without improving the efficiency of the food systems was the prevalent policy. This consideration increased generation of wasted food along supply chains. In the 21th century, escalating demands for processed foods have required identification of concrete directions to minimize energy demands and economic costs as well as reduce food losses and waste. Today, food wastes account as a source of valuable compounds and deal with the prospects of feeding fast growing population. Perspectives originate from the enormous amounts of food-related materials (food 'losses', 'wastes', 'by-products', or 'wasted by-products'), which are discharged worldwide and the existing technologies that promise not only the recovery but also the recycling and sustainability of valuable ingredients inside food chain. The prospect of recovering valuable compounds from food by-products is a story that started few decades ago. Citrus peel was one of the first by-products to be utilized for the recovery of essential oils and flavonoids, and their re-utilization as additives and flavourings in foods and fruit juices. Even earlier, solvent extraction had been applied to recover oil from olive kernels, which are one of the by-products derived from olive oil production. Nowadays, olive kernels are considered an established commodity similar to olive fruit, whereas researchers focus on the recovery of polyphenols from olive mill wastewater. Over the last decade, several companies have started commercializing the latest process and are ambitious to turn this waste into valuable compounds. In the field of animal-derived side streams, cheese whey constitutes the most intensely investigated food by-product and represents a successful reference of valorization. Protein concentrates and various sugar derivatives are the prominent compounds derived from this source, as reflected by the numerous processes and products existing in the market.

These commercially available applications inspired the scientific community to intensify its efforts for the valorization of all kind of food by-products for recovery purposes. Besides, the perpetual disposal of high nutrition proteins, antioxidants, or dietary fibres in the environment is a practice that could not be continuing for a long time within the sustainability and bioeconomy frame of the food industry. Indeed, the depletion of food sources, the fast growing population, and the increasing need for nutritionally proper diets do not allow considering other alternatives. As a result, a large number of projects have been initiated all over the globe and across scientific disciplines, whereas the existence of numerous scientific articles, patents, congresses, and industrialization efforts has emerged as a wealth of literature in the field.

Despite this plethora of information and the developed technologies, the respective shelf products remain rather limited. For instance, only few companies across the globe are activated in this field, whereas many of them are in an early stage and have not developed their process in a commercial manner yet. This is happening because the industrial implementation of food waste exploitation for the recovery of valuable components is a complex approach that needs careful consideration of numerous aspects. Waste emerges seasonally and often in large quantities and is prone to microbial spoilage. Therefore, drying or immediate processing is required. In addition, the concentration of target compounds varies significantly, which may be a challenge for subsequent standardization of extracts or products. Scale-up of processes developed on laboratory or pilot-plant scale needs to be appropriately designed to retain the functional properties of the target compounds. Finally, the product to be developed needs to meet the high expectations of the consumers in an increasingly competitive market.

Following these considerations, a commercially feasible product can be generated only if a certain degree of flexibility and alternative choices can be adapted in the developing methodology. Experience has shown that a project focused on the recovery technologies without investigating and establishing particular applications is doomed to fail because the final product might not be as beneficial as initially expected. In addition to these challenges, regulatory issues still exist and constitute a severe problem for the industry. Without doubt there is a market need for nutraceuticals and functional foods. However, marketing of such products using health claims requires comprehensive research activities to unambiguously demonstrate their health benefits. These studies are costly and constitute an impediment especially to small- and medium-sized as well as spin-off and spin-out companies. So far, only a small number of compounds and products have been approved, whereas the overwhelming majority of health claims has been declined by regulatory bodies (13).

1.4.1.4 Municipal Wastes

Municipal solid waste (MSW) is another potential feedstock. The municipal solid waste includes sewage sludge and industrial waste. Residential, commercial, and institutional post-consumer waste usually contains good amounts of plant-derived organic materials that can be used as potential source of biomass. The waste paper, cardboard, wood waste, and yard waste are examples of MSW. Typically MSW contains about 36% paper and paperboard products and 12 % yard trimmings. All these materials, after separation from other components such as metals and plastics can be converted to biofuels and bioproducts using similar processes used for conversion of lignocellulosic feedstocks. Municipal solid waste contains paper and paper products (37.8%), food waste (14.2%), yards waste (14.2%), wood wastes (3.0%) along with plastic (4.6%), rubber and leather (2.2%), textiles (3.3%), glass and ceramics (9.0%), metals (8.2%), and miscellaneous (3.1%) (14, 15).

1.4.1.5 Dedicated Crops (Terrestrial and Aquatic)

Both plants and residues have been identified as biomass sources. The most promising could be the following:

- A few fibrous dicotyledon plants, already known for their textile or cordage bastfibre, i.e., flax (*Linum usitatissimum*), hemp (*Cannabis sativa*), and kenaf (*Hibiscus cannabinus*);
- Some fibrous monocotyledon plants, which are presently part of the natural vegetation in the area, i.e., reeds, common (*Phragmites communis*) or giant (*Arundo donax*), and esparto (*Lygeum spartum*) or alfa (*Stipa tenacissima*) grasses;
- Several fibrous residues of either present or future crops: cereal straws, sorghum (*Sorghum vulgare*), Jerusalem artichoke (*Helianthus tuberosus*), and sunflower (*Helianthus annus*) stalks are the most prominent candidates in this class; other residues can also be considered on smallscale, local applications; *Camelina* grown on marginal lands. The assumption was that marginal crops for biofuel production could avoid competing with food crops for land and resources.

- Wood from short-rotation plantations, created on low-value agricultural or marginal land; most promising genera include softwoods like *Pinus*, *Picea*, *Pseudotsuga*, and *Larix*, and hardwood like *Eucalyptus* and *Populus*.

1.4.1.5.1 Energy Crops

The energy crops are normally densely planted, high-yielding, and short-rotation crops. The crops are usually low cost and need low maintenance. These crops are grown dedicatedly to supply huge quantities of consistent-quality biomass for biorefinery. The energy crops mainly include herbaceous energy crops, woody energy crops, agricultural crops, and aquatic crops.

Herbaceous energy crops are perennials that are harvested annually. It takes 2–3 years to reach full productivity. These crops include grasses such switchgrass, miscanthus, bamboo, sweet sorghum, tall fescue, kochia, wheatgrass, reed canary grass, coastal bermuda grass, alfalfa hay, thimothy grass, and others.

Wood energy crops are fast growing hardwood trees that are harvested within 5–8 years of plantation. These crops include hybrid poplar, hybrid willow, silver maple, eastern cottonwood, green ash, black walnut, sweetgum, sycamore, etc. The short-rotation woody energy crops are traditionally used for manufacture of pulp and paper (16).

Agricultural crops comprise of oil crops (e.g., jatropha, oilseed rape, linseed, field mustard, sunflower, castor oil, olive, palm, coconut, groundnut, etc.), cereals (e.g., barely, wheat, oats, maize, rye, etc.), and sugar and starchy crops (e.g., sweet sorghum, potato, sugar beet, sugarcane, etc.). The crops are generally grown to produce vegetable oils, sugars, and extractives. The crops have potentials to produce plastics, chemicals, and products as well.

1.4.1.5.2 Short-rotation Forestry

Compared to other annual biomass production systems, short-rotation forestry (10–15 years) can be assumed as an extensive and most eco-efficient land use. In contrast to annual crops the production can contribute to different international conventions and commitments simultaneously (soil erosion, biodiversity, climate protection, and desertification). To optimize short-rotation forestry as an ecological and socio-economical sound land use, the different utilization techniques for the energetic use of dendromass have to be assessed. To minimize land consumption, different land-use-management systems for biomass production have to be compared and optimized. This needs an interdisciplinary approach of agricultural and forestry institutions of industrialized and developing countries. To guarantee ecological and socio-economic sound land use management systems international standards for the production of biomass have to be developed.

Sweet sorghum is a promising alternative crop for bioethanol production. Moreover, it is a 'food fuel-energy/-industrial crop' which ranks fifth among the world's grain crops, requires low water/fertilizer input, has a high yield of grains and biomass (starch/sugars/lignocelluloses) for integrated multi-purpose processing, and grows well in marginal lands, in semi-arid and temperate regions, including Africa, India, Latin America, and Europe. A limiting factor for its widespread cultivation is the lack of varieties adapted to different growth conditions, including colder climate. Consequently, research should address the optimization of sweet sorghum as an energy crop through breeding. Besides biomass yield and relevant quality traits, genetic improvement/selection should concentrate on general agronomic traits (such as water and nutrient use efficiency) and, in particular, adaptation of sweet sorghum to colder climates. The project should also address agronomic practices and harvesting technologies leading to improved yield, quality, sustainability, and competitiveness of sweet sorghum production. Environmental and economic analysis of sweet sorghum cultivation, including energy balance and life cycle assessment, should also be carried out.

The accessible quantity of these resources could be increased by the following ways: (1) the improvement of existing or the development of new cultivation practices; (2) the development of new crop rotation; (3) the rational management of natural vegetation; (4) the development of the appropriate harvesting technology; (5) the genetic manipulation of plants for the removal of undesired properties and/or the acquisition of desired ones; (6) the generation of viable multi-product or multi-use agricultural systems (17).

1.4.1.5.3 Aquatic Crops

These include several varieties of aquatic biomass, for example, algae, giant kelp, other seaweed, marine microflora, etc.

Macroalgae: These (seaweeds) that occupy the littoral zone are a great source of compounds with diverse applications; their types and content greatly determine the potential applications and commercial values. Algal polysaccharides, namely the hydrocolloids: agar, alginate, and carrageenan, as well as other non-jellifying polysaccharides and oligosaccharides are valuable bioproducts. Likewise, pigments, proteins, amino acids, and phenolic compounds are also important exploitable compounds. For the longest time the dominant market for macroalgae has been the food industry. More recently, several other industries have increased their interest in algal-derived products, for example, cosmetics, nutraceutical/functional food, cosmetic and pharmaceutical industries, and more recently, as a source of feedstock for biorefinery applications. Thus, as world energy demands continue to rise and fossil fuel resources are increasingly reduced, macroalgae have attracted attention, as a possible renewable feedstock to biorefinery applications, for the production of multiple streams of commercial interest including biofuels such as bioethanol and biogas, particularly because they have considerable contents of carbohydrates. In this field, macroalgae have several advantages over terrestrial biomass, primarily because of their potentially high yields, no competition with food crops for the use arable land and fresh water resources, and utilization of carbon dioxide as the only carbon input. Today, seaweeds are used in many countries for different purposes, including their direct consumption as food or supplements (by animals and humans), as feedstock for the extraction of phycocolloids, or for their bioactive components and as biostimulators and biofertilizers. Notably, direct use as food has strong roots in the East Asia, whereas the West seems to be more committed to extraction of the hydrocolloids, namely carrageenan, agar, and alginate. In addition, many seaweeds are receiving increasing attention as a potential, renewable sources for the food industry, as feed for livestock and as food directly. The industry uses 7.5–8 million tons of wet seaweed annually. This is harvested either from naturally grown (wild) seaweed or from open-water, cultivated (marine agronomy, farmed) crops. The farming of seaweeds has expanded rapidly as demand has outstripped the supply available from natural resources. Commercial harvesting occurs in about 35 countries, spread between the Northern and Southern Hemispheres, in water ranging from cold, through temperate, to tropical (18, 19).

Microalgae: Climate, energy, and food security are three of the most important global challenges society faces during the 21st century. At the 2015 Paris Climate Conference, nations agreed to limit the rise in mean global temperature to not more than 2°C relative to pre-industrial levels and to pursue additional efforts to limit the rise to below 1.5°C.

At present, large-scale industrial cultivation of marine microalgae appears to be one of the most promising approaches for achieving these climate goals while simultaneously contributing to global energy and food security. Microalgae exhibit rates of primary production that are typically more than the most productive terrestrial energy crops. Thus they have the potential to produce an equivalent amount of bioenergy and/or food in less than one-tenth of the land area. Research and development investments during the next decade will be necessary to further improve the performance and reduce the cost and resource requirements associated with large-scale production of fuels, animal feeds, and human nutritional products from marine microalgae (20).

Microalgae biofuels belong to the third generation of biofuels, which are considered as an alternative energy source for fossil fuels without the disadvantages associated with first and second generation of biofuels. Generally, the first-generation biofuels are derived from crop plants such as soyabean, corn, maize, sugar beet, and sugar cane; palm oil; rapeseed oil; vegetable oil; and animal fats. These types of biofuels have created a lot of disputes due to their negative impacts on food security, global food markets, water scarcity, and deforestation. In addition, the second-generation biofuels derived from nonedible oils (*Jatropha curcas, Pongamia pinnata, Simarouba glauca*, etc.), lignocellulosic biomass, and forest residues require huge areas of land otherwise that could be used for food production.

Currently, the second-generation biofuel production also lacks efficient technologies for commercial exploitation of wastes as source for biofuel generation. Based on above-mentioned drawbacks associated with the first- and second-generation biofuels, microalgae biofuel seems to be a viable alternative source of energy to replace or supplement the fossil fuels. Microalgae biofuel production is commercially viable because it is cost competitive with fossil-based fuels, does not require extra lands, improves the air quality by absorbing atmospheric CO_2, and utilizes minimal water (21–23).

1.5 ENERGY AND BIOPRODUCTS

Biomass as renewable raw material can be used in an agroforestry/industrial integrated complex to capitalizing on all components. The hybrid technologies are environmentally sound, economically sustainable assuring social development, and the products are biodegradable and recyclable.

In these conditions we can talk about a new concept like green economy which encompasses the economic activities related to reducing the use of fossil fuels, decreasing pollution and greenhouse gas emissions, increasing efficiency of energy usage, recycling materials, and developing and adopting renewable sources of energy. The four main sectors of the green economy are as follows: *Environment* (preserving and protecting natural resources, managing their use in a sustainable way; using them more efficiently and productively; reducing or eliminating pollution and toxic waste), *Energy* (creating, storing, distributing, and saving energy), *Infrastructure* (reducing the impact of human development activities on our world), and *Support* (government and regulatory administration research, design, and consulting services).

Probably the areas of most interest to chemists are *green chemistry and sustainability*. Green chemistry is the utilization of a set of principles that reduce or eliminate the use or generation of hazardous substances in the design, manufacture, and application of chemical products (24).They are as follows: (1) prevention; (2) atom economy; (3) less hazardous chemical syntheses; (4) designing safer chemicals; (5) safer solvents and auxiliaries; (6) design energy efficiency: (7) use of renewable feedstocks; (8) reduce derivatives; (9) catalysis; (10) design for degradation; (11) real time analysis for pollution prevention; (12) inherently safer chemistry for accident prevention.

1.5.1 ENERGY

It estimates that if current energy consumption trends and structural changes in the global economy continue, the world's consumption will be doubled over the next 30 years. Having in mind that in 2030, 90% of energy consumption will be based on fossil resources, the use of biomass for energy can be one way to reduce the ever-increasing emissions of carbon dioxide, one of the main gases responsible for global warming.

By the photosynthesis process in Nature a huge quantity of biomass/phytomas is accumulated. Thus the photosynthesis process could be used to assure raw material for energy and chemicals. At the same time it is possible to reduce the negative influence of CO_2 on greenhouse effect or global warming by photosynthesis. The possibilities to increase the quantity of biomass accessible to be processed to assure sustainable development were mentioned above. Therefore, biomass from organic household, industrial and public wastes, garden and roadside chippings, clean residues of all sorts of wood and from forests, energy crops; agricultural residues, microalgae, and macroalgae are suitable for sustainable energy.

The world's total annual use of energy is only one tenth of annual photosynthetic energy storage, i.e., photosynthesis already stores ten times as much energy as the world needs. The problem is getting it to the people who need it. The energy content of stored biomass on the earth's surface today, which is 90% in trees, is equivalent to our proven fossil fuel reserves. In other words the energy content of the biomass is equivalent to commercially extractable oil, coal, and gas (Table 1.1).

The process of photosynthesis embodies the two most important reactions in life. The first one is water-splitting reaction which evolves oxygen as a by-product; all life depends on this reaction. The second is the fixation of carbon dioxide to organic compounds. All our food and fuel is derived from this CO_2 fixation from the atmosphere. When looking at an energy process we need to have some understanding of

TABLE 1.1
Fossil Fuel Reserves and Resources, Biomass Production, and CO_2 Balances (Adapted)

Proven Reserves	Tons Coal Equivalent
Coal	4×10^{11}
Oil	2×10^{11}
Gas	1×10^{11}
Estimated resources	7×10^{11}t $= 25 \times 10^{21}$J
Coal	85×10^{11}
Oil	5×10^{11}
Gas	3×10^{11}
Unconventional gas and oil	20×10^{11}
Total	113×10^{11}t $= 300 \times 10^{21}$J
Fossil fuels used energy use (1976)	2×10^{11}t carbon $= 6 \times 10^{21}$J
World's annual energy use	3×10^{20}J (5×10^9 t carbon from fossil fuel)
Annual photosynthesis	**8×10^{10}t carbon**
(a) net primary production	(2×10^{11}t organic matter)$= 3 \times 10^{21}$J
(b) cultivated land only	0.4×10^{11}t carbon
Stored in biomass	8×10^{11}t carbon$= 20 \times 10^{21}$J
(a) total (90% in trees)	0.06×10^{11} t carbon
(b) cultivated land only (standing biomass)	7×10^{11}t carbon
Atmospheric CO_2	6×10^{11}t carbon
CO_2 in ocean surface layers	$10–30 \times 10^{11}$t carbon
Soil organic matter	17×10^{11}t carbon
Ocean organic matter	

what the efficiency of this process will be; one needs to look at the efficiency over the entire cycle of the system, and in the process of photosynthesis we mean incoming solar radiation converted to stored endproducts. The practical maximum efficiency of photosynthesis is between 5 and 6 %, but this represents stored energy.

The photosynthetic efficiency will determine the biomass dry weight yields. Obviously if we can grow and adapt plants to increase the photosynthetic efficiency, the dry weight yields will increase and of course alter the economics of the crop. One of the very interesting areas of research is to try to understand what the limiting factors are in photosynthetic efficiency in plants both for agriculture and for biomass for energy and bioproducts.

There is another aspect of photosynthesis that we should all appreciate; i.e., the health of our biosphere and our atmosphere is totally dependent on the process of photosynthesis. Every three hundred years all CO_2 in the atmosphere is cycled through plants; every two thousand years all the oxygen; and every two million years all the water is cycled. Thus three ingredients in our atmosphere are dependent on cycling through the process of photosynthesis.

Biomass for fuels is a complex subject involving the growth, collection, densification, transport, conversion, and utilization of organic material. Often, biomass for fuels must compete with important alternative uses. The impact of biomass for fuels on food, feed, and fibre prices is not fully known. And the need to return organic material to the soil for erosion control and organic matter maintenance continues to be of concern. Also competition between food crops and fuel production from biomass is an unresolved issue and will need a great deal more attention. Certainly, a net energy gain from biomass fuels relative to the petroleum input is essential for a successful biomass fuels program. However, an overall net energy gain may not be necessary in the short run if a low-quality bulky fuel is upgraded to a high-quality clean burning fuel, especially a high energy density fuel to power existing mobile vehicles.

The potential of wood and forest biomass for making a greater contribution to our energy needs is gaining wider recognition. For this reason, energy farming has become conceptually feasible. In order to estimate energy yield of certain forest species, information on the quantification and characterization of tree biomass will be needed. The characterization of forest fuel for direct combustion normally involves the determinations of (1) heat of combustion, (2) density, and (3) moisture content (MC). Heat of combustion, calories per gram of moisture-free sample, is a measure of the basic thermochemical property of an organic material and is normally used for the comparison of fuels. Since the chief drawback of woody fuels is their bulkiness and natural MC, the economic value of a woody fuel also depends upon its density and MC. Hence, heavier woods like oak, hickory, and birch are sometimes considered being better fuels than the lighter weight softwoods such as pine or spruce that have a higher value in heat of combustion. Wood with low MC is also more valuable for fuel than that containing higher amount of moisture because wood moisture diminishes the usable heat produced by the fuel.

Dry biomass can be burnt to produce heat, steam, and/or electricity and can be converted to liquid or gaseous form for use in mobile vehicles by anaerobic fermentation, alcoholic fermentation, or gasification. The use of biomass for fuels and bioproducts raises complex and widely diversified issues and its impact must be assessed variously according to specific feedstocks, geographic areas, conversion technology, and use application.

Many processes or technologies exist for converting biomass to more useful form for fuel or industrial feedstocks. Most are classified as dry or wet processes. Dry processes include direct combustion and gasification; wet processes include anaerobic and alcohol fermentation.

If heat is needed and the biomass is relatively dry, direct combustion may be the most efficient and effective process. Heat may be used to produce steam and electricity if desired. Various gasification processes will produce a low or medium Btu gas (gas of low or medium calorific value, about ¼ and ½ of natural gas, respectively) if that energy form is desired for stationary engines or heat.

If the biomass is wet, for example, animal manure, anaerobic fermentation will yield a low Btu gas, primarily methane. For mobile vehicles requiring a high energy density fuel (most commonly a liquid), synthetic liquid fuel may be produced by direct fermentation of sugar crops or by hydrolysis of starches or cellulosic materials followed by fermentation (25).

1.5.1.1 Direct Combustion

Technology for direct combustion is old and highly developed. It is in wide use commercially accounting for most of the $1.9 \cdot 10^{18}$ J of energy presently generated from wood. Much research is under way to develop suitable combustion systems for wet biomass and to study optimum particle size, feeding systems, particles dimensions control, biomass mixtures with oil or coal, suspended burning systems, etc.

1.5.1.2 Burning in an Excess of Air

One purpose of burning is to eliminate unwanted waste: burning in an open pile and incineration are examples. Some type of furnace is required to collect and distribute the heat generated, and combustion may occur inside or outside tubes. Provisions must be made to: (a) introduce the organic particles or shredded material; (b) provide an adequate air flow to maintain an excess oxygen supply; (c) remove the residues or ash; and (d) control particles emission. Air flow may be by natural or forced draft. There are two types of air-suspended combustion systems: (a) those which suspend the burning fuel in the gas stream in the combustion enclosure and (b) those which suspend the fuel in the gas stream and in another medium, the fluidized bed. Advantages of flue-gas stream suspension include a more rapid response to automatic control, an initial cost saving due to lack of grate surface and mechanical stocking devices, and the ability to complete combustion with much smaller percentage of excess air in the furnace. Fluidized-bed suspension burning systems have all advantages of the flue-gas stream suspension, plus one that is important when a system must operate intermittently. Fluidized beds, usually sand, comprise a 'thermal flywheel' or large capacity. Once operating temperature is reached, they retain heat over a long period, losing only about 110°C during an overnight shutdown. The saving of auxiliary fuel for preheat on the next start-up are appreciable.

1.5.1.3 Burning in a Controlled Atmosphere

Gasification is the conversion of a solid or liquid to a gas. If the oxygen supply is restricted, incomplete combustion occurs, releasing combustible gases such as carbon monoxide, hydrogen, and methane. A solid residue or char remains.

1.5.1.4 Heating in the Absence of Air

Pyrolysis is the transformation of an organic material into another form by heating in the absence of air. If heat is applied slowly, the initial products are water vapour and volatile organic compounds. Increased heat leads to recombination of the organic materials into complex hydrocarbons and water. The principal products of pyrolysis are gases, oils, and char.

1.5.1.5 Producer Gas Generation

A producer gas generator produces a combustible gas from crop residues, wood chips, or charcoal. (During World War II, in Europe and Japan, producer gas generator or gasifiers were often used to operate tractors, automobiles, and buses, because petroleum was scarce.) The feedstock is heated to 1000° C and reacts with air, oxygen, steam, or various mixtures of these to produce a gas containing about 30% carbon monoxide, 15% hydrogen, and up to 3% methane.

There are two basic generator designs: the updraft and the downdraft. In an updraft generator, hot gases flow counter to the feedstock. Part of the fuel stock is pyrolysed and the resulting gas has a high tar content. In the downdraft system, pyrolysis products are broken down as they pass through the reaction zone before combining within the exiting gases. Since downdraft generators have the potential to eliminate tar from gas, they are probably better suited for burning crop residues as a fuel source (26, 27).

1.5.1.6 Anaerobic Digestion

Anaerobic digestion is a conversion process for wet biomass such as animal manure, municipal sewage, and certain industrial wastes (28).

Through this process complex organics are converted into methane and other gases. An effluent is also produced which can be used as fertilizer or animal feed. Anaerobic digestion is a biological process carried out by living microorganisms:

Organic matter + bacteria +water → methane + carbon dioxide + hydrogen sulfide + stabilized effluent.

This process occurs only in the absence of free oxygen. Methane-forming bacteria are sensitive to environmental conditions such as pH (6.6–7.6 optimum), temperature (35°C and 54°C are two preferred levels), and carbon/nitrogen ratio (30 optimum). Man-made digesters or containers that keep the feedstock isolated from air can be of either the batch or continuous flow type. Advantages of the batch type include the following: feedstock availability is often sporadic and comes in batches; daily management is minimal; and relative inexpensive. Disadvantages of the batch type are as follows: much labour is needed to load and unload digester; gas production is sporadic; not as efficient as continuous digester.

While early digesters were usually of the batch type, the continuous flow type is considered an improvement.

Biogas consists of 60–70% methane, 30–40% carbon dioxide, and a trace of hydrogen sulfide, ammonia gas, and water vapour. Biogas has an energy content around 22 MJ/m^3. Methane or biogas are 'permanent' gases and cannot be liquefied at any pressure at commonly occurring temperatures, seriously limiting their use in mobile vehicles.

These gases are better suited for use in high compression (13–14:1) stationary engines designed or modified to operate on methane. In biogas-powered stationary engines, waste heat can be recirculated in the digester coil and gas can be used as it is produced without a compressor storage unit. Full engine power is biogas is realized only if carbon dioxide is removed from the biogas mixture to increase the energy content of the gas. Longer engine life is attained if hydrogen sulfide is also eliminated from the gas before use. Biogas may be used to heat livestock buildings by scrubbing H_2S only, but the 30–40% CO_2 will necessitate additional venting and this requires more heating energy. However, the CO_2 would not present a problem in greenhouse heating.

Digester waste or sludge is an excellent fertilizer containing all the potassium and phosphorous and up to 99% of the nitrogen originally in the manure. In addition, trace elements such as B, Ca, Fe, Mg, S, and Zn remain unchanged. Sludge could also be used in livestock ratios if mixed with molasses, grains, and roughage. Water must be removed by centrifuge to concentrate the protein, but some of the protein will be dissolved in the water and lost.

1.5.2 Biofuels

Rapid depletion of finite petroleum reserves and the environmental concerns associated with their use are two major motivations that drive technology innovations towards sustainable technology (29–34).

This is why there is an increased interest for the biofuels in the last years. Thus, starting from 1975 first generation of the biofuel was developed.

First-generation biofuels are produced from food crops such as corn and sugarcane, and soyabeans and rapeseed. Ethanol produced from corn starch and biodiesel, as currently sold commercially, are examples of first-generation biofuels.

Second-generation biofuels are produced from non-food lignocellulosic materials such as wood, agricultural crop residues (e.g., corn stalks, bagasse), forest harvest residues, municipal solid waste, and dedicated biomass energy crops such as switchgrass. Waste vegetable oil is also being used as raw material. The end products can be ethanol, biodiesel, aviation fuel, or any one of a wide array of industrial biochemicals. Because production of such fuels is more technically challenging than for first-generation biofuels, requiring complex biochemical or thermochemical processing, it has taken decades to bring second-generation biofuels to commercialization. While the raw materials for production of second-generation biofuels are from non-food sources, food/fuel concerns largely become a non-issue. However, to the extent that dedicated fibre crops are used as a source of raw material, the potential for competition with food crops will continue. While fibre crops can be grown on marginal farm land, higher yields are obtainable on better sites, posing difficult land use decisions for farmers. Issues related to use of pesticides and fertilizers are also linked with dedicated fibre crops such as switchgrass.

Lignocellulosic biomass composition varies among species but generally consists of ~25% lignin and ~75% carbohydrate polymers in dry weight (cellulose and hemicelluloses) and it is the largest known renewable carbohydrate source. Major challenges for biological conversion are posed by biomass recalcitrance. The cellulose and hemicelluloses components of biomass can be separated from the lignin and depolymerized to obtain the constituent sugars, mainly glucose, xylose, and arabinose. To overcome the biomass recalcitrance, feedstock deconstruction is therefore required. For that purpose, there are different second-generation routes currently under development, for example, biomass gasification followed by syngas fermentation/catalysis or biomass cellulose enzymatic hydrolysis followed by sugar fermentation. Up to date, biomass enzymatic hydrolysis and microbial sugar conversion is the preferred technology for ongoing industrial projects; however, a myriad of technologies are competing in that space in biomass pretreatment, enzymes supply, microorganism capable of consumption of glucose, xylose and hydrolysis by-products, fermentation strategies, and downstream technologies. Depending on the feedstock choice and the cultivation technique, second-generation fuels production has the potential to provide benefits such as consuming waste residues and making use of marginal lands. This way, the new fuels can offer considerable potential to promote rural development and improve economic conditions in emerging and developing regions. However, while second-generation crops and production technologies are more efficient, their production could become unsustainable if they compete with food crops for available land. Thus, their sustainability will depend on whether producers comply with criteria like minimum lifecycle greenhouse gas (GHG) reductions, including land use change and social standards.

Third-generation biofuels, which at this point exist for the most part in research laboratories, early-stage development facilities, and a few small-scale enterprises, are produced from algae. The term algae encompasses a diverse group of organisms that include microalgae, macroalgae (seaweed), and cyanobacteria (formerly known as blue-green algae).There are more than 72,000 different species of algae, and possibly as many as 800,000, which occur either in the form of microalgae or seaweeds (35).

Previously, algae were lumped in with second-generation biofuels. However, when it became apparent that algae are capable of much higher yields with lower resource inputs than other feedstock, many suggested that they be moved to their own category. The diversity of fuels that algae can produce results from two characteristics of the organism. First, some algae produce an oil that can easily be refined into diesel or even certain components of gasoline. More importantly, however, is a second property in them that they can be genetically manipulated to produce everything from ethanol and butanol to even gasoline and diesel fuel directly. The list of fuels that can be derived from algae includes the following: biodiesel, biohydrogen, biogas, butanol, gasoline, methane, ethanol, vegetable oil, jet fuel.

Another favourable property of algae is the diversity of ways in which they can be cultivated. Algae can be grown in any of the following ways:

- *Open ponds* – These are the simplest systems in which algae is grown in a pond in the open air. They are simple and have low capital costs but are less efficient than other systems. They are also of concern because other organisms can contaminate the pond and potentially damage or kill the algae
- *Closed-loop systems* – These are similar to open ponds, but they are not exposed to the atmosphere and use a sterile source of carbon dioxide. Such systems have potential because they may be able to be directly connected to carbon dioxide sources (such as smoke stacks) and thus use the gas before it is released into the atmosphere.
- *Photobioreactors* – These are the most advanced and thus most difficult systems to implement, resulting in high capital costs. Their advantages in terms of yield and control, however, are unparalleled. They are closed-loop systems.

Note that all three systems mean that algae are able to be grown almost anywhere if the temperatures are warm enough. This means that no farm land need be threatened by algae. Closed-loop and photobioreactor systems have even been used in desert settings.

What is more, algae can be grown in wastewater, which means they can offer secondary benefits by helping to digest municipal waste while avoiding to take up any additional land. All of the factors above combine to make algae easier to cultivate than traditional biofuels.

The fourth-generation biofuels are envisioned as sustainable fuels, derived from engineered biological systems, to achieve high levels of energy efficiency and environmental performance which would involve the following: (1) metabolic engineering of plants and algae to achieve high biomass yield, improved feedstock quality, and high CO_2 fixation, (2) direct conversion of solar energy into fuel using designer microorganisms via application of emerging synthetic biology technologies as enabling pathways (36).

1.5.3 BIOPRODUCTS

Since 2007, the number of commercial-scale production capacities for bio-based chemicals has grown significantly and the number of intermediates introduced on the market has reached an average path of 2 per year. Over the past ten years, the production capacities for bio-based chemistry sector have grown faster than the ones for oil-based chemistry with more than 8% of annual growth. The integrated biorefinery is a processing facility allowing extraction of carbohydrates, oils, lignin, and other chemicals from biomass (extractives), and converting them into multiple products including fuels and high value chemicals and materials.

Bioproducts can be produced through a wide range of conversion platforms including biochemical (BC), thermochemical (TC), hybrid thermochemical/biochemical (TC/BC) lignocellulosics, and algal routes (37).

Examples of end-use markets derived from bioproducts include the following: construction and building materials, household and personal items, paints, human and animal health care products, furniture, fuels and oils, automotive supplies, cleaners, shipping and packaging materials (including biodegradable), non-woven and composites, landscaping supplies, compost/fertilizers, natural plant protection products, soil remediation, and health products. They are more environmentally benign than their petroleum-based counterparts and are usually biodegradable. Thus, biomass resources devoted to new industrial uses can be organized within four major biomass opportunity clusters: health, energy, biochemical, and environmental products:

- *Health* – pharmaceuticals, cosmetics, personal care, flavours, and fragrances;
- *Energy* – bio-diesel, ethanol, fuel additives, biomass combustion, two cycle oils, transmission fluids, and lubricants;
- *Biochemical* – coatings, film degradable polymers, plastics, dyes, paints, pigments, gums, adhesives, agrochemicals, soaps, detergents, specialized industrial organics, fine chemicals, biological control products, cleaning agents, solvents, surfactants, and ink;
- *Environmental Products* – paper and board, bio-composites, moulded fibre, non-woven fabric, textile fibres, filler, and insulation.

It is believed the emerging 'Life Science Revolution' will reshape the world's agriculture and agri-food industry in the 21st century. The boundaries between agriculture, industrial chemicals, energy, and health will converge developing what is envisioned as the largest industry in the world – the life science industry. It is estimated that biomass could provide 25% of industrial chemicals and chemical feedstocks. These efforts have led to development of a number of thermochemical and biochemical conversion processes for converting biomass components to industrial chemicals, fuels, and power. Biochemical technologies use enzymes or microorganisms to convert biomass feedstocks to desired products (e.g., through fermentation).These kinds of technologies may be used alone, or may augment more traditional thermochemical technologies such as those used to remove wood extractives or separate fibre. Fermentation processes are most commonly used for the manufacture of various products. Thermochemical technologies may utilize catalysts (acid, alkali, or a combination) and/or high pressure and temperature to convert biomass components to various biocompounds. Through biochemical conversion technologies, products are derived from constituent sugars representing the most abundant renewable resource available and there are many ways to transform sugars into bioproducts.

Several pretreatment methods can be employed to make polysaccharides more accessible for the saccharification process. Physical pretreatment methods are milling and grinding, extrusion and expansion, high-pressure steaming, and steam explosion. Chemical pretreatment methods are alkali treatment, acid treatment, gas treatment, oxidizing agents, cellulose solvents, solvent extraction of lignin, and swelling agents. Biological pretreatment methods use lignin-degrading organisms, polysaccharides-degrading organisms, or a combination of both. The most intensively studied pretreatment method is steam explosion with and without addition of catalysts.

The carried out analysis established that at present 12 building block chemicals could be produced from *sugars*. These can be subsequently converted to a number of high-value bio-based chemicals and materials (38).

The bioproducts proposed to be obtained from sugars resulted from polysaccharides of biomass are the following.

1.5.3.1 Butadiene (1,3-)

1,3-butadiene (BD) is the building block for the production of polybutadiene and styrene-butadiene rubbers, which are used in the production of tires for passenger cars and light-duty vehicles. Butadiene is currently produced from petroleum as a by-product of ethylene manufacturing. However, the shift to lighter feedstocks in ethylene production due to the abundance of shale gas has resulted in a constrained supply of butadiene and large price fluctuations over the past several years, thus providing an opportunity for bio-derived butadiene. There are multiple production routes to bio-derived butadiene that are under development. Many of these routes are through biological upgrading of sugars to intermediates including butanediol (1–4, or), butenes (normal or iso-), or alcohols (ethanol or butanol). These intermediates are converted catalytically to butadiene. A direct route for the production of butadiene via the fermentation of sugars is also being pursued; although this direct process is more complex and incurs increased capital and operating costs and lower overall yields.

1.5.3.2 Butanediol (1,4-)

1,4-butanediol (BDO) is a four-carbon primary alcohol that is a building block for the production of polymers, solvents, and specialty chemicals. Current demand for BDO is just under 2 million metric tons per year. With this large market bio-derived

BDO is being produced on a commercial scale utilizing commodity sugars. A promising conversion strategy currently being scaledup is the catalytic conversion of succinic acid to BDO.

1.5.3.3 Ethyl Lactate

Ethyl lactate is a biodegradable solvent produced by the esterification of ethanol and lactic acid. It was listed as a solvent that meets the ozone protection. It has been used in industrial applications to replace volatile organic petroleum-derived compounds. The primary use for ethyl lactate is as an industrial solvent, as its properties and performance have been found to meet or exceed those of traditional solvents like toluene, methyl ethyl ketone, and N-methyl-pyrrolidone in many applications. Ethyl lactate is expected to compete effectively in that market as economies of scale and process or supply chain improvements drive down costs. It has attracted attention because both of the starting materials used to make ethyl lactate, lactic acid and ethanol have a high potential to be made from lignocellulosic sugars.

1.5.3.4 Fatty Alcohols

Fatty alcohols, also called detergent alcohols, are linear alcohols of 12 or more carbons, used primarily to produce anionic and non-ionic surfactants for household cleaners, personal care products, and industrial applications. Shorter chain alcohols, in contrast, are used most often as plasticizers. Fatty alcohols accounted for 11% of current bio-based chemicals in commercial production. Fatty alcohols are derivatized by ethoxylation, sulfation, or sulfonation before use. They can be produced from tallow, vegetable oils, or petroleum. Each of these classes of feedstocks can be substituted for one another (i.e., as functional replacements) and the market for detergent alcohols is price sensitive. This has led to rapidly changing geographic shifts in detergent alcohol production and an overall move to price-competitive plant-derived feedstocks. Fatty alcohols have the potential to be produced from renewable sources by autotrophic and heterotrophic algae, or by the microbial fermentation of carbohydrates.

1.5.3.5 Furfural

Furfural is a heterocyclic aldehyde, produced by the dehydration of xylose, a monosaccharide often found in large quantities in the hemicelluloses fraction of lignocellulosic biomass. In theory, any material containing a large amount of pentose (five-carbon) sugars, such as arabinose and xylose, can serve as a raw material for furfural production. Most furfural is converted to furfuryl alcohol (FA), which is used for the production of foundry resins. The anti-corrosion properties of FA make it useful in the manufacture of furan fibre-reinforced plastics for piping. Furfural has a broad spectrum of industrial applications, such as the production of plastics, pharmaceuticals, agrochemical products, and non-petroleum-derived chemicals. Furfural is not produced from fossil feedstocks, therefore current production methods from biomass do not displace production from petroleum. However, recent research has reported a furfural-to-distillate fuel pathway that could potentially make furfural derivatives a drop-in replacement for petroleum-derived distillate fuels (jet and diesel).

1.5.3.6 Glycerin

Glycerin is a polyhydric alcohol and is a main component of triglycerides found in animal fats and vegetable oil. The word 'glycerin' generally applies to commercial products containing mostly glycerol, whereas the word 'glycerol' most often refers specifically to the chemical compound 1,2,3-propanetriol and to the anhydrous content in a glycerin product or in a formulation. Glycerin is the main by-product of biodiesel production. It is also generated in the oleochemical industry during soap production and is produced synthetically from propene. Biodiesel and soap production accounts for most current glycerin production; therefore, the overall supply of glycerin is driven primarily by the demand for these products. Glycerin from biomass is a drop-in replacement for synthetic glycerin from propene. The most common use is as a humectant (hygroscopic substance) in food and personal care products, but it has more than 1,500 uses. The glycerin market is currently saturated, which has resulted in stable and low glycerin prices. As a result, the primary market driver for producing glycerin may be leveraging the low, stable prices by using it as a feedstock for conversion to more valuable products, such as epichlorohydrin and succinic acid. The low, stable prices may also allow emerging uses, such as for animal feed and marine fuel.

1.5.3.7 Isoprene

Isoprene is the building block for polyisoprene rubber, styrene copolymers, and butyl rubber. Currently all commercially available isoprene is derived from petroleum. The majority of isoprene has typically been produced by separating the C5 stream from ethylene crackers fed with heavier feedstocks like naphtha or gas oil. This has created a favourable opportunity for bio-based isoprene to enter the market. Bio-based isoprene, produced by aerobic bioconversion of carbohydrates, is identical to petroleum-based isoprene and functions as a drop-in replacement molecule. Production of isoprene from biological sources is in the early stages of development but is accelerating with the backing from major tire manufacturers.

1.5.3.8 Lactic Acid

Lactic acid is the most frequently occurring carboxylic acid in nature. Most commercial production of lactic acid is by microbial fermentation of carbohydrates. It is an alpha-hydroxy acid with dual functional groups making it suitable for use in a variety of chemical transformations and products. Lactic acid is used globally for applications in food, pharmaceuticals, personal care products, industrial uses, and polymers. Growth in polylactic acid polymer production is expected to drive growth for lactic acid consumption. The biodegradable polylactic acid has gained popularity for use in food packaging, disposable tableware, shrink wrap, three-dimensional (3-D) printers, and elsewhere.

1.5.3.9 Acrylic Acid

Acrylic acid has a large market share of about 5 million metric tons per year with projected growth rates of 4%–5% per year. Acrylic acid is produced by the catalytic oxidation of propylene with air to form acrolein and then acrylic acid. With the availability of cheap shale gas utilized to produce ethylene, the market has seen a reduction in propylene production and is projecting higher costs for acrylic acid, opening opportunities for biomass-derived products. Lactic acid is also being pursued as a precursor to the production of bio-derived acrylic acid. Alternative routes to acrylic acid are through 3-hydroxypropionic acid (3-HPA), a fermentation intermediate, followed by catalytic dehydration to make acrylic acid. Acrylic acid production via sugars has also been demonstrated using fumaric acid, glycerin, or ethylene oxide as intermediary.

1.5.3.10 Propanediol (1,3-)

1,3-propanediol (PDO) is a linear aliphatic diol, which makes it a useful chemical building block. It has attracted investment for production from both petroleum and renewable feedstocks. PDO can be used for a variety of applications including polymers, personal care products, solvents, and lubricants. Production of PDO is stimulated because it is an additive in polytrimethylene terephthalate (PTT) polymer production. PTT is used in textiles and fibres due to its superior durability and stain resistance compared to nylon. PDO has been produced synthetically from acrolein or ethylene oxide, but today bio-based PDO is the primary source of production. Bio-based PDO uses 40% less energy than the typical petroleum-based route, giving the bio-based route a significant advantage.

1.5.3.11 Propylene Glycol

Propylene glycol (PG) is a diol (i.e., two hydroxyl groups) also known as 1,2-propanediol, propane-1,2-diol, and monopropylene glycol. It is a viscous, colourless, odourless liquid that does not evaporate (nonvolatile) at room temperature and is completely soluble in water. PG is safe for human consumption, and therefore is used in the production of consumer products such as antiperspirants, suntan lotions, eye drops, food flavourings, and bulking agent in oral and topical drugs. Industrial grade propylene glycol (PGI) is used in the production of unsaturated polyester resins for end-use markets such as residential and commercial construction, marine vessels, passenger vehicles, and consumer appliances. PG is also used as an engine coolant and antifreeze in place of ethylene glycol, and in the airline industry as an airplane and runway de-icer. In liquid detergents, PG serves as a solvent, enzyme stabilizer, clarifying agent, and diluent. Conventional PG has been produced by hydrating propylene oxide (PO), although bio-PG can be produced by hydrogenolysis of glycerin over mixed-metal catalysts, or hydrocracking of sorbitol. Bio-PG is a drop-in replacement for conventional PG. A primary driver of bio-PG production is profitable disposition of excess glycerin coproduct of biodiesel production.

1.5.3.12 Succinic Acid

Succinic acid is a dicarboxylic acid that can be produced from petroleum or biomass. The current market for petroleum-derived succinic acid is small with a primary focus on specialty chemicals. The potential for biomass-derived succinic acid is projected to be large, however, as a possible precursor for the synthesis of high-value products derived from renewable resources, including commodity chemicals, polymers, surfactants, and solvents. Due to the large market potential for succinic acid, production facilities have started operation utilizing biomass feedstocks.

1.5.3.13 Adipic Acid

Succinic acid is being pursued as a functional replacement for adipic acid in specific applications such as polyurethanes, resins, and plasticizers. However, traditionally petroleum-derived adipic acid is produced in a two-stage process that involves the oxidation of cyclohexane followed by nitric acid oxidation of the intermediate to adipic acid. The later steps of the process have a large impact on the sustainability of the overall process. Given the large market size of adipic acid and the potential to improve the sustainability of the process, a number of companies are pursuing renewable adipic acid.

There are several alternative pathways for adipic acid production from biomass at varying stages of development. The two processes are further studied to be developed. The first biologically converts fatty acids and plant-based oils directly to adipic acid. The second utilizes heterogeneous catalysis, which first produces glucaric acid through an aerobic oxidation of glucose followed by catalytic hydrogenation to adipic acid.

1.5.3.14 Xylene (Para)

Para-xylene (pX) is used to produce both terephthalic acid (TA) and dimethyl terephthalate (DMT). Both TA and DMT are raw materials for the production of polyethylene terephthalate (PET) bottles. Major consumers of PET bottles, including Coca-Cola and Pepsi Co., are supporting efforts to develop renewable bottles on a targeted timeline and by actively funding R&D efforts for the production of renewable PET. The renewable replacements that are currently being developed include a direct replacement of petroleum-derived pX from biomass or via functional replacements of PET (such as polyethylene furanoate). Renewable pX can be produced via the traditional biochemical fermentation process followed by upgrading, thermochemical pyrolysis routes, and hybrid thermochemical/biochemical strategies of catalytic upgrading of sugars.

1.5.3.15 Furan Dicarboxilic Acid

Not only has Coca-Cola been investing in routes towards drop-in replacements for polyethylene terephthalate, but it also has actively funded efforts to develop functional replacement options such as polyethylene furanoate (PEF). The production of PET and PEF are similar in that each plastic is produced via a reaction involving ethylene glycol. In contrast, however, PET is produced using a PTA monomer, whereas PEF is produced via furandicarboxylic acid (FDCA) and is a polyfuran.

It is a well-known fact that, in combination with cellulose and hemicelluloses, lignin constitutes the most abundant organic material in the phytomass on the earth surface. Generally, lignin is distributed together with hemicelluloses in the spaces of intercellulose microfibrils from the primary and secondary walls, as well as in the middle lamellae, as an adhesive component to connect cells and reinforce the cell walls of xylem tissues. The total amount of wood in the world's forests has been estimated to about 1.5 billion metric tons. On a worldwide level 50 million tons of lignin are produced annually, but only 1% is sold. At the same time, about 100 million tons of polymers are produced from oil. The question is *why, at an industrial level, no representation for lignin production corresponding to its percent occurrence in biomass can be established* (39).

Lignin is accessible as by-product from biorefining of biomass, especially from pulp manufacture, and the production will be increased in the future when the biomass will be used to obtain biofuels and chemicals from polysaccharides components. In these conditions the possibilities to use lignin as source of biocompounds were analysed.

Thus the applications and uses of lignin are categorized into, near-term, medium-term, and long-term opportunities. Lignin-based opportunities could readily be divided into three categories: (1) power, green fuels, and syngas; (2) macromolecules; and (3) aromatics and other chemicals. The first category represents the use of lignin purely as a carbon source using aggressive means to break down its polymeric structure. The second category is the opposite extreme and seeks to take advantage of the macromolecular structure imparted by nature in high-molecular weight applications. Somewhere between the two extremes come technologies that would break up lignin's macromolecular structure but maintain the aromatic nature of the building block molecules – the third category.

Lignin could have a significant impact on increasing the amount of fuels available from biomass over the carbohydrate fraction alone by up to 20%. Thus lignin could be used for combustion, gasification, pyrolysis, and hydroliquefaction.

Macromolecules are primarily applications targeted at dispersants, emulsifiers, binders, and sequestrants. In fact, nearly three-quarters of commercial lignin products are believed to lie within these applications. Other, smaller applications include adhesives and fillers. Generally, lignin is used in these applications with little or no modification other than sulfonation or thiohydroxymethylation.

Lignin's commercial applications as a polyelectrolyte and polymeric material could be greatly expanded into higher valued macromonomer and polymer applications with the development of appropriate chemical and catalytic processes. This will require a better fundamental understanding of lignin reactivity relative to the source and the process by which it was isolated. This knowledge will help lead to the development of appropriate technology to modify, control, and amplify lignin's polyelectrolyte, chemical reactivity, including copolymerization, and compatibility properties with other monomers and polymers.

The main directions proposed to be developed to use lignin as raw material for bioproducts are the following (40).

1.5.3.16 Carbon Fibres

Lignin represents a potential low-cost source of carbon suitable for displacing synthetic polymers such as polyacrylonitrile (PAN) in the production of carbon fibre. Using lignin in the carbon fibre manufacturing process improves raw material availability, decreases raw material sensitivity to petroleum cost, and decreases environmental impacts. The goal of replacing steel panels with lightweight, yet strong, carbon fibre-reinforced plastics is to significantly reduce vehicle weight and improve fuel economy.

1.5.3.17 Polymer Modifiers

Polymer modifiers can be simple, low-cost fillers or may be high-value additives that improve various polymer physical or performance properties. Currently, lignin use concentrates on the former; future research should concentrate on the latter by creating technologies that improve polymer alloying, mutual solubility, cross-linking, and control of colour. Relevant technologies include predictable molecular weight control, facile introduction of reactive groups and polyelectrolytic functionality. Examples of reactive functionality might include the addition of ethyoxy, epoxy, vinyl, and carbonyl moieties. Molecular weight control could include polydispersity, depolymerization, molecular weight increase, intermolecular cross-linking, or increasing phenolic functionality. The technical challenges surrounding polymer modifiers include understanding how to make these modifications economically, effectively, and predictively with lignins of different sources. The modified lignins need to be validated in a variety of high-value applications. Applications may include high-strength engineering plastics, heat-resistant polymers, under-the-hood uses, antibacterial surfaces, high-strength and formaldehyde-free adhesives, and light- and ultraviolet-resistant polymers. Another technology challenge for some of these applications is control of lignin colour.

1.5.3.18 Resins/Adhesives/Binders

Resins and adhesives offer a large opportunity, especially for formaldehyde-free applications. Formaldehyde is currently considered a carcinogen and its banish from consumer and packaging goods and building products is highly likely in the near term. Technical needs and challenges for lignin in this area centre on effective, practical means for molecular weight and viscosity control, functional group enhancement (e.g., carbonylation, carboxylation, amination, epoxidation, and de-etherification, i.e., methoxy conversion to phenolic) to improve oxidative and thermal stability, provide consistent mechanical processing properties, control lignin colour, and provide precise control of cure kinetics. Product consistency in these application targets will also be a technical challenge.

Lignin is the only renewable source of an important and high-volume class of compounds—the aromatics (BTX, phenol, terephthalic acid). It is easy to conclude that direct and efficient conversion of lignin to discrete molecules or classes of high-volume, low-molecular weight aromatic molecules is an attractive goal. As petroleum resources diminish and prices increase, this goal is very desirable, and is perhaps the most challenging and complex of the lignin technology barriers. Bringing high-volume aromatics efficiently from a material as structurally complex and diverse as lignin becomes a challenging but viable long-term opportunity.

1.5.3.19 BTX Chemicals

When considering the different kinds of structural motifs present in lignin from various biomass sources, it is easy to conclude that technology developments may lead to two sets of compound classes. One of these, which would arise from aggressive (i.e., non-selective) depolymerization in the form of C–C and C–O bond rupture, is aromatics in the form of BTX plus phenol and includes aliphatics in the form of C1 to C3 fractions. Of course, there is the possibility of forming some C6-C7 cycloaliphatics as well.

1.5.3.20 Monomeric Lignin Molecules

Another view is that very selective depolymerization, also invoking C–C and C–O bond rupture, could yield a plethora of complex aromatics that are difficult to make via conventional petrochemical route. These compounds are closely related to the basic building blocks of lignin and may be highly desirable if they can be produced in reasonable commercial quantity. However, two barriers would need to be overcome. First, technology would need to be developed that would allow highly selective bond-scission to capture the monomeric lignin building block structures. Development of this technology will be more difficult than the more aggressive processes that would yield BTX or phenols. Second, markets and applications for monomeric lignin building blocks would need to be developed. For the reasons listed above, this technology is longestterm and currently has unknown market pull for large-scale use. Since most of the chemical industry is used to single, pure-molecule raw materials, using mixtures of products in a chemical raw material feed, as would arise from lignin processing, constitutes a challenge. New/improved separation techniques for aromatic lignin monomers constitute a related challenge.

1.5.3.21 Low Molecular Weight By-products

In lignin processing, low-molecular weight aliphatics (C1-C3 etc.) also will be produced along with aromatics. Residual lights may include formic or acetic acids as well as aliphatics and olefins. Such low-molecular weight materials could be applied to syngas (reforming), alkylated gasoline or propane (LP) fuels. Alternatively, this material could be dehydrogenated to provide low-molecular weight olefins and residual lights.

1.5.3.22 Fermentation Products

Few fermentation routes are available today that use lignin as a nutrition source other than some routes that are of academic interest. In one such route, 2-keto-adipic acid is produced. Commercially viable fermentation technologies, possibly based on modifying 'white-rot' fungi, represent a higher-risk area of research.

1.5.3.23 Socio-economic Benefits

Renewable sources of biomass offer an answer to maintaining sustainable development of economically and ecologically attractive technology. The innovations in the development of materials from bioresources, the preservation of fossil-based raw materials, complete biological degradability, the reduction in the volume of garbage and compostability in the natural cycle, protection of the climate through reduction of carbon dioxide released, and the application possibilities of agricultural and aquacultural resources for the production of bio/green materials are some of the reasons why such materials have attracted the public interest.

We need a cultural revolution and we need to work together in synergy in order to enable our economies to work towards sustainability enabling people to live better with less. This is a structural and behavioural change that we need to instil in society if we really want to preserve our planet for a better more sustainable future.

The change of paradigm leading to a post petroleum and resource-efficient society cannot be only by science or technology pushed. It needs to be market driven and sustainable from an environmental and social perspective. The interaction of public authorities, farmers, waste management companies, and civil society will be also recommended.

Therefore, we appreciate that the use of biomass to obtain biofuels and bioproducts could have the following advantages: reduced dependence on imported fossil oil, reductions in greenhouse emissions, building on the existing innovation base to support new development, a bio-industry that is globally competitive, the development of processes that use biotechnology to reduce energy consumption, the use of renewable materials, jobs and wealth creation, the development of new, renewable materials, new markets for the agriculture and forestry sectors, including access to high-value markets, underpinning a sustainable rural economy and infrastructure, sustainable development along the supply chain from feedstocks to products and their end-of-life disposal.

1.6 CONCLUDING REMARKS

The use of renewable carbon sources in place of non-renewable carbon sources can have significant beneficial impact on the chemical industry. Shifting society's dependence away from petroleum to renewable biomass resources is generally viewed as an important contributor to the development of a sustainable industrial society and effective management of greenhouse gas emission. Due to its accessibility from different sources it is expected that biomass will become a significant part of the worldwide shift from a fossil-based economy to one that is biobased.

In view of changing world energy and bioproducts needs, a research and road map for the biorefinery of 21st century is vital. This biorefinery vision will contribute to sustainability not only by its inherent dependence on sustainable bioresources but also by recycling waste with the entire process becoming carbon neutral. By developing biorefinery concepts, the potential exists to explore higher value options for product streams, therefore increasing the productivity, profitability, efficiency, and sustainability of manufacturing plants. It leverages our knowledge in plant genetics, biochemistry, biotechnology, biomass chemistry, separation, and process engineering to have a positive impact on the economic, technical, and environmental well-being society. The goal of research is to modify plant species for use in sustainable biomass production. Interrelated plant traits such as higher yield, altered stature, resielience to biotic and abiotic challenge, and biomass composition will increase industrial crop value in terms of biofuels and biomaterials. The challenge is to weave these different strands of research into an integrated production strategy. The grand challenge for biomass production is to develop crops with a suite of desirable physical and chemical traits while increasing biomass yields by a factor of 2 or more. Advances in plant sciences and genetics are providing researchers with the tools to develop the next generation of agroenergy/material crops having increased yield and utility tailored for modern biorefinery operations.

Having in mind the chemical composition of biomass and the possibilities offered by biorefining, it was already demonstrated that there are a lot of products with energy and chemical value that could be obtained to substitute fossil raw materials and to assure a sustainable development.

REFERENCES

1. United Nations Official Document. *www.un.org. Source: United Nations. Sustainable Development Goals: 17 Goals to Transform our World*. Available at http://www.un.org/sustainabledevelopment/sustainable-development-goals/ Accessed Jan. 26, 2020; E. B. Barbier, J. C. Burgess, The sustainable development goals and the systems approach to sustainability, *Economics*, 11, 2017–2028 (2017), http://dx.doi.org/10.5018/economics-ejournal.ja.2017-28.

2. V. I. Popa, Biorefining and the pulp and paper industry, in *Pulp production and processing: From paper making to high-tech products*, p.1–33, Editor Valentin I. Popa, Smithers Rapra Technology Ltd, Shropshire, 2013.

3. National Renewable Energy Laboratory (NREL), http://www.nrel.gov/biomass/biorefinery.html.

4. C. I. Simionescu, V. Rusan, V. I. Popa, Options concerning phytomass valorification, *Cellulose Chemistry and Technology*, 21, 3–16 (1987).

5. V. I. Popa, Chapter 1, Biomass for fuels and biomaterials in *Biomass as Renewable Raw Material to Obtain Bioproducts of High-tech Value*, Editors: V. Popa, I. Volf, Elsevier, 1–37, 2018.

6. I. Volf, V. I. Popa, Chapter 4 Integrated processing of biomass resources for fine chemical obtaining: polyphenols, in *Biomass as Renewable Raw Material to Obtain Bioproducts of High-tech Value*, Editors: V. Popa, I. Volf, Elsevier, 113–160, 2018.

7. A. Koopmans, J. Koppejan, Agricultural and forest residues – generation utilization and availability, Paper presented at *the Regional Consultation on Modern Applications of Biomass Energy*, January, 6–10, 1997, Kuala Lumpur, Malaysia (see FAO, 1998).

8. F. Rosillo-Calle, P. de Groot, S. L. Hemstock, Chapter 3 Assessment methods for woody biomass supply, in *The Biomass Assessment Handbook, Bioenergy for a Sustainable Environment*, Editors: F. Rosillo-Calle, P. de Groot, S. L. Hemstock, J. Woods, Earthscan, London (2007).

9. V. I. Popa, Wood bark as valuable raw material for compounds with biological activity. *Celuloză şi Hârtie*, 64, 5–17 (2015).

10. E. Kantarelis, A. Zabaniotou, Valorization of cotton stalks by fast pyrolysis and fixed bed air gasification for syngas production as precursor of second generation biofuels and sustainable agriculture. *Bioresource Technology*, 100, 942–947 (2009).

11. M. Adla, K. Sheng, A. Gharibib, Technical assessment of bioenergy recovery from cotton stalks through anaerobic digestion process and the effects of inexpensive pre-treatments. *Applied Energy*, 93, 251–260 (2012).

12. F. Rosillo-Calle, P. de Groot, S. L. Hemstock. Chapter 4 Non-woody biomass and secondary fuels, in *The Biomass Assessment Handbook, Bioenergy for a sustainable Environment*, Editors: F. Rosillo-Calle, P. de Groot, S. L. Hemstock, J. Woods, Earthscan, London, (2007).

13. C. Galankis, Editor, *Food Waste Recovery: Processing Technologies and Industrial Technique* Academic Press, 412 p.. (2015).

14. A. Louwrier. Industrial products-return to carbohydrate-based industries, *Biotechnology and Applied Biochemistry*, 27, 1–8 (1998).

15. C. M. Drapcho, Nghiem PhuNhuan, T. H. Walker, *Biofuels: Engineering and Process Technology*, McGraw Hill, New York etc. Part 2, Biofuels, Chapter 4. Biofuel Feedstocks, 69–103 (2008).

16. C. Simionescu, M. Grigoras, A. Cernatescu-Asandei, G. Rozmarin, *Chimia lemnului din România: plopul si salcia (Wood Chemistry in Romania-poplar and Willow)*, Editura Academiei Republicii Socialiste România, p. 297 (1973).

17. V. I. Popa, Un model pentru valorificarea biomasei ca sursa de energie si produse chimice (A model for valorization of biomass as energy and chemical products), *Celuloză și Hârtie*, 65, 14–20 (2016).

18. S. M. Cardoso, L. G. Carvalho, P. J. Silva, M. S. Rodrigues, O. R. Pereira. L. Pereira, Bioproducts from seaweeds: A review with special focus on the Iberian Peninsula, *Current Organic Chemistry*, 18, 896–917 (2014).

19. C.I. Simionescu, V. Rusan, V. I. Popa, *Chimia algelor marine (Seaweeds Chemistry)*, Editura Academiei Republicii Socialiste România, p. 211 (1974).

20. C. H. Greene, M. E. Huntley, I. Archibald, L.N. Gerber, et al., Marine microalgae: climate, energy, and food security from sea, *Oceanography*, 29, 10–15, (2016). doi:10.5670/oceanog.2016.91

21. S. Reddy Medipally, F. Md. Yusoff, S. Banerjee, M. Shariff, Microalgae as sustainable renewable energy feedstock for biofuel production. *BioMed Research International*, Article ID 519513, 13 (2015). doi:10.1155/2015/519513

22. M. Frac, S. Jezierska-Tys, J. Tys, Microalgae for biofuels production and environmental applications: A review. *African Journal of Biotechnology*, 9 (54), 9227–9236 (2010), 27 December, 2010, Available online at http://www.cademicjournals.org/AJB, ISSN 1684-5315© 2010 Academic Journals.

23. X. Zhang, J. Rong, H. Chen, C. He, Q. Wang, Current status and outlook in the application of microalgae in biodiesel production and environmental protection, *Frontiers in Energy Research Bioenergy and Biofuels*, 2 Article 32,1–15 (2014), doi:10.3389/fenrg.2014.00032

24. P. T. Anastas, John C. Warner. *Green Chemistry: Theory and Practice*, 148, Oxford University Press (2000). ISBN-10: 9780198506980, ISBN-13: 978-0198506980.

25. B. A. Stout, Agricultural biomass for fuel, *Experientia*, 38, 145–151(1982).

26. N. Canabarro, J. F. Soares, C. G. Anchieta, C. S. Kelling, M. A. Mazutti, Thermochemical processes for biofuels production from biomass, *Sustainable Chemical Processes*, 1, 22 (2013), doi:10.1186/2043-7129-1-22.

27. M. Verma, S. Godbout, S. K. Brar, O. Solomatnikova, S. P. Lemay, J. P. Larouche, Biofuels production from biomass by thermochemical conversion technologies, *International Journal of Chemical Engineering*, Article ID 542426, 18 (2012), doi:10.1155/2012/542426.

28. J. N. Meegoda , B. Li, K. Patel, L. B. Wang, A review of the processes, parameters, and optimization of anaerobic digestion, *International Journal of Environmental Research and Public Health*, 15, 2224–2240 (2018); doi:10.3390/ijerph15102224.

29. A. J. Ragauskas, C. K. Williams, B. H. Davison, G. Bitovsek, J. Caurney et al, The path forward for biofuels and biomaterials, *Science*, 311, 484–489 (2006).

30. G. Eggleston, Chapter 1 Future sustainability of the sugar and sugar–ethanol industries, In *Sustainability of the Sugar and Sugar Ethanol Industries*; Eggleston, G.; ACS Symposium Series; American Chemical Society: Washington, DC, 1–19, (2010).

31. C. I. Santos, C. C. Silva, S. I. Mussatto, P. Osseweijer, L. A. M. van der Wielen, J. A. Posada, Integrated 1st and 2nd generation sugarcane bio-refinery for jet fuel production in Brazil: Techno-economic and greenhouse gas emissions assessment, *Renewable Energy*,129, Part B, 733–747 (2018) doi:10.1016/j.renene.2017.05.011

32. Y. Pu, D. Zhang, P. M. Singh, A. J. Ragauskas, The new forestry biofuels sector *Biofuels, Bioproducts and Biorefining*,2, 58–73 (2008); doi:10.1002/bb

33. J.-H. Ng, H. K. Ng, S. Gan, Recent trends in policies, socio-economy and future directions of the biodiesel industry, *Clean Technology and Environmental Policy*,12, 213–238 (2010). doi:10.1007/s10098-009-0235-2

34. C. Dragisic, E. Ashkenazi, L. Bede, M. Honzák, T. Killeen, A. Paglia, B. Semroc, C. Savy, Tools and methodologies to support more sustainable biofuel feedstock production *Journal of Industrial Microbiology and Biotechnology*, 38, 371–374 (2011). doi:10.1007/s10295-010-0858-7.

35. M. Guiry, How many species of algae are there? *Journal of Phycology* 48, 1057–1063 (2012). doi:10.1111/j.1529-8817.2012.01222.x. 2.

36. Y. Yan,· J. C. Liao, Engineering metabolic systems for production of advanced fuels, *Journal of Industrial Microbiology and Biotechnology*, 36, 471–479 (2009). doi:10.1007/s10295-009-0532-0

37. V. Rusan, V. I. Popa, Interferences between chemical components of wood and synthetic copolymers, *Cellulose Chemistry and Technology*, 26, 591–606 (1992).

38. T. Werpy, G. Petersen, *Top Value Added Chemicals from Biomass. Volume I-Results of Screening for Potential Candidates from Sugars and Synthesis Gas* (2004). https://dx.doi.org/10.2172/15008859.https://www.nrel.gov/docs/fy04osti/35523.pdf.

39. V. I. Popa, Lignin and sustainable development, *Cellulose Chemistry and Technology*, 41, 591–593 (2007).

40. J. J. Bozell, J.E. Holladay, D. Johnson, J.F. White, *Top Value Added Chemicals from Biomass. Volume II-Results of Screening for potential Candidates from Biorefinery Lignin*, Pacific Northwest National Laboratory, Richland, WA, PNNL-1683, 2007. https://dx.doi.org/10.2172/921839/https://www1,eere.enewrgy.gov/bioenergy/pdfs/pnnl-16983.pdf.

2 Selectively Transformation of Lignin into Value-added Chemicals

Xiaojun Shen and Runcang Sun

CONTENTS

2.1 INTRODUCTION

The transformation of biomass into chemicals has a long history. At the beginning of the 20th century, a large number of industrial products were manufactured from wood-based resources and crops (1). During the past 100 years, the interest in the utilization of bio-based resources was inversely correlated with the oil supply and price (2). However, petroleum resources are limited. Developing inexhaustible renewable natural resource as the alternative is urgently desired in the long term, which is the necessary requirement for sustainable development (2, 3).

As one of the most abundant renewable carbon resource on earth, lignocellulosic biomass attracts most of the interest (4). Lignocellulose is mainly composed of the semi-crystalline polysaccharide cellulose, amorphous multicomponent polysaccharide hemicelluloses, and amorphous aromatic polymer lignin (5, 6). Lignin in plant fills in the gaps between cellulose and hemicelluloses, which acts like adhesives that keep the lignocellulose matrix together and add strength and rigidity to cell walls (6, 7). The lignin content depends on the plant taxonomy, about 30% in conifer wood, 20–25% in broadleaf wood, and 10–15% in grass, respectively (6, 8). A number of technologies have been developed for lignin fractionation from lignocellulosic biomass (9), and pulp industry yields large amount of Kraft lignin annually. Among the three components of lignocellulose, lignin is the most complex, and its utilization for value-added chemicals is a severer challenge compared to cellulose and hemicelluloses (10). However, great attentions should be paid to lignin, because lignin is the only abundant renewable natural aromatic biopolymer, which can be a sustainable candidate feedstock for aromatic chemicals (6, 11). Although some aliphatic and cycloaliphatic compounds are industrially available from cellulose (12), starch (12), or triglycerides (13), many key chemicals are aromatic compounds and ultimately derived from petroleum (14). However, currently almost 98 wt.% of lignin as waste stream is combusted to produce heat and power for the biorefinery in the pulp and paper industry (15, 16). It is not only harmful to the environment but also a waste of resource (17). The rest lignin is used in low-value commercial applications, such as a low-grade fuel for heat and power (18), water reducing dispersing (19), emulsifying agents (20), phenolic resin (21), and polyurethane foams (22). The potential of lignin is far from being fully developed. The good news is that more and more researchers are recognizing the great value of lignin, and researches on lignin conversion into platform chemicals and alternative fuels are booming in the last decade (Figure 2.1).

The basic premise for lignin utilization is to depolymerize it into small molecules, and there have been several systematic reviews (6, 23, 24). The lignin deconstruction strategies can be broadly classified into acidolysis (25–28), base depolymerization (29–32), pyrolysis (33–36), gasification (37–39), oxidation depolymerization (40), hydrotreating (including hydrogenation (41–43), hydrodeoxygenation (44–46) and hydrogenolysis (47–52)), and the combined processes (53–56). Based on the types of reactions involved, the above strategies can be simplified into three modes: (1) oxidizing reaction (nitrobenzene, metal oxides, O_2, air, H_2O_2, or other peroxide was used as an oxidant), (2) reducing reaction (H_2, hydride, or hydrogen donor solvent was used a reductant), (3) non-redox reaction. Oxidation reaction generally occurs at relatively low temperatures (below 250°C) (6, 24). In the depolymerization system, oxidizing reaction involves the cleavage of the aryl ether bonds (β-O-4, 4-O-5, α-O-4), C–C bonds (β-5, β-β, β-1 or 5-5), or other linkages in the lignin. Meanwhile, oxidation of the functional groups (hydroxyl, aromatic ring, or others) in lignin also occurs and produces polyfunctional aromatic compounds. It favours the production of various target fine chemicals, such as aromatic alcohols, aldehydes, and acids. However, the oxidation process cannot be easily controlled, and the alcohols, aldehydes, and acids were usually formed simultaneously in one-pot, which increased the cost of products separation in industry. Reducing reaction involves thermal

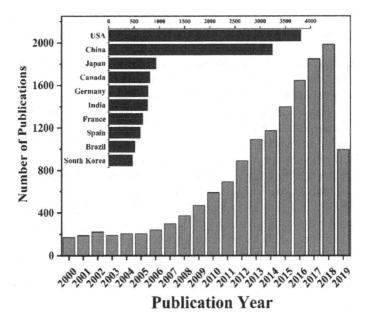

FIGURE 2.1 Publications on lignin transformation from the beginning of this century (derived from Web of Science using 'lignin utilization/ transformation/ conversion/depolymerization' as the key words).

hydroprocessing reductions of lignin in hydrogen atmosphere at high temperatures ranging from 100 to 350°C (6, 57, 58), including hydrogenolysis, hydroalkylation, hydrodeoxygenation, hydrogenation, and integrated hydrogen-related reactions (6). In reducing depolymerization system, not only the cleavage of ether linkages in lignin occurs, but also the functional groups (hydroxyl, aldehyde and carboxyl group, carbon–carbon double bonds, and the aromatic rings) in lignin can be reduced. However, as compared to oxidizing reaction, the selectivity of these reactions is easier to be controlled, which is benefit for highly selective target fine chemicals. It is by far one of the most popular and efficient technologies applied in depolymerization of lignin into oligomers, phenols, and further upgrading productions such as aromatic and aliphatic hydrocarbons. Non-redox reaction mainly contains acidolysis, base depolymerization, pyrolysis, and gasification. Both acid (typically at 0–200°C) and base (100–300°C) can catalytically cleavage the ether linkage and C–C bond in lignin into low depolymerized lignin and monomeric phenols. Acid environment is liable to condensation reaction of the lignin fragments, which suppress further depolymerization of lignin, while base can efficiently break ether linkage and C–C bond to form various depolymerized productions. Pyrolysis (typically at 450–700°C) is a process that directly produces bio-oil from lignocellulosic biomass. Gasification involves two different commercial processes: Fischer–Tropsch synthesis and methanol/dimethyl ether synthesis to produce syngas (hydrogen and carbon monoxide) from a range of real lignin feedstocks.

Although much effort has been made for transformation of lignin into value-added chemicals, almost all of the current technologies provide a mixture of numerous products, such as oligomers, phenols, aldehydes, ketones, acids, esters et al. Isolating each of these compounds is currently not possible and even less economically viable (14). Even if harvesting the molecules of interest, the separation and purification process can also be extremely energy intensive, and sometimes it is challenging (59). For example, the separation of vanillin from lignin oxidation mixture is rather complex, which can generate huge amount of caustic soda wastes (160 kg/kg vanillin) (60)! Separation of vanillin and syringaldehyde from the lignin oxidation products was a subject of intense researches, and the separation problem has been preventing the preparation of syringaldehyde from lignin for decades (61). The problem of complex mixed products may be one of the main limiting factors for the scaling high-value utilization of lignin in industry. The realization of lignin industrialization presupposes not only high conversion of lignin but also high selectivity of depolymerization products, which can reduce energy consumption and cost. Thus, it is desperately desired to develop highly selective approaches to lignin transformation and avoid the formation of side products as less as possible at the very beginning. Some researchers have been aware of the significance of 'selectivity', and great efforts are being made to harvest product with high selectivity from lignin. However, selective production of target fine chemicals or platform chemicals from lignin is a severe challenge because of the complex structure. Up to now, there is only one report that claimed to achieve one single product from lignin (62). The term 'selective' in other literatures about lignin utilization actually means that a specific class of chemicals sharing the same functional groups is obtained. Here in this minireview, we follow this convention. Although there have been numerous comprehensive reviews on lignin valorization, selective transformation of lignin has not been focused. Thus, we prepare this minireview to highlight this key issue.

The thesis of this chapter focused on the selective transformation of lignin into value-added chemicals. Contributions from model compounds are not included, considering the selectivity differs a lot from real lignin due to its much more complicated structures. For the researches on model molecules, the readers are suggested to consult other reviews (26, 63). In the first section, we give a brief introduction to the structure of lignin. Then in the second section, we summarize typical researches focused on selectivity utilization of lignin to produce target chemicals, organized by the types of products (phenols, phenolic aldehydes, carboxylic acids, alkanes, arenes, and other compounds). Finally, we make a brief discussion on the challenges and opportunities for selective transformation of lignin.

2.2 STRUCTURE OF LIGNIN

In nature, lignin presents a complex three-dimensional amorphous structure, arising from enzymatic dehydrogenative polymerization of three different cinnamyl alcohol monomers: sinapyl alcohol (syringyl (S) units), coniferyl alcohol (guaiacyl (G) units), and p-coumaryl alcohol (p-hydroxyphenyl (H) units) (64), as shown in Figure 2.2. The content and the ratio of monomers in lignin vary in different types of plants (65). Generally, lignin from conifer wood has more guaiacyl (G) units than the other two

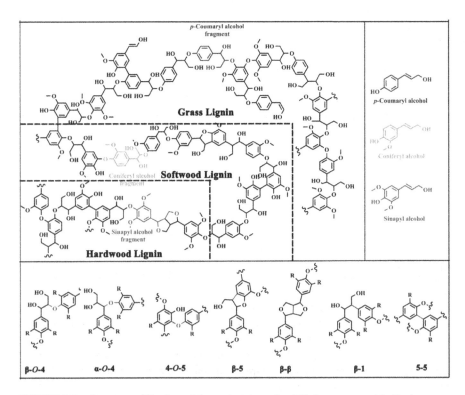

FIGURE 2.2 Structure of lignin and the main inter-subunit linkages present in lignin.

TABLE 2.1
Abundance of the Three Basic Lignin Units in Different Types of Plants (6)

Monolignol	Grass	Conifer Wood	Broadleaf Wood
p-Coumaryl alcohol (H)	10–25%	0.5–3.4%	Trace
Coniferyl alcohol (G)	25–50%	90–95%	25–50%
Sinapyl alcohol (S)	25–50%	0–1%	50–75%

monolignols, broadleaf wood only contains guaiacyl (G) and syringyl (S) units, and grass lignin presents a mixture of all three monolignols (8). Table 2.1 shows the typical content of the three primary lignin units in different plants. Based on composition and abundance, lignin can be classified into three categories: G-lignin, G-S-lignin, and G-S-H-lignin (6). However, going deeper into the research work, researchers found that some other aromatic species also can involve in the formation of the lignin, which includes coniferaldehyde, sinapaldehyde, ferulic acid, 5-hydroxyconiferyl alcohol, and the above monolignols containing acylated groups (acetate, p-hydroxybenzoate, or p-coumarate) (66–68), as shown in Figure 2.3. One of the most significant obstacles to high-value utilization of lignin is its heterogeneity of structure, which depends on the biomass sources, wood type, and even some growing parameters (69).

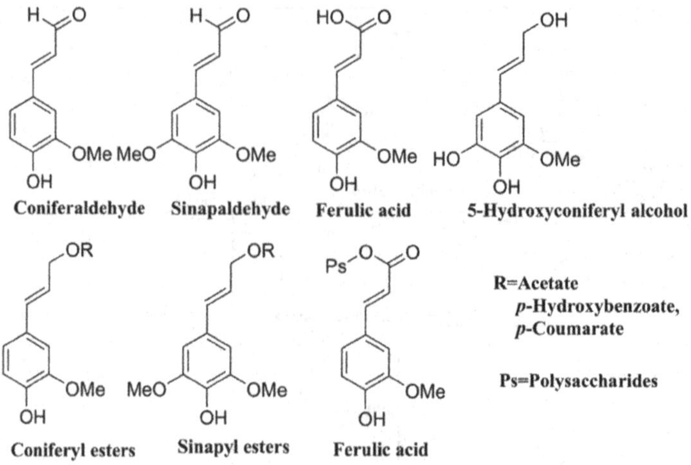

FIGURE 2.3 Other lignin precursors found in lignin in recent years (64).

These building units of lignin are predominantly bonded by a series of C–O and C–C linkages, such as β-O-4 (β-aryl ether), 4-O-5 (diaryl ether), α-O-4 (α-aryl ether), β-5 (phenylcoumaran), β–β (resinol), β-1 (spirodienone), or 5-5, to form a very complex polymer (21b,31b). Figure 2.2 presents the representative structures of these linkages in lignin, and Table 2.1 lists the typical content of the major linkages in different types of plants. Although the contents of these linkages are related to the type of lignocellulosic biomass, generally more than 50% of the linkages in lignin are ether linkages. Importantly, among these ether linkages, the β-O-4 (β-aryl ether) linkage is dominant in both softwood and hardwood, representing from *ca.* 43% (G-lignin) to *ca.* 65% (G-S-lignin) as indicated in Table 2.2 (6, 61, 70–72). Moreover, it is reported that proportion of resistant linkages (β-β, β-5, β-1 and 5-5) in softwood lignin is higher than that in hardwood lignin, because the C_5 positions of guaiacyl (G) units in

---

TABLE 2.2

Content Distribution and Bond Dissociation Energies of the Lignin Linkages in Hardwood and Softwood (6, 61, 70–72)

| Linkage | Number/100 ppu | | Bond Dissociation Energies (kcal/mol) |
	Softwood (%)	Hardwood (%)	
β-O-4	43–50	50–65	56.54–72.30
α-O-4	6–8	4–8	48.45–57.28
4-O-5	4	6–7	77.74–82.54
β-β	2–4	3–7	–
β-5	9–12	4–6	125.2–127.6
β-1	3–7	5–7	64.7–165.80
5-5	10–25	4–10	114.9–118.4
Others	16	7–8	–

softwood lignin tends to commit coupling reaction to form resistant linkages whereas those in syringyl (S) unit is sterically inhibited due to addition of methoxy group (6, 61). Therefore, G-type lignin (softwood lignin) contains much higher condensation degree than hardwood lignin. Besides, due to steric effect, the additional methoxy groups in syringyl (S) unit could inhibit formation of other linkages, leading to more linear structures in the hardwood lignin as compared with softwood lignin. Grass lignin involves other noncanonical monolignols (Figure 2.3) in its biosynthesis process and its special structure (70). The representative model structure of lignin is illustrated in Figure 2.2. It should be noted that these model structures of lignin are not the real structure in raw lignocellulosic biomass, but it can represent the linkages between adjacent building units, as well as the contents in lignin.

2.3 SELECTIVE TRANSFORMATION OF LIGNIN

The basis of lignin transformation is the cleavage of the linkages (intermolecular C–O and C–C bonds) between the monomers. As shown in Table 2.1, the bond energies of C–O bonds are much lower than those of the C–C bonds. Thus, C–O bond is considered as cleavable bond, while C–C bonds are non-cleavable linkage (73). Fortunately, the cleavable C–O bonds, especially β-O-4 type, are the major linkage in lignin (Table 2.2), which indicates that lignin can be depolymerized into small molecules under appropriate conditions. Numerous methods have been proposed for the depolymerization of lignin, and various products can be obtained, heavily depending on the treatment methods, reaction conditions, and lignin source (24). Phenols, aldehydes, carboxylic acids, alkanes, and alkenes can be derived from lignin with high selectivity.

2.3.1 PHENOLS

Phenols were important platform chemicals with applications in many fields, such as biofuel (74–76) and key precursors of plastics/cosmetics/pharmaceutics (77). Substituted phenols are the most commonly reported derivatives from lignin. Numerous methods have been developed for the selective transformation of lignin into phenols.

2.3.1.1 Hydrogenolysis

Hydrogenolysis is one of the most employed methods for the cleavage of ether bonds (78–83). The most widely used catalysts are supported metal (Pt, Ru, Pd, Rh) particles or metal organic complex (78). Although the cleavage of different types of ether bonds can be well controlled in model compounds, there are only a few reports on hydrogenolysis of lignin into phenols with a high selectivity. For example, Wang, Zhang, and co-workers (77) achieved the selective hydrogenolysis of Kraft lignin into phenols over Pd/C using choline-derived ionic liquids as the reaction media. Under optimum reaction condition, the conversion of Kraft lignin and the selectivity to phenol (PL) and catechol (COL) reached 20.3%, 18.4%, and 18.1%, respectively (77). One of the challenges to obtain phenols from lignin by hydrogenolysis is that the aromatic ring also can be hydrogenated. Although noble metals have demonstrated efficient catalytic hydrogenation activity, it is sometimes difficult to achieve high selectivity. As alternatives, non-noble metal catalysts are attracting researchers' interest since it is easier to control their selectivity. For example, the hydrogenolysis product of lignin over Ni-based catalyst can be significantly tuned by the solvent. Rinaldi et al. (84) found that the Lewis basicity of the solvent is an important parameter in the reaction performance with Raney Ni. Basic solvents can prevent the hydrogenation of the aromatic ring while have no effect on the hydrogenolysis of the ether bond, resulting in a high selectivity to phenol products. The hydrogenolysis of lignin in methanol was found to produce mostly phenols over Raney Ni. Cu-doped porous metal oxides were also found to be efficient for promoting the selective hydrogenolysis of lignin (48). Temperature affected the products significantly, and C_9 catechols were obtained with high selectivity at 140–220°C. Especially, 4-(3-hydroxypropyl)-catechol was the major product and could be isolated in a high purity at 140°C.

A commonly existing problem in lignin depolymerization is that the depolymerized fragments can also condense to form non-cleavable C–C bonds and form insoluble materials (85), which hampers the conversion of lignin into well-defined aromatic chemicals (26). Luterbacher et al. (47, 86) put forward a solution to facilitate lignin monomer production by pretreating the lignin with aldehyde, which prevented lignin condensation by forming 1,3-dioxane structures with lignin side-chain hydroxyl groups. Via this pretreatment, the soluble lignin fraction can be converted to guaiacyl and syringyl monomers at near theoretical yields during subsequent hydrogenolysis. The yield of the monomers reached up to 47 mole% and 78 mole% for Klason lignin of beech and high-syringyl transgenic poplar, respectively. In addition, Ralph et al. (87) discovered that C-lignin from vanilla seed coat can maintain its native structure after treatment under even strongly acidic conditions, which prevent condensation and the generation of undesired new C–C bonds. More important, C-lignin can be completely depolymerized by a hydrogenolytic method to produce simple monomeric catechols near-quantitatively and, by selecting the catalyst, with a single monomer accounting for 90% of the monomer product.

As lignin is symbiotic with polysaccharide (cellulose and hemicelluloses) in nature, researchers also make efforts to directly utilize lignin using lignocellulose or even unpretreated wood as starting material (73, 88–90). This alternative strategy is known as lignin-first approach (91), which is considered to have the potential to yield nearly theoretical amounts of phenolic monomers by performing solvolytic delignification and lignin depolymerization (88). For example, Sels et al. (92) presented a catalytic lignocellulose biorefinery process, where birch sawdust was efficiently delignified through simultaneous solvolysis and catalytic hydrogenolysis of Ru/C in methanol. The yield of phenolic monomers was above 50% in the lignin oil, mainly 4-n-propylguaiacol and 4-n-propylsyringol. Hensen and co-workers (91) reported an approach to efficiently release lignin fragments from lignocellulose by Lewis acid metal triflates. Combined with metal-catalysed (Pd/C) hydrogenolysis, the lignin in lignocellulose can be converted to C9 phenol derivatives in 46 wt.% yield, leaving cellulose largely untouched. Ni-based catalysts are also widely used in the lignin-first strategy. Abu-Omar and co-workers (93) reported the catalytic conversion of lignin in *Miscanthus* over Ni/C catalyst. By modification of the reaction conditions (temperature and pressure of H_2), saturated or unsaturated branched products can be obtained selectively. Optimal conditions gave 69% yield of phenol derivatives. Zhang et al. developed a carbon supported Ni-W_2C catalyst which can directly catalyse the hydrogenolysis of raw woody biomass into two groups of chemicals (51). The reaction was conducted at 235°C and 6 MPa H_2 for 4 h in water media. Cellulose and hemicelluloses were converted to ethylene glycol and other diols, while lignin can be selectively converted into phenols with a total yield of 46.5% (mainly guaiacylpropanol, syringylpropanol, guaiacylpropane, and syringylpropane). Recently, Sels et al. (88) investigated the reductive catalytic fractionation of wood over Ni/Al_2O_3, producing a lignin oil that contained over 40% phenolic monomers, of which 70% consisted of 4-n-propanolguaiacol and -syringol. The role of the Ni/Al_2O_3 catalyst was also well addressed. Bimetal catalysts were also employed, for example, Ni-Fe alloy catalyst supported on activated carbon was also used for hydrogenolysis of lignin, which achieved 23.2 wt.% total yield of monomers at 225°C under 2 MPa H_2, mainly propylguaiacol and propylsyringol (88).

2.3.1.2 Acidolysis

Acidolysis is one of the most prevalent methods for the fractionation of lignocellulose into its main components (24). Both Brønsted acid and Lewis acid can cleavage the C–O bond in lignin. For example, Stahl et al. (27) proposed a method for the depolymerization of oxidized lignin under mild conditions (110°C) in aqueous formic acid. More than 52% of the original oxidized lignin was converted to well-defined phenol derivatives, among which syringyl and guaiacyl-derived diketones are two of the major depolymerization products (19.8%). However, this method is not as efficient for normal unoxidized lignin, and the yield of low-molecular-mass aromatics was only 7.2 wt.%. Catalytic amount of appropriate Lewis acid can be effective for lignin depolymerization. Metal triflates were found to be an excellent promoter. For example, Bruijnincx and co-workers reported a route to depolymerize lignin by a combination of Lewis acid metal triflates and rhodium complexes. The selectivity towards either 4-methylphenols or 4-(1-propenyl)phenols can be controlled by varying the amount and strength of the Lewis acid catalyst (26). Barta and co-workers (26, 94) also proved that Fe(OTf)$_3$ can effectively catalyse the conversion of lignin into phenolic C_2-acetal products.

The recondensation phenomenon is also severe in acidolysis process. Barta and co-workers (95) revealed the cause of this problem is that the aldehyde intermediates are very reactive and easily form C–C bond with other species. They proposed a strategy to increase the yield of aromatic monomers by in situ conversion of reactive intermediates in the acid-catalysed depolymerization of lignin (95). Three specific methods were developed, i.e., in situ catalytic hydrogenation, in situ catalytic decarbonylation, and in situ conversion of the aldehyde intermediates to acetals. The efficiency and mechanism was verified by model compounds, and the methods were demonstrated to be viable for real lignin (95).

2.3.1.3 Non-hydrogen Reductive Depolymerization

Strictly, hydrogenolysis also belongs to reductive depolymerization. Since it is extremely important, we have discussed it independently, and here in this section we just review selective reductive depolymerization of lignin using non-hydrogen reductant. It has been demonstrated in previous paragraphs that Ni-based catalyst is efficient and selective for hydrogenolysis of ether bonds (78, 80). Actually, it can also promote the non-hydrogen reductive depolymerization of lignin and has been widely used. For example, Luque and Labid proposed an approach to reductive depolymerize organosolv lignin over Ni/AlSBA-15 catalyst using formic acid as the reductive reagent as well as the solvent under microwave irradiation. It yielded simple phenolic monomeric products, such as desaspidinol, syringaldehyde, and syringol (96). Alcohols are used frequently as solvent in lignin treatment, and they can act as hydrogen source at appropriate reaction condition. For example, Wang and Xu et al. (49) showed that Ni/C catalyst can selectively promote the transformation of lignin into monomeric phenols, propylguaiacol, and propylsyringol. The selectivity of the phenol products can reach up to more than 90% and 50% conversion of birch wood lignin. Alcohols, such as methanol, ethanol, and ethylene glycol are suitable solvents for the depolymerization, and it was demonstrated that alcohol played dual role in the cleavage of the linkage of lignin. It acted as the nucleophilic reagent for C–O–C cleavage in alcoholysis reaction and functioned as hydrogen source in the reduction cleavage steps (49). The reaction mechanism was shown in Figure 2.4. Besides organic reductant, metal Zn was also used for selectively reductive depolymerization of lignin. Westwood et al. (97) developed a one-pot method which involves an initial catalytic oxidation (using DDQ and O_2) of the β-O-4 linkages in lignin followed by zinc-mediated cleavage of the C–O–Aryl bond. Application of this depolymerization method to birch lignin followed by extraction and purification gave isolated phenolic monomers in 6.0 wt.% yield, among which 80% is 3-hydroxy-1-(4-hydroxy-3,5-dimethoxyphenyl) propan-1-one.

In general, harsh reaction conditions (high temperature, high pressure, strong acid, or base) are required for the cleavage of lignin, which lead to problematic condensation of the depolymerized fragment via C–C bond formation (98). In 2015, Cantat and co-workers (98) reported a route to reductive depolymerization of lignin under ambient conditions using hydrosilanes as reductants and $B(C_6F_5)_3$ as a Lewis acid catalyst. This is the first example of metal-free catalysis for the reductive depolymerization of lignin. Different lignin preparations derived from 15 gymnosperms and angiosperms species were successfully depolymerized using this method, which yielded a narrow distribution of phenol derivatives. The isolated yield of the phenol monomers (e.g., 5-(3-hydroxypropyl)benzene-1,2,3-triol and 4-propylbenzene-1,2-diol) were 7–24 wt.% from lignin and 0.5–2.4 wt.% from wood. Interestingly, by tuning the catalytic conditions and selecting the wood source and the lignin extraction method, several different aromatic products can be isolated selectively from wood (98). Together with the work of Westwood's, these two contributions are the rare examples that lignin is depolymerized under mild conditions. However, the reagent in the reactions are a bit expensive, more economically viable approach are yet to be developed.

2.3.1.4 Other Methods

Other methods such as thermolysis and solvolysis are also developed for selective transformation of lignin into phenols. For example, Li and Hu et al. (99) achieved the

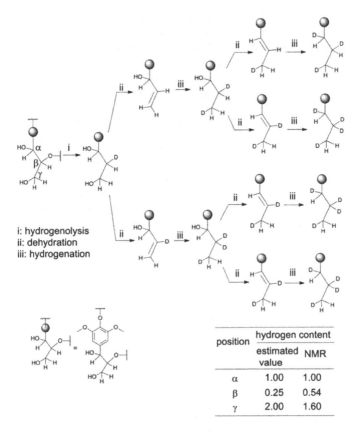

FIGURE 2.4 Mechanism of the formation of monomeric phenols by hydrotreating lignin using alcohols as hydrogen source over Ni/C catalyst (49).

selective conversion of lignin in corncob residue to monophenols via a two-step process without significant degradation of cellulose. Lignin was thermotreated in THF at 300°C, and the yield of total monophenols reached up to 24.3 wt.%, among which 86.8% were the predominant products, 4-ethylphenol (10.5 wt.%), 2,6-dimethoxyphenol (6.6 wt.%), and 4-ethylguaiacol (4.0 wt.%). The authors also treated lignin in *Phyllostachys heterocycla cv. pubescens* in an ethanol solvothermal system at 220°C, and 45.3% of the lignin was degraded for 2 h. A maximum yield of 10.6 wt.% of 4-ethyl phenols was obtained under optimized conditions (100). Subsequently, they extended the solvent to Na_2CO_3-H_2O-THF solution, and the total yield of monophenols reached up to 26.9 wt.%. Therein, the reaction mechanism of the cleavage of inter- and intramolecular linkages was investigated in details (74). Chornet and coworkers (59) performed hydrolytic based-catalysed depolymerization of a steam exploded aspen lignin; the results showed that the yield of the monomers is 28 wt.%, and the selectivity of phenol derivatives is 89.4%. The authors also investigated the separation methods of these monomers.

Oxidation is also an important method for ether bond cleavage (40, 101, 102). However, due to over-oxidation, it is challenging to get phenol derivatives with high selectivity. For example, Zhang et al. (103) reported a highly efficient lignin-to-monomeric phenolic compounds conversion method based on peracetic acid (PAA) treatment. Diluted acid pretreated corn stover lignin and steam exploded spruce lignin were treated with this method, and the highest yield of phenolic compounds reached up to 47%. However, the 'phenolic compounds' contain quantities of acid products, such as p-hydroxybenzoic acid, vanillic acid, syringic acid, and 3,4-dihydroxybenzoic acid. Nevertheless, selective oxidation of the β-O-4 linkages in lignin has been demonstrated to be feasible (104), which weakens the C–O and C–C bond energy (101). The resulted oxidized lignin is easier to be depolymerized with other method. Thus, a two-step oxidation depolymerization process is an efficient approach to the selective transformation of lignin, such as the work of Stahl's group (27) and Westwoods' group (97) which has been reviewed in previous paragraphs.

2.3.2 PHENOLIC ALDEHYDES

Phenolic aldehydes are also frequently reported as major depolymerization products of lignin, which are mainly derived from p-hydroxyphenyl (H), guaiacyl (G), or syringyl (S) units (14).

The most important phenolic aldehyde derived from lignin is vanillin, which is currently one of the only molecular phenolic compounds manufactured on an industrial scale from biomass (14). The lignin-to-vanillin process can be traced back to the year of 1936 (14), and nowadays 15% of the overall vanillin are produced from lignin (105). However, only lignin from the sulfite pulping process is used to prepare vanillin industrially (14), and the price of lignin-based vanillin is consistently ten times higher than that of guaiacol-based vanillin (106). Recent years, lignin-to-vanillin researches focused on extending the feedstock to all kinds of lignin, improving the selectivity and understanding the corresponding mechanism (60, 107). For more details about vanillin production from lignin, please refer to the corresponding reviews (14, 61).

Lignin is not naturally rich of carbonyl groups in raw lignocellulose. The carbonyl groups mainly come from the oxidation of the hydroxyls (63, 108) or the rearrangement of the enolic structures derived from hydrolysis (107). Thus, oxidation depolymerization and hydrolysis are the main methods to obtain phenolic aldehydes from lignin (109). Usually oxidation depolymerization also involves hydrolysis (60, 107). The production of vanillin from lignin represents the typical mechanism of the generation of aldehyde group (Figure 2.5) (110). Sun et al. (111) studied the oxidation degradation of wheat straw lignin by alkaline nitrobenzene and cupric compound (CuO, Cu(OH)$_2$, CuSO$_4$) in detail, and the major products were the phenolic aldehydes of vanillin and syringaldehyde using both of the oxidizing agents. With nitrobenzene and NaOH at 170°C, the selectivity of vanillin and syringaldehyde reached 39% and 37%, respectively. It has now been recognized that nitrobenzene is the oxidant that gives the highest yields of the depolymerized phenolic products (14). However, it is highly toxic and the reduction product (aniline) is difficult to separate. Thus, O$_2$ as a green oxidant is more attractive (61) and has been widely reported in

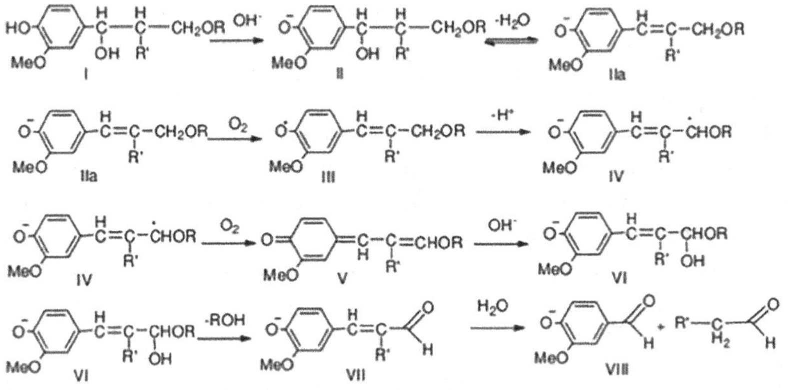

R and R' could be H, OH, or other linked ppu

FIGURE 2.5 Proposed mechanism for lignin oxidation to generate vanillin by Tarabanko et al. (61, 110).

recent years. For example, Chornet et al. (112) conducted the alkaline oxidation of a steam-explosion hardwood lignin in the presence of Cu^{2+} and Fe^{2+} using O_2 as the oxidant, and the yield of combined aldehydes (vanillin + syringaldehyde + hydroxy-benzaldehyde) can reach 14.6 wt.% with only 1.3 wt.% yield of ketone products. Zhu et al. (60) carried out the lignin oxidation reaction in phenol media with a catalyst of γ-Al_2O_3 supported ReO_x nanoparticles. The yield of vanillin was 7.3 wt.% with selectivity up to more than 95.0% under moderate reaction conditions (120°C and 2 bar O_2). Therein, phenol was used as reaction media to enhance the dissolution of lignin. The pulsed corona discharge (PCD) was applied to lignin oxidation in aqueous solution, which can remove lignin from wastewater and transfer it to aldehydes, and the selectivity to aldehydes can be tuned by oxygen concentration (113). Overall, it is a challenge to obtain complete selectivity to aldehydes by oxidation depolymerization. There's a balance between aldehydes production, aldehydes degradation, and competing reactions, which determines the maximum yield and selectivity (61). Because of the different reactivities of varies types of lignin and particular structures, the oxidation conditions should be carefully tuned to move the balance for the aldehydes production (61).

2.3.3 CARBOXYLIC ACIDS

Carboxylic acids are widely reported as the depolymerization product of lignin, but usually co-exist with other products, such as phenols and aldehydes (101, 114). They are often considered as 'side products', and few works reported them as the main product with a high selectivity. Lignin derived carboxylic acid are mainly aromatic acids. The most famous are vanillic acid, syringic acid, and their derivatives, which are the oxidation products of the monomers. However, vanillin and syringaldehyde are more fundamentally important, so the formation of the corresponding acids is not preferred and considered as over oxidation.

Recently, some researchers attempted to transform lignin into aliphatic acids, such as acetic, muconic, maleic, and succinic acids. The market for this kind of aliphatic acids is much larger than that for aromatic acids and aldehydes (115). They are important and highly valuable industrial chemicals and intermediates used in biopolymer, pharmaceutical, and food additives industries (116). Currently, they are mainly produced from petroleum-based feedstocks (117). Producing these important compounds from renewable lignin provides a new alternative and meets the requirement of sustainable development. Zhang et al. (118) reported the selective production of dicarboxylic acids through chalcopyrite ($CuFeS_2$)-catalysed oxidation of biorefinery lignin including diluted-acid corn stover lignin and steam-exploded spruce lignin using hydrogen peroxide (H_2O_2). The total yields of dicarboxylic acids (malonic acid, succinic acid, maleic acid, fumaric acid, and malic acid) were 14% and 11% for these two lignin samples, respectively, and the selectivity to the dicarboxylic acids was up to 91%. The reaction mechanism is shown in Figure 2.6, which involves the depolymerization of lignin, aromatic ring opening and the formation of

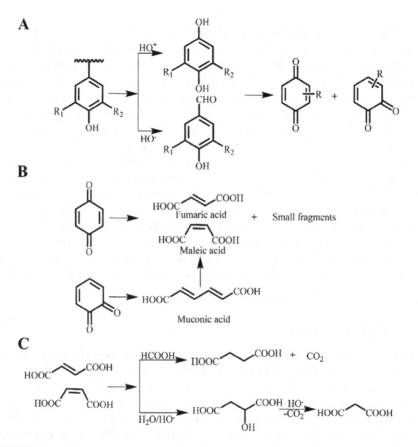

FIGURE 2.6 Mechanisms for (A) lignin depolymerization and aromatic nuclei oxidation, (B) aromatic ring cleavage, and (C) formation of carboxylic acids (118).

the final products (118). Patience and co-workers (115) developed a two stage gas–solid conversion method that transformed lignin into carboxylic acids. Lignin was thermo-oxidatively steam cracked in the first step to volatile compounds which contact a catalyst bed in the second step to produce carboxylic acids. The products are mostly C_4 carboxylic acids such as maleic acid and butyric acid, and the selectivity to different carboxylic acids can be turned by the metal catalyst as well as the support (115). Currently, Li et al. (119) devised a new catalytic system to promote selectively oxidative lignin in air for producing carboxylic acid derivatives diethyl maleate (DEM), an important, versatile, and widely used bulk chemical currently derived from fossil fuel, by using polyoxometalate ionic liquid of $[BSmim]CuPW_{12}O_{40}$ (POM-IL). In their work, DEM was formed at impressively high yield of 404.8 mg g^{-1} and selectivity of 72.7% over the POM-IL catalysts. This high catalytic activity is ascribed to a five-coordinated Cu^+ species through the formation of end-on dioxygen species in vacant orbitals which facilitates the selective oxidation of basic lignin aromatic units (phenylpropane C_9 units) (Figure 2.7).

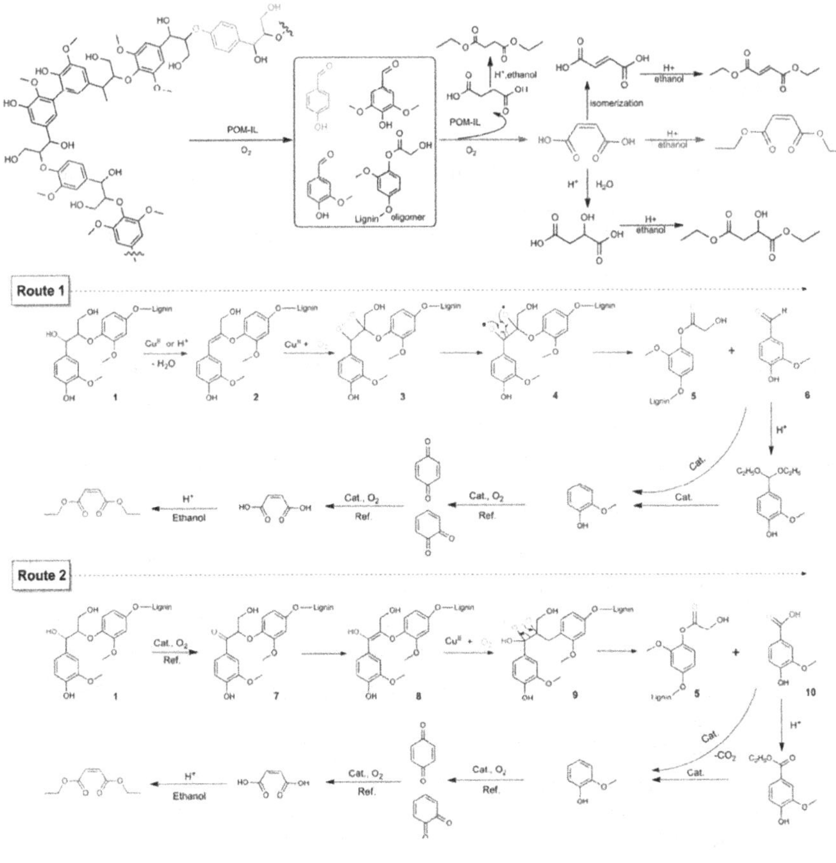

FIGURE 2.7 The pathway for lignin depolymerization and in situ aromatic ring oxidation to efficiently and selectively produce diethyl maleate (DEM) (119).

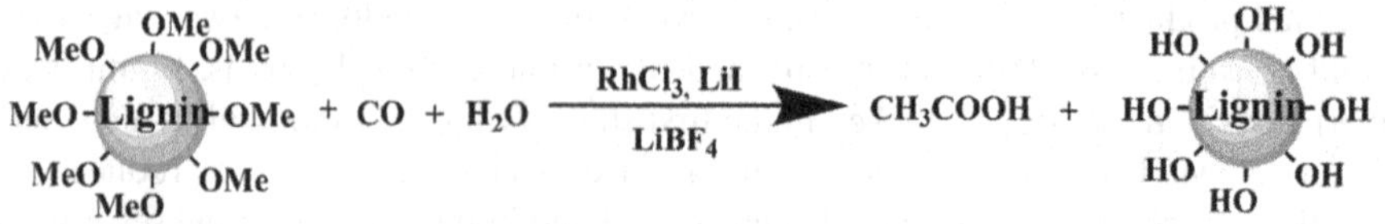

FIGURE 2.8   Highly selective transformation of lignin into acetic acid (62).

More recently, Han et al. (62) reported a striking result that acetic acid can be selectively obtained from lignin with 100% selectivity (Figure 2.8). This is the first time that a single product is harvested from lignin. They proposed a strategy that selectively transformed a specific group in lignin to value-added chemicals (62). Applying this strategy, they converted the methoxy groups in lignin into acetic acid as the single product in the liquid phase, while the molecular weight distribution of the residue lignin didn't change significantly according to the GPC trace. This method works for both Kraft and organosolv lignin, and the utilization of methoxy group reached up to 87.5%. NMR spectra and HPLC trace verifies that acetic acid is the only product.

2.3.4 ALKANES

One of the most significant value of lignin is that it can be converted into bio-oil which has the potential to be an alternative to fossil fuels (1). Despite extensive research, lignin still has a low commercial value and usually is combusted as a low-grade fuel (120). One of the main distinct differences between bio-oil and fossil fuels is their higher oxygen content (121), which has many drawbacks such as instability, causticity, high viscosity and low energy density (4, 76, 122). Therefore, it is desperate to develop methods to decrease oxygen content while increase the H/C ratio in the depolymerized liquid product for use of bio-fuel. Because of the similarities to hydrocarbons derived from petroleum, the lignin-derived alkanes could well be refined into drop-in fuels by conventional oil refinery processes. For the catalytic hydrodeoxygenation (HDO) of lignin-derived phenols or bio-oils, please refer to the existing reviews (76, 123–125). Here we focus on directly depolymerizing lignin into hydrocarbons with high selectivity.

To transform lignin into alkanes, all types of C–O bonds and aromatic rings should be via hydrogenolysis and hydrogenation, as presented in Figure 2.9. This requires a very efficient hydrogenolysis and hydrogenation catalyst, and noble metal catalysts are good candidates. Using a two-step process, Kou et al. (126) successfully degraded wood lignin into alkanes over noble metal (Pd/C, Ru/C, Pt/C) catalysed hydrogenolysis. Lignin was first depolymerized into monomers (200°C), which were further hydrogenated into alkanes at higher temperature (250°C) in the second step. This approach yielded about 42 wt.% C_8-C_9 alkanes, 10 wt.% C_{14}-C_{18} alkanes, and 11 wt.% methanol from lignin, which are close to the empirical maximum values. Ru/Al_2O_3 combined with acidic zeolite (H$^+$-Y) was also reported to catalyse the HDO of dilute alkali extracted corn stover lignin into jet fuel range hydrocarbons (primarily C_{12}-C_{18} cyclic structures) (127). Earth abundant Ni-based

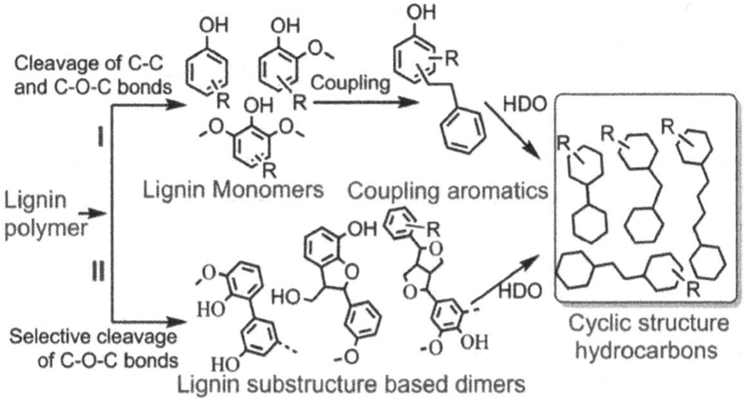

FIGURE 2.9 Routes of depolymerization and hydrodeoxygenation of lignin to cyclic structure hydrocarbons (127).

catalyst was also applied for the HDO of lignin. For example, Zhao et al. (128) developed an efficient method for one-pot hydrodeoxygenation of enzymatic lignin to C_6-C_9 cycloalkanes in liquid dodecane over supported Ni catalyst. The method enables 80 wt.% lignin conversion by using amorphous silica-alumina supported Ni particles at 300°C in the presence of 6 MPa H_2 with the yield of cycloalkanes approaching 50 wt.%, and almost no aromatic compounds generate. The solvent and the acid sites in the support are found to be crucial for the high selectivity. Ni/Al-SBA-15 is also demonstrated to be capable of hydrodeoxygenation organosolv lignin with selectivity to cycloalkanes higher than 99% (129). Although good achievement has been made on transformation of lignin into liquid alkanes, some researchers are devoted to directly converting raw lignocellulose into liquid fuels. Without chemical pretreatment of raw wood, this strategy has a great potential to tackle tremendous workload and energy saving (130). Xia, Chen and co-workers (130) designed a multifunctional Pt/NbOPO$_4$ catalyst that can promote the direct HDO of raw woods into liquid alkanes with mass yields up to 28.1wt.%. Cellulose, hemicelluloses and, more significantly, lignin fractions in the wood sawdust are simultaneously converted into hexane, pentane, and alkylcyclohexanes, respectively, which paved an energy-efficient route for converting raw lignocellulose into valuable alkanes.

2.3.5 ARENES

Lignin is the only abundant renewable source for aromatic chemicals. Converting it into liquid alkanes as fuel does not take full use of lignin, because this kind of compounds can be derived from other renewable resource such as cellulose and hemicelluloses. Depolymerization and deoxygenation of lignin while retaining the aromatic ring are very interesting and of great significance, because it provides an inexhaustible supply for arenes which are difficult to be derived from other renewable resources. To achieve this goal, great care should be taken to promote hydrogenolysis of C–O bonds while preventing the hydrogenation of the aromatic rings during the lignin hydrodeoxygenation. However, it is extremely challenging because the aryl–O bonds in aryl ethers are strong, especially that in phenol (414 KJ mol^{-1}) (131), and the hydrogenation of the aromatic ring is more thermodynamically favoured (41).

Up to now, there are only a few reports on the highly selective transformation of lignin into aromatic hydrocarbons. Yang et al. (46) developed an aqueous phase catalytic process to selectively depolymerize lignin polymeric framework and remove oxygen via hydrodeoxygenation (HDO) reactions over supported noble metals (Ru, Rh and Pt) in the presence of solid acid zeolites. The reaction mechanism is shown in Figure 2.10. Under various reaction conditions, the conversion of lignin was 35%–60% and the selectivity for aromatic hydrocarbons was 65%–70% (46). More recently, Yang and Wang et al. (131) reported a one-pot lignin conversion over a Ru/Nb$_2$O$_5$ catalyst in water media. The yield of liquid hydrocarbons was 35.5 wt.%, which was a near-quantitative carbon yield based on lignin monomers obtained via nitrobenzene oxidation method. The selectivity for monomer arenes approached 71%. The authors also performed DFT calculations to reveal reaction mechanism, and the results showed that the synergistic effect between Ru and NbO$_x$ species contributed to the high selectivity to arenes (131).

2.3.6 OTHER COMPOUNDS

Other compounds such as ethers (132), esters(133) also appeared as major depolymerization products of lignin in some reports, but the authors did not aim to produce this kind of compounds and there's no data of the selectivity. Substituted cyclohexanols

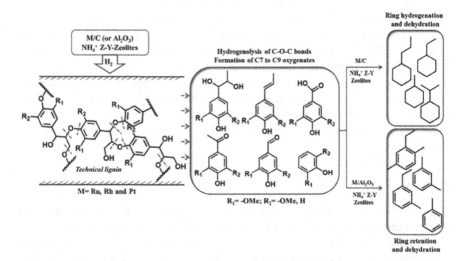

FIGURE 2.10 Proposed pathway for formation of aromatic and cyclic aliphatic hydrocarbons from lignins with supported (C and/or Al$_2$O$_3$) noble metals integrated with zeolite NH$_4^+$ Z-Y 57277-14-1 catalyst matrix (46).

can be obtained from phenolic model compounds with a high selectivity (82), but there are no reports for real lignin. Quinones widely exist in oxidation depolymerization products, and there are a few works that yielded it as major product. For example, Bösmann and Wasserscheid et al. (134) found that $Mn(NO_3)_2$ can effectively catalyse the oxidation depolymerization of lignin in ionic liquid media, especially in [EMIM] [CF_3SO_3] (1-ethyl-3-methylimidazolium trifluoromethanesulfonate). By adjusting the reaction conditions and catalyst loading, 2,6-dimethoxy-1,4-benzoquinone (DMBQ) was formed as the main product, which could be isolated from the reaction mixture as a pure substance in 21.0% selectivity and 11.5 wt.% overall yield with respect to the amount of lignin (134).

2.4 CONCLUSIONS AND OUTLOOK

In this chapter, we summarized the selective transformation of lignin into value-added chemicals. Phenols, aldehydes, carboxylic acids, alkanes, arenes, etc. can be derived from lignin with relatively high selectivity. However, it is far less than enough. Because we still get mixed products even if the selectivity to a specific class of compounds is 100%! What's worse, the separation of congeners is usually more difficult than different classes of species. There's a lot of work yet to do, because it is the high selective to one single compound that is the ultimate goal for highly efficiently utilization of lignin. This is extremely challenging because of the complex structure of lignin. Although there are only several main types of linkages in lignin, selective cleavage one specific type of the linkages is a rough ride. Nevertheless, selective cleavage of the linkages does not guarantee the selectivity of the product. Future opportunities for highly selective utilization of lignin are as follows:

1. Design more efficient and robust catalyst taking advantages of confinement effects to achieve higher selectivity;
2. Take advantages of the condensation. Although the condensation of the fragments in lignin is usually considered as side reaction which prevents the depolymerization of lignin, it may be helpful for higher selectivity of the depolymerization product. If most of the undesired products could be condensed into larger molecules by carefully controlling the reaction condition, the selectivity to the desired product would be very high and the separation would be easier.
3. Think out of the box, and make the selectivity even worse! If lignin can be depolymerized into different classes of compounds and only one product in each class, the separation would be relatively easy and energy saving;
4. Multilevel utilization of lignin. Convert lignin in multisteps, and one specific group is selectively utilized in each step. We only need to achieve high selectivity in each step, and the product can be different from different steps. This will make the separation relatively simple. What's more, different catalysts can be used in each step and they don't need to be multifunctional, which also reduce the challenge of catalyst design.

REFERENCES

1. M. Stöcker, Biofuels and biomass-to-liquid fuels in the biorefinery: Catalytic conversion of lignocellulosic biomass using porous materials. *Angewandte Chemie International Edition* 47, 9200–9211 (2008).

2. A. J. Ragauskas et al., The path forward for biofuels and biomaterials. *Science* 311, 484–489 (2006).

3. K. K. Lange, E. I. Tellgren, M. R. Hoffmann, T. Helgaker, A paramagnetic bonding mechanism for diatomics in strong magnetic fields. *Science* 337, 327–331 (2012).

4. D. M. Alonso, S. G. Wettstein, J. A. Dumesic, Bimetallic catalysts for upgrading of biomass to fuels and chemicals. *Chemical Society Reviews* 41, 8075–8098 (2012).

5. A. J. Ragauskas et al., Lignin valorization: Improving lignin processing in the biorefinery. *Science* 344, 1246843 (2014).

6. C. Li, X. Zhao, A. Wang, G. W. Huber, T. Zhang, Catalytic transformation of lignin for the production of chemicals and fuels. *Chemical Reviews* 115, 11559–11624 (2015).

7. R. Rinaldi et al., Paving the way for lignin valorisation: Recent advances in bioengineering, biorefining and catalysis. *Angewandte Chemie International Edition* 55, 8164–8215 (2016).

8. F. G. Calvo-Flores, J. A. Dobado, Lignin as renewable raw material. *ChemSusChem* 3, 1227–1235 (2010).

9. M. P. Pandey, C. S. Kim, Lignin depolymerization and conversion: A review of thermochemical methods. *Chemical Engineering & Technology* 34, 29–41 (2011).

10. X.-J. Shen et al., Efficient and product-controlled depolymerization of lignin oriented by raney Ni cooperated with CsxH3-xPW12O40. *BioEnergy Research* 10, 1155–1162 (2017).

11. C. O. Tuck, E. Pérez, I. T. Horváth, R. A. Sheldon, M. Poliakoff, Valorization of biomass: Deriving more value from waste. *Science* 337, 695–699 (2012).

12. M. N. Belgacem, A. Gandini, *Monomers, Polymers and Composites from Renewable Resources*. Amsterdam: Elsevier Ltd. (2008).

13. M. A. R. Meier, J. O. Metzger, U. S. Schubert, Plant oil renewable resources as green alternatives in polymer science. *Chemical Society Reviews* 36, 1788–1802 (2007).

14. M. Fache, B. Boutevin, S. Caillol, Vanillin production from lignin and its use as a renewable chemical. *ACS Sustainable Chemistry & Engineering* 4, 35–46 (2016).

15. J. H. Lora, W. G. Glasser, Recent industrial applications of lignin: A sustainable alternative to nonrenewable materials. *Journal of Polymers and the Environment* 10, 39–48 (2002).

16. W. Thielemans, E. Can, S. S. Morye, R. P. Wool, Novel applications of lignin in composite materials. *Journal of Applied Polymer Science* 83, 323–331 (2002).

17. P. C. A. Bruijnincx, B. M. Weckhuysen, Lignin up for break-down. *Nature Chemistry* 6, 1035 (2014).

18. D. Stewart, Lignin as a base material for materials applications: Chemistry, application and economics. *Industrial Crops and Products* 27, 202–207 (2008).

19. W. O. S. Doherty, P. Mousavioun, C. M. Fellows, Value-adding to cellulosic ethanol: Lignin polymers. *Industrial Crops and Products* 33, 259–276 (2011).

20. G. Xu, J.-h. Yang, H.-h. Mao, Z. Yun, Pulping black liquor used directly as a green and effective source for neat oil and as an emulsifier of catalytic cracking heavy oil. *Chemistry and Technology of Fuels and Oils* 47, 283 (2011).

21. N. S. Çetin, N. Özmen, Use of organosolv lignin in phenol-formaldehyde resins for particleboard production: II Particleboard production and properties. *International Journal of Adhesion and Adhesives* 22, 481–486 (2002).

22. B.-L. Xue, J.-L. Wen, R.-C. Sun, Lignin-based rigid polyurethane foam reinforced with pulp fiber: Synthesis and characterization. *ACS Sustainable Chemistry & Engineering* 2, 1474–1480 (2014).

23. C. Xu, R. A. D. Arancon, J. Labidi, R. Luque, Lignin depolymerisation strategies: Towards valuable chemicals and fuels. *Chemical Society Reviews* 43, 7485–7500 (2014).

24. J. Zakzeski, P. C. A. Bruijnincx, A. L. Jongerius, B. M. Weckhuysen, The catalytic valorization of lignin for the production of renewable chemicals. *Chemical Reviews* 110, 3552–3599 (2010).

25. R. Jastrzebski et al., Tandem catalytic depolymerization of lignin by water-tolerant lewis acids and rhodium complexes. *ChemSusChem* 9, 2074–2079 (2016).

26. P. J. Deuss, K. Barta, From models to lignin: Transition metal catalysis for selective bond cleavage reactions. *Coordination Chemistry Reviews* 306, 510–532 (2016).

27. A. Rahimi, A. Ulbrich, J. J. Coon, S. S. Stahl, Formic-acid-induced depolymerization of oxidized lignin to aromatics. *Nature* 515, 249 (2014).

28. A. K. Deepa, P. L. Dhepe, Lignin depolymerization into aromatic monomers over solid acid catalysts. *ACS Catalysis* 5, 365–379 (2015).

29. V. Roberts, S. Fendt, A. A. Lemonidou, X. Li, J. A. Lercher, Influence of alkali carbonates on benzyl phenyl ether cleavage pathways in superheated water. *Applied Catalysis B: Environmental* 95, 71–77 (2010).

30. V. M. Roberts et al., Towards quantitative catalytic lignin depolymerization. *Chemistry – A European Journal* 17, 5939–5948 (2011).

31. A. Toledano, L. Serrano, J. Labidi, Improving base catalyzed lignin depolymerization by avoiding lignin repolymerization. *Fuel* 116, 617–624 (2014).

32. J. Long et al., Efficient base-catalyzed decomposition and in situ hydrogenolysis process for lignin depolymerization and char elimination. *Applied Energy* 141, 70–79 (2015).

33. D. Mohan, Pittman, Charles U., P. H. Steele, Pyrolysis of wood/biomass for bio-oil:? A critical review. *Energy & Fuels* 20, 848–889 (2006).

34. M. Kosa, H. Ben, H. Theliander, A. J. Ragauskas, Pyrolysis oils from CO_2 precipitated Kraft lignin. *Green Chemistry* 13, 3196–3202 (2011).

35. J. Cho, S. Chu, P. J. Dauenhauer, G. W. Huber, Kinetics and reaction chemistry for slow pyrolysis of enzymatic hydrolysis lignin and organosolv extracted lignin derived from maplewood. *Green Chemistry* 14, 428–439 (2012).

36. P. F. Britt, A. C. Buchanan, K. B. Thomas, S. K. Lee, Pyrolysis mechanisms of lignin: Surface-immobilized model compound investigation of acid-catalyzed and free-radical reaction pathways. *Journal of Analytical & Applied Pyrolysis* 33, 1–19 (1995).

37. F. L. P. Resende, S. A. Fraley, M. J. Berger, P. E. Savage, Noncatalytic gasification of lignin in supercritical water. *Energy & Fuels* 22, 1328–1334 (2008).

38. F. L. P. Resende, P. E. Savage, Kinetic model for noncatalytic supercritical water gasification of cellulose and lignin. *AIChE Journal* 56, 2412–2420 (2010).

39. A. Yamaguchi, N. Hiyoshi, O. Sato, M. Shirai, Gasification of organosolv-lignin over charcoal supported noble metal salt catalysts in supercritical water. *Topics in Catalysis* 55, 889–896 (2012).

40. G. Chatel, R. D. Rogers, Review: Oxidation of lignin using ionic liquids—An innovative strategy to produce renewable chemicals. *ACS Sustainable Chemistry & Engineering* 2, 322–339 (2014).

41. X. Wang, R. Rinaldi, A route for lignin and bio-oil conversion: Dehydroxylation of phenols into arenes by catalytic tandem reactions. *Angewandte Chemie International Edition* 52, 11499–11503 (2013).

42. X. Wang, R. Rinaldi, Exploiting H-transfer reactions with RANEY® Ni for upgrade of phenolic and aromatic biorefinery feeds under unusual, low-severity conditions. *Energy & Environmental Science* 5, 8244–8260 (2012).

43. H. Ben, W. Mu, Y. Deng, A. J. Ragauskas, Production of renewable gasoline from aqueous phase hydrogenation of lignin pyrolysis oil. *Fuel* 103, 1148–1153 (2013).

44. R. Ma, W. Hao, X. Ma, Y. Tian, Y. Li, Catalytic ethanolysis of kraft lignin into high-value small-molecular chemicals over a nanostructured α-molybdenum carbide catalyst. *Angewandte Chemie International Edition* 53, 7310–7315 (2014).

45. X. Ma, Y. Tian, W. Hao, R. Ma, Y. Li, Production of phenols from catalytic conversion of lignin over a tungsten phosphide catalyst. *Applied Catalysis A: General* 481, 64–70 (2014).

46. D. D. Laskar, M. P. Tucker, X. Chen, G. L. Helms, B. Yang, Noble-metal catalyzed hydrodeoxygenation of biomass-derived lignin to aromatic hydrocarbons. *Green Chemistry* 16, 897–910 (2014).

47. L. Shuai et al., Formaldehyde stabilization facilitates lignin monomer production during biomass depolymerization. *Science* 354, 329–333 (2016).

48. K. Barta, G. R. Warner, E. S. Beach, P. T. Anastas, Depolymerization of organosolv lignin to aromatic compounds over Cu-doped porous metal oxides. *Green Chemistry* 16, 191–196 (2014).

49. Q. Song et al., Lignin depolymerization (LDP) in alcohol over nickel-based catalysts via a fragmentation–hydrogenolysis process. *Energy & Environmental Science* 6, 994–1007 (2013).

50. Q. Song, F. Wang, J. Xu, Hydrogenolysis of lignosulfonate into phenols over heterogeneous nickel catalysts. *Chemical Communications* 48, 7019–7021 (2012).

51. C. Li, M. Zheng, A. Wang, T. Zhang, One-pot catalytic hydrocracking of raw woody biomass into chemicals over supported carbide catalysts: Simultaneous conversion of cellulose, hemicellulose and lignin. *Energy & Environmental Science* 5, 6383–6390 (2012).

52. J. Zhang et al., Highly efficient, NiAu-catalyzed hydrogenolysis of lignin into phenolic chemicals, *Green Chemistry* 16, 2432–2437 (2014).

53. C. Zhao, D. M. Camaioni, J. A. Lercher, Selective catalytic hydroalkylation and deoxygenation of substituted phenols to bicycloalkanes. *Journal of Catalysis* 288, 92–103 (2012).

54. S. M. Leckie, G. J. Harkness, M. L. Clarke, Catalytic constructive deoxygenation of lignin-derived phenols: New C–C bond formation processes from imidazole-sulfonates and ether cleavage reactions. *Chemical Communications* 50, 11511–11513 (2014).

55. T. L. Marker et al., Integrated hydropyrolysis and hydroconversion (IH2®) for the direct production of gasoline and diesel fuels or blending components from biomass, Part 2: Continuous testing. *Environmental Progress & Sustainable Energy* 33, 762–768 (2014).

56. T. L. Marker, L. G. Felix, M. B. Linck, M. J. Roberts, Integrated hydropyrolysis and hydroconversion (IH2) for the direct production of gasoline and diesel fuels or blending components from biomass, part 1: Proof of principle testing. *Environmental Progress & Sustainable Energy* 31, 191–199 (2012).

57. R. Ma, Y. Xu, X. Zhang, Catalytic oxidation of biorefinery lignin to value-added chemicals to support sustainable biofuel production. *ChemSusChem* 8, 24–51 (2015).

58. F. S. Chakar, A. J. Ragauskas, Review of current and future softwood kraft lignin process chemistry. *Industrial Crops and Products* 20, 131–141 (2004).

59. A. Vigneault, D. K. Johnson, E. Chornet, Base-catalyzed depolymerization of lignin: Separation of monomers. *The Canadian Journal of Chemical Engineering* 85, 906–916 (2007).

60. J. Luo, P. Melissa, W. Zhao, Z. Wang, Y. Zhu, Selective lignin oxidation towards vanillin in phenol media. *ChemistrySelect* 1, 4596–4601 (2016).

61. P. C. R. Pinto, E. A. B. D. Silva, A. E. Rodrigues, *Lignin as Source of Fine Chemicals: Vanillin and Syringaldehyde*. Berlin: Springer Berlin Heidelberg (2012).

62. Q. Mei et al., Selective utilization of the methoxy group in lignin to produce acetic acid. *Angewandte Chemie International Edition* 56, 14868–14872 (2017).

63. S. K. Hanson, R. T. Baker, Knocking on wood: Base metal complexes as catalysts for selective oxidation of lignin models and extracts. *Accounts of Chemical Research* 48, 2037–2048 (2015).

64. J. Ralph et al., Lignins: Natural polymers from oxidative coupling of 4-hydroxyphenyl-propanoids. *Phytochemistry Reviews* 3, 29–60 (2004).

65. W. Boerjan, J. Ralph, M. Baucher, Lignin biosynthesis. *Annual Review of Plant Biology* 54, 519–546 (2003).

66. R. C. Sun, Cereal straw as a resource for sustainable biomaterials and biofuels. *Cereal Straw as a Resource for Sustainable Biomaterials & Biofuels*, 289–292 (2010).

67. R. Vanholme, K. Morreel, J. Ralph, W. Boerjan, Lignin engineering. *Current Opinion in Plant Biology* 11, 278–285 (2008).

68. S. D. Mansfield, H. Kim, F. Lu, J. Ralph, Whole plant cell wall characterization using solution-state 2D NMR. *Nature Protocols* 7, 1579 (2012).

69. J.-L. Wen, S.-L. Sun, B.-L. Xue, R.-C. Sun, Recent advances in characterization of lignin polymer by solution-state nuclear magnetic resonance (NMR) methodology. *Materials* 6, 359–391 (2013).

70. A. Zhang, F. Lu, R. Sun, J. Ralph, Ferulate–coniferyl alcohol cross-coupled products formed by radical coupling reactions. *Planta* 229, 1099–1108 (2009).

71. X. Shen, P. Huang, J. Wen, R. Sun, Research status of lignin oxidative and reductive depolymerization. *Progress in Chemistry* 29, 162–178 (2017).

72. R. Parthasarathi, R. A. Romero, A. Redondo, S. Gnanakaran, Theoretical study of the remarkably diverse linkages in lignin. *The Journal of Physical Chemistry Letters* 2, 2660–2666 (2011).

73. M. V. Galkin, J. S. M. Samec, Lignin valorization through catalytic lignocellulose fractionation: A fundamental platform for the future biorefinery. *Chem Sus Chem* 9, 1544–1558 (2016).

74. Z. Jiang et al., Understanding the cleavage of inter- and intramolecular linkages in corncob residue for utilization of lignin to produce monophenols. *Green Chemistry* 18, 4109–4115 (2016).

75. C. Zhao, Y. Kou, A. A. Lemonidou, X. Li, J. A. Lercher, Highly selective catalytic conversion of phenolic bio-oil to alkanes. *Angewandte Chemie International Edition* 48, 3987–3990 (2009).

76. Q. Bu et al., A review of catalytic hydrodeoxygenation of lignin-derived phenols from biomass pyrolysis. *Bioresource Technology* 124, 470–477 (2012).

77. F. Liu, Q. Liu, A. Wang, T. Zhang, Direct catalytic hydrogenolysis of kraft lignin to phenols in choline-derived ionic liquids. *ACS Sustainable Chemistry & Engineering* 4, 3850–3856 (2016).

78. A. G. Sergeev, J. F. Hartwig, Selective, nickel-catalyzed hydrogenolysis of aryl ethers. *Science* 332, 439–443 (2011).

79. F. Gao, J. D. Webb, J. F. Hartwig, Chemo- and regioselective hydrogenolysis of diaryl ether C–O bonds by a robust heterogeneous Ni/C catalyst: Applications to the cleavage of complex lignin-related fragments. *Angewandte Chemie International Edition* 55, 1474–1478 (2016).

80. J. He, C. Zhao, J. A. Lercher, Ni-catalyzed cleavage of aryl ethers in the aqueous phase. *Journal of the American Chemical Society* 134, 20768–20775 (2012).

81. P. Kelley, S. Lin, G. Edouard, M. W. Day, T. Agapie, Nickel-mediated hydrogenolysis of C–O bonds of aryl ethers: What is the source of the hydrogen? *Journal of the American Chemical Society* 134, 5480–5483 (2012).

82. V. Molinari, C. Giordano, M. Antonietti, D. Esposito, Titanium nitride-nickel nanocomposite as heterogeneous catalyst for the hydrogenolysis of aryl ethers. *Journal of the American Chemical Society* 136, 1758–1761 (2014).

83. A. G. Sergeev, J. D. Webb, J. F. Hartwig, A heterogeneous nickel catalyst for the hydrogenolysis of aryl ethers without arene hydrogenation. *Journal of the American Chemical Society* 134, 20226–20229 (2012).

84. X. Wang, R. Rinaldi, Solvent effects on the hydrogenolysis of diphenyl ether with raney nickel and their implications for the conversion of lignin. *ChemSusChem* 5, 1455–1466 (2012).

85. E. Adler, Lignin chemistry—past, present and future. *Wood Science and Technology* 11, 169–218 (1977).

86. W. Lan, M. T. Amiri, C. M. Hunston, J. S. Luterbacher, Protection group effects during α, γ-diol lignin stabilization promote high-selectivity monomer production. *Angewandte Chemie* 130, 1370–1374 (2018).

87. Y. Li et al., An "ideal lignin" facilitates full biomass utilization. *Science Advances* 4, eaau2968 (2018).

88. S. Van den Bosch et al., Integrating lignin valorization and bio-ethanol production: On the role of Ni-Al2 O3 catalyst pellets during lignin-first fractionation. *Green Chemistry* 19, 3313–3326 (2017).

89. P. Ferrini, R. Rinaldi, Catalytic biorefining of plant biomass to non-pyrolytic lignin bio-oil and carbohydrates through hydrogen transfer reactions. *Angewandte Chemie International Edition* 53, 8634–8639 (2014).

90. X. Huang et al., Reductive fractionation of woody biomass into lignin monomers and cellulose by tandem metal triflate and Pd/C catalysis. *Green Chemistry* 19, 175–187 (2017).

91. X. Huang, J. Zhu, T. I. Korányi, M. D. Boot, E. J. M. Hensen, Effective release of lignin fragments from lignocellulose by Lewis acid metal triflates in the lignin-first approach. *ChemSusChem* 9, 3262–3267 (2016).

92. S. Van den Bosch et al., Reductive lignocellulose fractionation into soluble lignin-derived phenolic monomers and dimers and processable carbohydrate pulps. *Energy & Environmental Science* 8, 1748–1763 (2015).

93. H. Luo et al., Total utilization of Miscanthus biomass, lignin and carbohydrates, using earth abundant nickel catalyst. *ACS Sustainable Chemistry & Engineering* 4, 2316–2322 (2016).

94. P. J. Deuss et al., Phenolic acetals from lignins of varying compositions via iron(iii) triflate catalysed depolymerisation. *Green Chemistry* 19, 2774–2782 (2017).

95. P. J. Deuss et al., Aromatic monomers by in situ conversion of reactive intermediates in the acid-catalyzed depolymerization of lignin. *Journal of the American Chemical Society* 137, 7456–7467 (2015).

96. A. Toledano et al., Fractionation of organosolv lignin from olive tree clippings and its valorization to simple phenolic compounds. *ChemSusChem* 6, 529–536 (2013).

97. C. S. Lancefield, O. S. Ojo, F. Tran, N. J. Westwood, Isolation of functionalized phenolic monomers through selective oxidation and C–O bond cleavage of the β-O-4 linkages in lignin. *Angewandte Chemie International Edition* 54, 258–262 (2015).

98. E. Feghali, G. Carrot, P. Thuery, C. Genre, T. Cantat, Convergent reductive depolymerization of wood lignin to isolated phenol derivatives by metal-free catalytic hydrosilylation. *Energy & Environmental Science* 8, 2734–2743 (2015).

99. Z. Jiang, T. He, J. Li, C. Hu, Selective conversion of lignin in corncob residue to monophenols with high yield and selectivity. *Green Chemistry* 16, 4257–4265 (2014).

100. L. Hu et al., The degradation of the lignin in *Phyllostachys heterocycla* cv. pubescens in an ethanol solvothermal system. *Green Chemistry* 16, 3107–3116 (2014).

101. M. Wang et al., Two-step, catalytic C–C bond oxidative cleavage process converts lignin models and extracts to aromatic acids. *ACS Catalysis* 6, 6086–6090 (2016).

102. Y. Yang et al., Ionic liquid [OMIm][OAc] directly inducing oxidation cleavage of the β-O-4 bond of lignin model compounds. *Chemical Communications* 53, 8850–8853 (2017).

103. R. Ma et al., Peracetic acid depolymerization of biorefinery lignin for production of selective monomeric phenolic compounds. *Chemistry–A European Journal* 22, 10884–10891 (2016).

104. A. Rahimi, A. Azarpira, H. Kim, J. Ralph, S. S. Stahl, Chemoselective metal-free aerobic alcohol oxidation in lignin. *Journal of the American Chemical Society* 135, 6415–6418 (2013).

105. E. B. da Silva et al., An integrated process to produce vanillin and lignin-based polyurethanes from Kraft lignin. *Chemical Engineering Research and Design* 87, 1276–1292 (2009).

106. M. B. Hocking, Vanillin: Synthetic flavoring from spent sulfite liquor. *Journal of Chemical Education* 74, 1055 (1997).

107. A. W. Pacek, P. Ding, M. Garrett, G. Sheldrake, A. W. Nienow, Catalytic conversion of sodium lignosulfonate to vanillin: Engineering aspects. Part 1. Effects of processing conditions on vanillin yield and selectivity. *Industrial & Engineering Chemistry Research* 52, 8361–8372 (2013).

108. A. Azarpira, J. Ralph, F. Lu, Catalytic alkaline oxidation of lignin and its model compounds: A pathway to aromatic biochemicals. *BioEnergy Research* 7, 78–86 (2014).

109. C. Díaz-Urrutia et al., Towards lignin valorisation: Comparing homogeneous catalysts for the aerobic oxidation and depolymerisation of organosolv lignin. *RSC Advances* 5, 70502–70511 (2015).

110. V. Tarabanko, D. Petukhov, G. Selyutin, New mechanism for the catalytic oxidation of lignin to vanillin. *Kinetics and Catalysis* 45, 569–577 (2004).

111. R. Sun, J. M. Lawther, W. Banks, The effect of alkaline nitrobenzene oxidation conditions on the yield and components of phenolic monomers in wheat straw lignin and compared to cupric (II) oxidation. *Industrial Crops and Products* 4, 241–254 (1995).

112. G. Wu, M. Heitz, E. Chornet, Improved alkaline oxidation process for the production of aldehydes (vanillin and syringaldehyde) from steam-explosion hardwood lignin. *Industrial & Engineering Chemistry Research* 33, 718–723 (1994).

113. I. Panorel et al., Pulsed corona discharge oxidation of aqueous lignin: Decomposition and aldehydes formation. *Environmental Technology* 35, 171–176 (2014).

114. J. Zeng et al., Biomimetic fenton-catalyzed lignin depolymerization to high-value aromatics and dicarboxylic acids. *ChemSusChem* 8, 861–871 (2015).

115. S. Lotfi, D. C. Boffito, G. S. Patience, Gas–solid conversion of lignin to carboxylic acids. *Reaction Chemistry & Engineering* 1, 397–408 (2016).

116. S. Y. Lee, S. H. Hong, S. H. Lee, S. J. Park, Fermentative production of chemicals that can be used for polymer synthesis. *Macromolecular Bioscience* 4, 157–164 (2004).

117. A. Castellan, J. C. J. Bart, S. Cavallaro, Industrial production and use of adipic acid. *Catalysis Today* 9, 237–254 (1991).

118. R. Ma, M. Guo, X. Zhang, Selective conversion of biorefinery lignin into dicarboxylic acids. *ChemSusChem* 7, 412–415 (2014).

119. Z. Cai et al., Selective production of diethyl maleate via oxidative cleavage of lignin aromatic unit. *Chem*, (2019).

120. M. Nagy, K. David, G. J. Britovsek, A. J. Ragauskas, Catalytic hydrogenolysis of ethanol organosolv lignin. *Holzforschung* 63, 513–520 (2009).

121. M. Schlaf, Selective deoxygenation of sugar polyols to α, ω-diols and other oxygen content reduced materials—A new challenge to homogeneous ionic hydrogenation and hydrogenolysis catalysis. *Dalton Transactions*, 4645–4653 (2006).

122. C. Zhao, J. A. Lercher, Upgrading pyrolysis oil over Ni/HZSM-5 by cascade reactions. *Angewandte Chemie International Edition* 51, 5935–5940 (2012).
123. D. D. Laskar, B. Yang, H. Wang, J. Lee, Pathways for biomass-derived lignin to hydrocarbon fuels. *Biofuels, Bioproducts and Biorefining* 7, 602–626 (2013).
124. M. Saidi et al., Upgrading of lignin-derived bio-oils by catalytic hydrodeoxygenation. *Energy & Environmental Science* 7, 103–129 (2014).
125. P. M. Mortensen, J.-D. Grunwaldt, P. A. Jensen, K. Knudsen, A. D. Jensen, A review of catalytic upgrading of bio-oil to engine fuels. *Applied Catalysis A: General* 407, 1–19 (2011).
126. N. Yan et al., Selective degradation of wood lignin over noble-metal catalysts in a two-step process. *ChemSusChem: Chemistry & Sustainability Energy & Materials* 1, 626–629 (2008).
127. H. Wang et al., Biomass-derived lignin to jet fuel range hydrocarbons via aqueous phase hydrodeoxygenation. *Green Chemistry* 17, 5131–5135 (2015).
128. J. Kong, B. Li, C. Zhao, Tuning Ni nanoparticles and the acid sites of silica-alumina for liquefaction and hydrodeoxygenation of lignin to cyclic alkanes. *RSC Advances* 6, 71940–71951 (2016).
129. X. Wang, R. Rinaldi, Bifunctional Ni catalysts for the one-pot conversion of Organosolv lignin into cycloalkanes. *Catalysis Today* 269, 48–55 (2016).
130. Q. Xia et al., Direct hydrodeoxygenation of raw woody biomass into liquid alkanes. *Nature Communications* 7, 11162 (2016).
131. Y. Shao et al., Selective production of arenes via direct lignin upgrading over a niobium-based catalyst. *Nature Communications* 8, 16104 (2017).
132. X. Huang, T. I. Korányi, M. D. Boot, E. J. Hensen, Ethanol as capping agent and formaldehyde scavenger for efficient depolymerization of lignin to aromatics. *Green Chemistry* 17, 4941–4950 (2015).
133. J. S. Luterbacher et al., Lignin monomer production integrated into the γ-valerolactone sugar platform. *Energy & Environmental Science* 8, 2657–2663 (2015).
134. K. Stärk, N. Taccardi, A. Bösmann, P. Wasserscheid, Oxidative depolymerization of lignin in ionic liquids. *Chem Sus Chem* 3, 719–723 (2010).

3 Nanocellulose-based Materials for the Solar Cell, Wearable Sensors, and Supercapacitors

Sheng Chen, Meng Wang, Changyou Shao, and Feng Xu

CONTENTS

3.1 INTRODUCTION

With the development of world economy and the population explosion, we have been facing the issues of energy shortage and environmental pollution in the past decades. The traditional fossil resources have limited reserves and are not renewable, which are not conductive to the sustainable development of the world. To solve this problem, recently the 'green' comprehensive utilization of bioresources has attracted great attention from government and researchers (Gaurav et al. 2017, Zhang et al. 2018). Some conventional materials are expected to be replaced by the natural bioresources for the eco-efficient and environment-friendly materials, products, and devices.

Cellulose, composed of β-1,4-anhydro-D-glucopyranose units, is the most abundant organic polymer in nature (Fu et al. 2020, Cannon and Anderson 1991). It is the main component of most biomass like wood, cotton, and other plant fibres (Kotelev et al. 2017). Cellulose holds the merits of renewability, sustainability, and biodegradability etc.; therefore, it has been regarded as the potential material for replacing the conventional fossil resource in widespread applications, such as energy conversion, energy storage, and sensing.

Recently, a nanostructured cellulose, also known as nanocellulose, has drawn great attention from researchers (Xue et al. 2017, Chen, Yu, et al. 2018b). Nanocellulose is isolated from the cellulose fibres at nanoscale via mechanical and/ or chemical methods; however, nanocellulose has extraordinary physicochemical properties that are absent in the original cellulose fibres. Therefore, nanocellulose and nanocellulose-based composites are becoming the study focus in the fields of 'green' energy and sensing (Sabo et al. 2016).

In this chapter, the production, characterization, and applications of the nanocellulose and/or nanocellulose-based composite are briefly introduced. Furthermore, our group's recent works on the various nanocellulose-based functional materials (i.e., nanopaper, hydrogel, and carbon aerogel) for solar cells (Chen, Song, et al. 2018a), wearable sensors (Shao et al. 2018), and supercapacitors (Wang et al. 2019) are discussed in detail.

3.2 NANOCELLULOSE

Nanocellulose is a form of nanostructured cellulose fibre and has lots of advantages, such as large specific area (100–200 gm^{-2}), adjustable aspect ratios (100–150), high tensile strength (7.5–7.7 GPa), high Young's modulus (110–220 GPa), low coefficient of thermal expansion, superior gas-barrier, high chemical resistance, and easy surface functionalization (Chen and Hu 2018, Chen, Yu, et al. 2018b, Thomas et al. 2018, Fang et al. 2019). Typical sources of nanocellulose are wood, cotton, wheat/ rice straw, bamboo, algae, and bacteria, etc. The obtained cellulose fibres from these materials can be further disintegrated into nanocellulose through mechanical and/or chemical treatment. Generally, nanocellulose can be classified into three types: cellulose nanofibrils (CNFs), cellulose nanocrystals (CNCs), and bacterial nanocelluloses (BNCs). The synonyms, formation processes, and average size of these nanocellulose materials are compared and presented in Table 3.1. The nanocellulose

TABLE 3.1

Three Types of Nanocellulose Materials (Klemm *et al.* 2011)

Type of Nanocellulose	Synonyms	Formation	Average Size
Cellulose nanofibrils (CNFs)	Nanofibrillated cellulose, cellulose nanofibres, nanofibrils, microfibrils	Delamination of wood pulp by mechanical pressure before and/or after chemical or enzymatic treatment	Diameter, 5–60 nm; length, several micrometres
Cellulose nanocrystals (CNCs)	Nanocrystalline cellulose, crystallites, whiskers, nanowhiskers, rod-like cellulose microcrystals	Acid hydrolysis of cellulose from various sources	Diameter, 5–70 nm; length, 100–250 nm (from plant), 100 nm to several micrometres (from cellulose of tunicates, algae, and bacteria)
Bacterial nanocelluloses (BNCs)	Bacterial cellulose, microbial cellulose, biocellulose	Bacterial synthesis	Diameter: 20–100 nm, different types of nanofibre networks

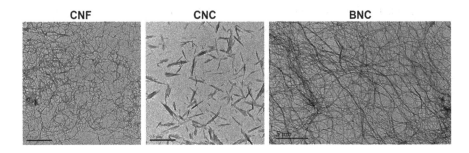

FIGURE 3.1   TEM images of three types of nanocellulose.

materials have different names, and CNF, CNC, and BNC are most widely used in literature. The size of these nanocellulose materials is also very different. As can be seen in Figure 3.1, TEM images present the morphology and especially dimensions of the cellulose-based nanostructured fibres. CNFs and CNCs have the similar diameter, but the rod-like CNCs are dramatically shorter than CNFs. BNCs also have the high aspect ratio like CNFs but different types of nanofibre networks. Besides, various technologies were utilized to prepare these nanocellulose materials.

CNFs are prepared mainly by mechanical fibrillation (Veigel et al. 2011). The equipment that has been commonly used includes high-pressure homogenizer, micro-jet machine, ball miller, freezing grinder, and ultrasonic grinder, etc. In 1983, Turbak produced the stable suspension of CNFs in water by a physical treatment of wood cellulose pulps for the first time. Although the obtained novel cellulose-based material was called microfibrillated cellulose (MFC), the treated cellulose fibres were in nanoscale. Mechanical fibrillation is an effective method to prepare CNFs.

However, the energy consumption is very high during mechanical processing, which limits the cost-effective productions of CNFs. To solve this problem, the chemical and/or biological technologies can be used to pretreat the cellulose pulps to destroy the original structure of cellulose fibres. This can decrease the energy consumption of post-mechanical treatment and thus the whole cost of CNFs production. For example, the 2, 2, 6, 6-tetramethylpiperidine-1-oxyl (TEMPO) mediated oxidation is able to break the interfibril hydrogen bonds in cellulose and convert C6 primary hydroxyls to carboxyls, which can reduce the energy needed for defibrillating cellulose by at least 2 orders of magnitude (Jiang and Hsieh 2016, Isogai et al. 2011).

Cellulose nanocrystals (CNCs) can be extracted from native cellulose fibres by acid hydrolysis. The most commonly used reagents are sulfuric acid and hydrochloric acid. Besides, phosphoric acid and hydrobromic acid can also be used for the preparation of CNCs. The Figure 3.2 illustrates the crystalline structure of native cellulose and its change during the fabrication of CNC and CNF (Hubbe et al. 2017). Cellulose is composed of crystalline domain and disordered (amorphous) regions. Strong acid hydrolysis can destroy and eliminate the amorphous regions of the cellulose but maintain its crystal domains. However, both crystalline and amorphous regions were contained for CNFs. The obtained CNCs are highly crystalline and relatively rigid rod-like nanoparticles, which have lower aspect ratios and flexibility when compared with CNFs.

As demonstrated by the schemes in Figure 3.3, bacterial nanocelluloses (BNCs) are purely synthesized by bacteria like *Acetobacter xylinum* in aqueous culture media in several days. During the production of BNCs, the bacterial body can generate glucose chains and extrude them out through tiny pore on the cell envelope. The glucan chain aggregated and formed the cellulose microfibre, which continued to aggregate as ribbons, i.e., BNCs. Compared to the CNFs and CNCs extracted from plant sources, the pure biosynthesis BNCs do not have the contaminants such as lignin and hemicellulose (Lin et al. 2013, Lin and Dufresne 2014). Therefore, the BNCs have excellent biocompatibility and exhibit potential applications in the fields of biomedicine and functional biomaterials.

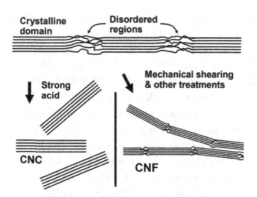

FIGURE 3.2 Schemes showing the crystalline changes for the fabrication of CNC and CNF (Hubbe *et al.* 2017).

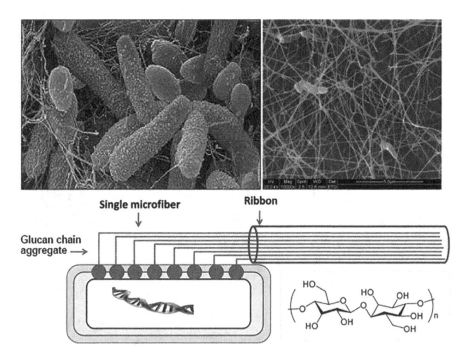

FIGURE 3.3 Schemes and SEM images of BNCs (Shi *et al.* 2014).

3.3 NANOCELLULOSE-BASED FUNCTIONAL MATERIALS

3.3.1 Nanocellulose-based Films (Nanopaper)

The common method for paper production used in paper-making industry can also be used to prepare nanocellulose-based films. In brief, the nanocellulose suspension is filtered through a porous substrate (e.g., microporous membrane) and then the obtained wet cellulose film is dried and/or pressed to fabricate nanopaper (Wei et al. 2014). Since the pure CNF-based film was firstly reported (Nogi et al. 2009) in 2009, various pure nanocellulose-based films and functional composite films have been developed. As can be seen in Figure 3.4, the nanocellulose-based films with fascinating optical, thermal, electrical, and mechanical properties exhibit potential applications in emerging fields, such as flexible electronics, energy devices, and water treatment, etc.

The pure nanocellulose films made of nanocellulose like CNFs present excellent mechanical properties. This nanopaper with a tensile strength of 100–300 MPa and a modulus of 5–30 GPa is much stronger than the widely used printing paper and packaging paper (Zhu et al. 2014). To further improve the mechanical strength of nanocellulose films, lots of researchers prepared the films made of cellulose nanofibres with preferred alignment rather than random distribution (Zhu et al. 2017, Wang et al. 2018, Tang et al. 2015, Walther et al. 2011, Sehaqui et al. 2012). The as-fabricated nanocellulose film composed with aligned BNCs exhibited extremely high strength (~1 GPa) and modulus (48.1 GPa) (Wang et al. 2018).

FIGURE 3.4 Schematic illustration of the nanocellulose-based films with excellent properties for various applications (Fang et al. 2019.)

The well-developed methods to prepare these anisotropic nanocellulose films include wet stretching (Tang et al. 2015), wet extrusion (Walther et al. 2011), and cold drawing (Sehaqui et al. 2012), etc. However, these down-top technologies have relatively high cost and complexity, which have limited the scalable production of the strong nanocellulose films. Recently, the top-down approach, combined with delignification and hot-press, to prepare wood-derived films with aligned cellulose nanofibres was developed by Hu group. This method is cost-effective, facile, and efficient, but the obtained nanocellulose films were a little weaker than the aforementioned films made by down-top technologies, which may have contributed to the degradation of cellulose during the delignification process (Fang et al. 2019). The nanocellulose films with excellent mechanical properties including high strength and toughness have been used as the substrate for various flexible electronics, such as transistors (Park *et al.* 2018, Huang et al. 2013), solar cells (Hu et al. 2013, Nogi et al. 2015), and organic light-emitting diodes (OLEDs) (Ummartyotin et al. 2012, Purandare et al. 2014), etc.

Unlike the common porous paper with ultralow transparency, the dense nanocellulose film always exhibited great optical properties, i.e., high transparency and/or high haze. The nanostructured cellulose fibres like CNFs have the diameter and interstice much smaller than the visible wavelength, which would cause less light scattering compared with common paper. Furthermore, most of the light diffusions are forward scattering rather than back scattering. Therefore, the as-prepared nanocellulose films have both high transparency and high haze. As light-management layers, these nanopapers could significantly improve the working efficiency of solar cells (with an enhancement of 10.1–23.9%) by facile lamination (Jia et al. 2017, Fang et al. 2014, Ha et al. 2014). Another important nanocellulose films with advanced optical properties are CNC-based photonic films with structural colours. Compared to pigmentary colour, the structural colour generated from light interference is much more stable under cyclic stimulation. The prepared CNCs by acid hydrolysis are promising chiral nematic crystals for photonic films for various applications (Fernandes et al. 2017, Yao et al. 2017, Guidetti et al. 2016). For example, the self-assembled CNC films with tunable photonic properties and barrier capabilities can be utilized in the fields of sustainable consumer packaging products, as well as effective templates for photonic and optoelectronic materials and structures (Guidetti et al. 2016).

The nanocellulose-based composite films combined with functional materials, such as metal nanoparticles, advanced carbon materials, and conductive polymers, have also been extensively developed for applications in energy-storage devices and flexible electronics, etc. As can be seen in Figure 3.5a, a freestanding conductive film is fabricated by polypyrrole (PPy)/BNCs combined with graphene (RGO) through insitu polymerization and filtration (Ma et al. 2016). Based on this film electrode, the assembled flexible supercapacitor exhibited excellent performances, including high areal capacitance of 1.67 F cm^{-2}, high areal energy density of 0.23 mWh cm^{-2}, and a maximum power density of 23.5 mW cm^{-2}. Besides, through vacuum filtration, an ultrathin and flexible MXene/CNFs composite film with a nacre-like lamellar structure has also been fabricated (Figure 3.5b) (Cao et al. 2018). This nanocellulose-based composite film exhibited high electrical conductivity (up to 739.4 S m^{-1}) and excellent specific electromagnetic interference (EMI), shielding efficiency (up to 2647 dB cm^2 g^{-1}), demonstrating its potential applications in various fields such as flexible electronics and smart robots.

3.3.2 NANOCELLULOSE-BASED HYDROGEL

Hydrogel is composed with heterogeneous mixtures of two (or more) phases; generally, the dispersed phase is water and another is a solid 3D network (Hoare and Kohane 2008). Since the hydrogel was first proposed in 1960 (Wichterle and Lim 1960), it has drawn widespread attentions from researchers and engineers (Ahmed 2015). Numerous hydrogel materials with various outstanding properties have been developed in the past decades. However, the mechanical properties (such as stiffness, tensile strength, and toughness) of hydrogels can be further improved for applications in a wider range. Recently, the nanocellulose materials with excellent mechanical property have been used to prepare hydrogels as reinforcing agents (fillers) and/or building

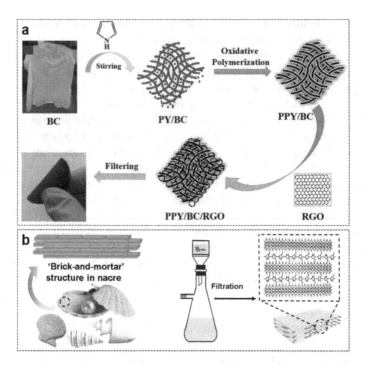

FIGURE 3.5   (a) Schematic illustration of the fabrication process of PPy/BNC/RGO film electrode (Ma *et al.* 2016). (b) Schematic illustration of the MXene/CNFs composite film with a nacre-like structure ('brick-and-mortar') by a vacuum filtration method (Cao *et al.* 2018).

blocks. Different from most conventional hydrogels made of water-soluble polymers, nanocellulose-based hydrogel is always fabricated from a colloidal suspension of nanocellulose materials, such as CNC, CNF, and BNC (Curvello et al. 2019, Mendoza et al. 2018).

CNCs, which have great rigidity and thus limited ability to entangle with each other, are typically utilized to prepare the hydrogel networks as effective reinforcing agents (or fillers) *via* physical and/or chemical interacting forces (Du et al. 2019, De France et al. 2017). For the physical method, the common technologies include straightforward mixing (homogenization) (McKee et al. 2014), free radical polymerization (Yang et al. 2014a), and freeze-thaw processing (Gonzalez et al. 2014), etc.; the widely used network polymers include poly(vinyl alcohol) (PVA), poly(acrylamide) (PAM), poly(ethylene glycol) (PEG), poly(N,N-dimethylacrylamide) (PDMA), etc. Besides, chemical incorporation of CNCs into hydrogel networks is also an effective and beneficial method to produce the hydrogels with excellent mechanical strength, which is mainly contributed to the emerging covalent bonds between CNCs and the other polymer phase (Chau et al. 2016). In these hydrogels, CNCs were cross-linked by themselves or the polymer networks. Our group fabricated a series of CNC-based hydrogels and investigated their properties. For example, as shown in Figure 3.6a, we fabricated a tough, stretchable, and hysteretic isotropic nanocomposite hydrogel through a simple approach (Yang et al. 2014a).

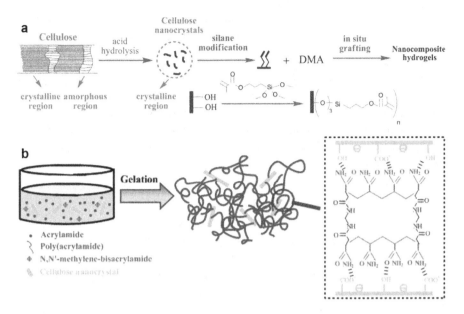

FIGURE 3.6 (a) Schematic illustration showing the preparation of CNC-PDMA nanocomposite hydrogels (Yang et al. 2014a). (b) Schematic illustration and network microscopic structure of CNC-PAM hydrogels (Yang et al. 2014b).

In this proposed hydrogel, the rod-like CNCs are encapsulated by flexible polymer chains of PDMA. In addition, as can be seen in Figure 3.6b, CNCs can also be combined with PAM to prepare the tough and stretchable composite hydrogels (Yang et al. 2014b). The attractive physical interactions in the network increased the fracture strength of the hydrogels by reversible adsorption–desorption processes on the CNC surface.

The CNFs with higher aspect ratio compared to CNCs have significant flexibility and propensity of entanglement. Therefore, the obtained hydrogels based on CNFs were mechanically strong and stable. CNFs have been widely used as building blocks to prepare the pure cellulose hydrogels. These hydrogels typically contain the CNFs of 0.05–6 wt%; the gelation can be induced by adding different metal salts and/or lowering the pH of suspensions (Fall et al. 2011, Dong et al. 2013). Besides, CNFs can also be used as reinforcing agents (or fillers) to fabricate the composite hydrogels. Compared with CNCs, lower content of CNFs was required for producing hydrogels, because CNFs have dramatic tendency to become the entangled networks. In 2017, our group prepared a composite hydrogel reinforced with CNFs, where there was an integration of reversible hydrogen bonds into a lightly covalently cross-linked poly(acrylamide) matrix (PAAm) (Yang and Feng 2017). The synergistic reinforcing mechanisms, i.e., interfacial dynamics, energy dissipation, and damage resistance, in the CNF-based composite hydrogels were also extensively demonstrated.

The as-proposed nanocellulose-based hydrogels have exhibited significant potentials for applications in various fields, such as biomedicine, tissue engineering scaffolds, strain/pressure sensors, electronic skins, and bioelectrodes (Heidarian et al. 2019). Specifically, the excellent biocompatibility, biodegradability, and adjustable surface chemistry of nanocellulose make the hydrogels to be the promising materials for drug delivery (Lin and Dufresne 2014, Xue et al. 2017). The BNC-based hydrogels were able to accelerate the wound healing (Lin et al. 2017, Mohamad et al. 2014). The electro-conductive hydrogels based on CNFs, PVA, and graphene have been used as stretchable strain sensor (Zheng et al. 2019). This sensor exhibited high flexibility, stretchability (up to 1000%), and sensitivity (GF = ~3.8). Furthermore, it had rapid self-healing ability and high healing efficiency, which endow the sensor with promising application in electronic skins.

3.3.3 Nanocellulose-based Aerogel

Aerogel is a synthetic porous ultralight material formed by replacement of liquid in a gel with gas. It was first created by Kistler (1931). The conventional aerogels mainly composed with silica were always fragile, which limited their applications in various fields. The mechanically flexible, strong, and compressible aerogel made of polymers were also prepared for various utilizations. Recently, the nanocellulose-based aerogels that combine the merits of cellulose (e.g., renewable, sustainable, biocompatible, and biodegradable) and porous architecture (e.g., ultralow density, high porosity, and excellent mechanical properties) have drawn increasing attentions from researchers (Lavoine and Bergström 2017).

The fabrication processes of nanocellulose-based aerogels include two steps: preparing a wet nanocellulose-based gel (i.e., hydrogel) and then removing the solvent without framework collapse. The recent progress on the fabrication of nanocellulose-based hydrogels has been discussed in Section 3.3.2. Herein, the methods to remove the solvent in hydrogel, i.e., drying technologies for production of aerogels are summarized. The critical point during solvent removal is to maintain the porous structure. However, the conventional drying process in atmospheric pressure would cause the collapsing of pores due to water evaporation (Scherer 1986). In the past decades, the main solvent-removal technologies to prepare aerogels are supercritical drying and freeze-drying. For the supercritical drying, the original solvent like water in hydrogels was replaced by a supercritical fluid (e.g., carbon dioxide), which avoided the formation of a liquid/vapour interface and thus the pores' collapse. Freeze-drying is always combined with ice-templating method to prepare the aerogels with iso- or anisotropic structures. Homogeneous freezing, no matter fast freezing by liquid nitrogen or slow freezing by refrigerator (−55°C), would produce the isotropic ice crystals and thus isotropic pores after freeze-drying. However, the directional freezing endowed the ice crystals with high alignment; the obtained aerogels hold the anisotropic structure, i.e., aligned pore channels. Importantly, the liquid/vapour interface is replaced by the solid/vapour interface during freeze-drying, which is critical to the production of porous aerogels.

The nanocellulose-based aerogels showed a wide range of applications, such as biomedical scaffolds, thermal insulation, pressure sensor, and energy-storage devices. The nanocellulose-based aerogels with interconnected porous network and the merits of biocompatibility and biodegradability were promising material for cell cultures. Besides, the composite aerogels made of nanocellulose and other conductive materials, such as carbon nanotube, graphene, and conductive polymers have been applied in the preparation of pressure sensors and/or energy devices. In addition, the carbon aerogels, which were fabricated by high-temperature treating nanocellulose-based aerogels under inert gases protected condition, have also demonstrated the potentials for applications in sensor and energy devices such as supercapacitors.

Our group has proposed a series of CNF-based aerogels for various applications. For example, as shown in Figure 3.7a, novel CNF-based aerogels with high lipophilicity, ultralow density, superior porosity were successfully prepared via freeze-drying CNFs suspension and salinization reaction in liquid phase (Zhou et al. 2016). Thus, as-prepared aerogels with excellent superhydrophobicity (water contact angle as high as 151.8°) exhibited excellent oil/water selective absorption capacity and reusability. As can be seen in Figure 3.7b, the obtained CNF-based aerogels by freeze-drying were used as templates for the synthesis of CNF/PPy/Ag hybrid aerogels by a simple dip-coating method (Zhou et al. 2015). These aerogels exhibited great antimicrobial, compressible, and electrically conductive properties, which endowed the aerogels with potentials for applications in wound healing, energy-storage devices, and pressure sensors. Besides, another superhydrophobic

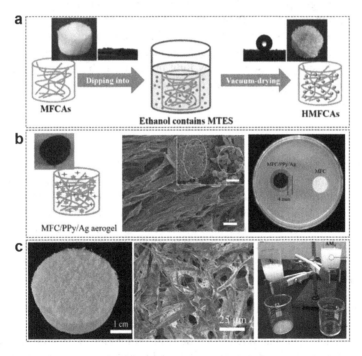

FIGURE 3.7 (a) Schematic illustration of the fabrication of superhydrophobic aerogels based on CNFs (Zhou et al. 2016). (b) Schemes and SEM image of the CNF/PPy/Ag composite aerogels with antimicrobial activities (Zhou et al. 2015). (c) Superhydrophobic CNF-assembled aerogels for highly efficient water-in-oil emulsions separation (Zhou et al. 2018).

composite aerogel made of CNFs, SiO_2, and methyltrimethoxysilane (MTMS) was created by a facile strategy (Figure 3.7c) (Zhou et al. 2018). The as-prepared aerogels had a hierarchical porous structure with high roughness and low surface energy, which endowed the resultant aerogels with superhydrophobicity (water contact angle = 168.4°). The composite aerogels could separate surfactant-stabilized water-in-oil emulsions without external pressure, with high separation efficiency (>99%) and high flux (1910 ± 60 L m^{-2} h^{-1}).

3.4 CELLULOSE NANOPAPER FOR SOLAR CELLS

As mentioned before, the cellulose nanopapers with both high transparency and haze have been applied in the field of solar cells. These nanopapers with the single function, i.e., light management, achieved a dramatic efficiency enhancement of solar cells by simple coating when compared with the bare cells (Fang et al. 2014, Ha et al. 2014, Jia et al. 2017). However, the accumulation of dust on the surface of solar cells outside would decrease light flux into the active layer and thus working efficiency of solar cells. To address this problem, the superhydrophobic surface with water contact angles (CAs) of larger than 150° and sliding angles (SAs) of less than 10°can be introduced to the solar cells, and the dust accumulation can be avoided *via* self-cleaning process.

In 2018, our group fabricated a superhydrophobic, highly transparent, and hazy nanopaper composed of TEMPO-oxidized CNFs and 3D nanostructured polysiloxanes via a facile process: vacuum filtration and in situ siloxanes growth. This nanopaper can not only lead to a significant enhancement in short circuit density and conversion efficiency of a solar cell by light management but also recover most of the photovoltaic performance losses due to dust accumulation by a self-cleaning process.

3.4.1 PREPARATION OF THE CELLULOSE NANOPAPER

As mentioned before, we obtained three types of nanocellulose, i.e., CNCs, CNFs, and 2,2,6,6-tetramethylpiperidine-1-loxy oxidized CNFs (TOCNFs), by chemical processes and/or mechanical treatment. Herein, the CNC, CNF, and TOCNF dispersions (0.1 wt %, 50 g) after stirring for 1 h at 800 rpm were vacuum filtrated with a glass suction filtration kit using a mixed cellulose ester membrane filter (pore size: 0.22 μm). Then, the wet sheets were carefully placed between filter papers and dried under pressure (0.01 MPa) at 105 °C for 10 min. Three types of cellulose nanopapers with ~40 μm thickness were fabricated and coded as CNC-P, CNF-P, and TOCNF-P, respectively.

To construct the superhydrophobic surface on the cellulose nanopaper, the nanopaper was further modified by the methyltrichlorosilane. In detail, the nanopapers were immersed into the 0.5 M methyltrichlorosilane solution of anhydrous toluene at room temperature for different time, as reported previously (Khoo and Tseng 2008, Gao and McCarthy 2006). Vessels were closed to the air but exposed to the chamber environment during the solution and sample introductions. After reaction, the samples were then removed and rinsed with various solvents in the following order:

toluene, ethanol, ethanol/deionized water (1:1), and DI water. Finally, the samples were dried in an oven at 55°C for 4 h. As an example, the obtained hydrophobic TOCNF-Ps after different modification time (10, 20, 30, 40 min) were coded as H10-TOCNF-P, H20-TOCNF-P, H30-TOCNF-P, and H40-TOCNF-P, respectively.

3.4.2 Characterization of the Cellulose Nanopaper

The surface morphology and structure of different nanopapers play a critical role in their properties. Herein, as can be seen in Figure 3.8, we compared the morphology of regular paper and three types of nanopapers that are made of different-sized cellulose nanofibres. The regular paper built up by the original cellulose fibres had a significantly rough surface and porous structure. These cavities would cause light scattering, thereby limiting the optical transparency of the regular cellulose paper. The various cellulose nanopapers illustrated different surface morphologies. The CNC-based nanopaper (CNC-P) had a significantly flattening and smooth surface; the CNF-P had a rough and uneven surface. For the TOCNF-P, its surface roughness was in between the CNC-P and CNF-P. While, all the cellulose nanopapers exhibited the dense structure rather than the porous structure in the regular paper. The surface morphology of the H40-TOCNF-P with different magnifications was also demonstrated. After methyltrichlorosilane (MTCS) modification for 40 min, the fibrous siloxanes network was constructed on the TOCNF-P surface. The pearl-necklace-like fibres were composed of interconnected spherical siloxanes particles, and the obtained siloxanes fibres were irregularly bent and randomly cross-linked with each other. However, the siloxanes were homogeneously distributed on the H40-TOCNF-P surface, as illustrated by the elemental mapping of Si.

A possible mechanism of polysiloxanes formation was proposed based on the literature. The trace water in the toluene solution led to the hydrolysis of MTCS, then the obtained silanols would react with the isolated hydroxyl groups on the surface of

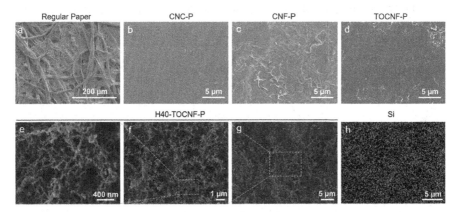

FIGURE 3.8 (a–d) The surface morphology of regular paper and nanopapers before modification under SEM. (e–h) The surface morphology of H40-TOCNF-P under SEM with different magnifications (left three) and elemental mapping of Si by energy dispersive X-ray spectroscopy (EDS) (rightmost).

TOCNF-P or other silanols. During this reaction, the Si-O-Si linkages were established on the substrates. Because the silanols with both hydrophilic groups (–OH) and hydrophobic groups (–CH₃) have the propensity to self-assemble, the nanospheres or nanofibres of siloxanes continued to grow in three dimensions and react with excessive Si-OH groups. Therefore, the 3D fibrous siloxanes network that consisted of rough pearl-necklace-like fibres was formed on the nanopaper surface. Besides, the successful silanization of the nanopaper was further confirmed by the Fourier transform infrared (FTIR) spectra. Compared with all the unmodified nanopapers, two new peaks were observed in the H40-TOCNF-P sample: the absorption bands at 1273 and 781 cm^{-1}, which are assigned to the asymmetric stretching vibrations of Si-CH₃ and the characteristic vibrations of Si-O-Si, respectively (Duan *et al.* 2014). These groups and linkages were consistent with the predicted chemical structures of polysiloxanes.

The optical properties of nanopapers are critical for substrates towards widespread applications. Both the total transmittance and transmittance haze of nanopapers were measured by the UV–Vis spectrometer and are shown in Figure 3.9. The CNC-P presented the highest transmittance of 91.3% but the lowest haze of only 20.1% at 550 nm. This may be due to the rod-like CNCs with small length constructed densely, allowing more light to propagate through and suppressing light scatter behaviour. Conversely, CNF-P showed the lowest transmittance (69.7% at 550 nm) but the highest haze (61.4% at 550 nm), which may be due to their rough surface caused by the large dimensions of CNFs. Interestingly, TOCNF-P not only had a high optical transmittance (90.4% at a wavelength of 550 nm), which was close to that of CNC-P, but it also exhibited a higher transmission haze (49.3% at 550 nm) than that of CNC-P. A possible explanation for this could be that each individual TOCNF led to small forward scattering rather than significant back scattering due to its nanoscale diameter and appropriate length, and thus the obtained densely laminated TOCNF-P allowed most of light to propagate through and retained an appropriate light-scattering effect, as reported in the literature (Hu et al. 2013). Furthermore, the obtained H40-TOCNF-P after modification still retained a high transmittance (90.2% at 550 nm) and high haze (46.5% at 550 nm).

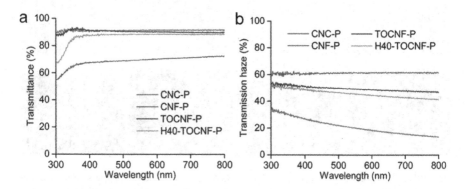

FIGURE 3.9 (a) Total transmittance and (b) transmittance haze of unmodified nanopapers and H40-TOCNF-P.

The mechanical properties like strength and toughness of nanopapers are critical for their wide range applications. Both the regular paper and CNC-P exhibited a very low strength of around 49 MPa. This may be attributed to that the native wood fibres with large dimensions in regular paper cannot bind tightly and densely, and the rod-like and rigid CNCs have limited contact area between fibres in the CNC-P. However, the CNF-P, TOCNF-P, and H40-TOCNF-P were much stronger (with the tensile strength of 92.8–103.7 MPa) and much tougher (with the toughness of 1.12–2.45 J M^{-3}) than the regular paper and CNC-P. This improvement in mechanical properties may be related to the enhanced cohesion, such as dispersion force and hydrogen bonds, and increased contact area between CNFs or TOCNFs (Nishiyama 2018). Interestingly, H40-TOCNF-P exhibited the highest toughness of 2.45 J M^{-3}, which was probably due to the formation of 3D polysiloxanes networks that increased the ability of H40-TOCNF-P to absorb energy or sustain deformation without breaking.

The wettability of regular paper and different cellulose nanopapers was characterized by the static CAs, which is illustrated in Figure 3.10. Due to outstanding hydrophilicity of original wood pulp and lots of cavities, the regular paper exhibited a completely hydrophilic property, as indicated by the zero CA and the wetted paper by water droplet. However, the TOCNF-P with CA of 93.5° presented hydrophobicity; this may be due to its dense structure and tight binding of TOCNFs inside. As expected, after modification by MTCS for 40 min, the obtained H40-TOCNF-P had

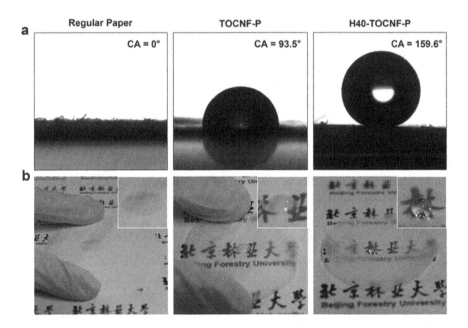

FIGURE 3.10 (a) Water contact angle of regular paper, TOCNF-P, and H40-TOCNF-P. (b) Photos of regular paper, TOCNF-P, and H40- TOCNF-P with the water droplet on the surface.

a significant water-repellency, as demonstrated by the sphere-like water droplet that settled on its surface. The H40-TOCNF-P has achieved the superhydrophilicity with the CA of 159.6°. Furthermore, the slide angle (SA) for H40-TOCNF-P is only 5.8°, indicating that the adhesion force between water droplet and nanopaper surface is ultralow. Therefore, the as-prepared superhydrophobic nanopaper was endowed with the self-cleaning function. The dust particles on the surface of H40-TOCNF-P could be easily collected and taken away by rolling water. This important characteristic reveals the potential applications of H40-TOCNF-P in many interdisciplinary technological fields.

3.4.3 Enhancement of Solar Cell Efficiency

The as-prepared superhydrophobic, transparent, and hazy cellulose nanopaper with both light-management and self-cleaning functions was used to improve the efficiency of solar cells. The photovoltaic performances of the bare cell and the cells coated with various nanopapers were evaluated by open circuit voltage (V_{oc}), short circuit density (J_{sc}), fill factor (FF), and the overall conversion efficiency. As shown in Table 3.2, both the TOCNF-P and H40-TOCNF-P significantly enhanced the conversion efficiency of the solar cell. This was mainly due to the optical properties, i.e., high transmittance and haze of nanopapers. High transmittance allowed most light to propagate through; high haze made the light diffusive, which increased the travelling path length and thus time of photos in the solar cell's active layer.

However, after contaminated with dust, the solar cell coated with nanopaper exhibited obvious decrease in photovoltaic performances, which was mainly contributed to the dust particles which blocked the incident light significantly. After self-cleaning process, the conversion efficiency of solar cell was dramatically increased. Compared to that of solar cell with native H40-TOCNF-P, the Jsc and the corresponding efficiency had recovered 94.80% and 96.76%, respectively. This photovoltaic performance was still slightly better than that of the bare solar cell.

TABLE 3.2

Photovoltaic Characteristics of the Bare Solar Cell and the Solar Cell with Different Nanopapers

	V_{oc} (V)	J_{sc} (mA cm^{-2})	FF (%)	Efficiency (%)
Bare cell	2.494	5.282	76.47	10.07
Cell + TOCNF-P	2.508	5.915	77.37	11.48
Cell + H40-TOCNF-P	2.531	5.692	77.21	11.12
Cell + contaminated H40-TOCNF-P	2.469	3.493	60.95	6.00
Cell + self-cleaned H40-TOCNF-P	2.546	5.396	78.35	10.76

3.5 NANOCELLULOSE-BASED HYDROGEL FOR STRAIN SENSOR

The wearable strain sensor with self-healing property has the ability to restore its structure and sensing function from damage. Therefore, this kind of self-healing strain sensor exhibits excellent safety, durability, and reliability, which have raised more attention of researchers (Li et al. 2017). Recently, the self-healing conductive gels have been extensively investigated and applied as wearable strain sensors in various fields. However, the conventional self-healing conductive gels always have a relatively low mechanical strength due to the dynamic cross-links, which has limited their application in strain sensing with stable performances (Chen et al. 2012). To endow the conductive gels with both great self-healing and mechanical properties, incorporating nanoscale fillers (e.g., CNCs) into the gel matrix that could dissipate energy is a promising approach, because the added fillers would tune the interactions between themselves and the surrounding polymer phase in conductive gels (De France et al. 2017).

In addition, self-adhesive function is also significant for the practical application of conductive gels as wearable strain sensors. The self-adhesive strain sensors can be directly attached onto the surface of the object to be tested (e.g., skin of human body) without the assistance of additional adhesive tapes and/or gels. Compared to the adhesive polydopamine (PDA), the colourless tannic acid (TA) with the merits of low cost and biocompatibility is regarded as a better material for the preparation of 'green' and safe self-adhesive hydrogels (Sileika et al. 2013, Krogsgaard et al. 2016). In this chapter, novel, tough, self-healing, and self-adhesive CNC-based hydrogels were designed and fabricated; their potential applications as wearable strain sensors were also demonstrated.

3.5.1 PREPARATION OF THE NANOCELLULOSE-BASED HYDROGEL

The CNCs coated with TA (TA@CNC) were firstly fabricated. As shown in Figure 3.11a, the CNCs colloidal suspension was stable with light blue colour. After mixing with TA, the obtained TA@CNC solution remained stable and with the colloidal dispersion, but the colour was changed into slightly yellow. Figure 3.11b presents the atomic force microscope (AFM) images of CNC and TA@CNC. As mentioned before, CNCs were rod-like crystals with low aspect ratio, while the TA@CNCs had dramatically thickened walls due to the deposition of TA onto the CNCs surface. The introduction of TA into the TA@CNC can be further proved by the dimension distribution and FTIR spectra.

The fabrication process of CNC-based hydrogel is illustrated in Figure 3.11c. The acrylic acid (AA, monomer), ammonium persulfate (APS, initiator), N,N′-methylenebis(acrylamide) (MBA, chemical cross-linker), and TA@CNC were added together in aqueous solution to prepare cross-linked composite gels by free radical polymerization. Then, the as-fabricated composite gels were soaked in 0.1 mol L^{-1} AlCl$_3$ aqueous solution to generate ionically cross-linked domains. Finally, the gels were soaked in DI water for more than 24 h to remove superfluous cations. The obtained CNC-based hydrogels had a hierarchically porous structure and coordination bonds among TA@CNC, PAA, and Al^{3+}. The possible coordination modes were

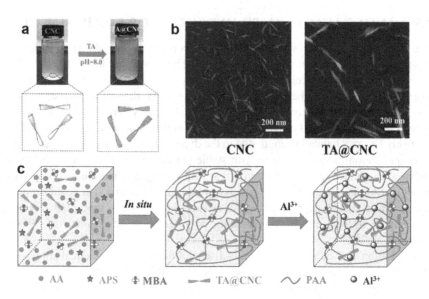

FIGURE 3.11 (a) Photographs of the CNC (1.02 wt%) and TA@CNC (1.2 wt%) suspension. (b) AFM images of CNC and TA@CNC. (c) Schematic illustration of the fabrication of the CNC-based hydrogel.

metal-phenolic coordination between TA@CNCs, metal-carboxylate coordination between PAA chains, as well as hybrid bridging between TA@CNCs and PAA chains.

3.5.2 Characterization of the Nanocellulose-based Hydrogel

The mechanical properties of CNC-based hydrogels were demonstrated by the tensile measurement. Figure 3.12a presents the stress–strain curves of the as-prepared hydrogels with different TA@CNC contents. When compared to the pure PAA, the hydrogels with TA@CNC exhibited significantly enhanced mechanical properties. In particular, the hydrogel with 0.6 wt% TA@CNC has the highest elasticity (strain up to 2900%), which is also illustrated by the photograph in Figure 3.12a. The hydrogel with 0.8 wt% TA@CNC exhibits the highest toughness (5.60 MJ m^{-3}), which is extremely higher than that for the pure PAA. The enhancement of mechanical properties by TA@CNC for hydrogels may be contributed to the resistance against crack propagation in bunting and energy dissipation. The hydrogels have great self-recovery ability and could dissipate energy with pronounced hysteresis loops. The highest recovery ratio for the CNC-based hydrogels was 92.5% with 12% residual strain, indicating their excellent resilience and fatigue resistance. Furthermore, the hydrogels also exhibited brilliant recovery properties against compressing; the great compressive toughness endowed the hydrogel with the ability to withstand 85% deformation and almost full recovery.

Self-healing function is critical to the CNC-based hydrogels with a stable performance of strain sensing in practical application. Figure 3.12b illustrates the

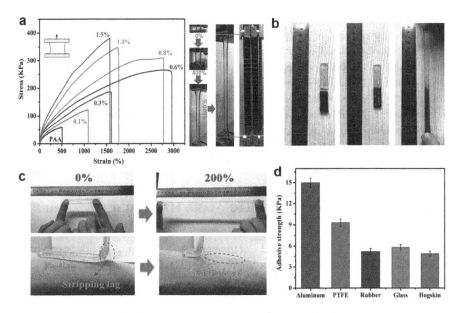

FIGURE 3.12 (a) Stress–strain curves for the hydrogels with different TA@CNC contents and the photograph of the hydrogel under stretching state. (b) Photograph of the hydrogels before and after self-healing. (c) Photograph showing the excellent adhesive properties of the hydrogels. (d) Adhesion strength of the hydrogels with different substrates tested by tensile adhesion tests.

self-healing behaviour of rectangle-shaped hydrogel specimens. The specimen was cut into two parts and one of them was stained by methylene blue to make it better visible. As can be seen, two parts of the hydrogel specimen could be recombined together when they were placed in contact without any additional stress. Furthermore, the self-healed hydrogel still exhibited great mechanical strength and toughness during stretching. Besides, the effect of time on the self-healing efficiency of hydrogels was also investigated. By increasing the self-healing time, the fracture stress of the healed hydrogels dramatically increased, which indicates that enlarging processing time would improve the self-healing properties of CNC-based hydrogels. In addition, increasing the TA@CNC contents in hydrogels was also beneficial for their self-healing behaviours, because the dynamic TA@CNC motifs play a crucial role for the reversible rearrangement in the self-healing process.

The CNC-based hydrogels exhibited an excellent self-adhesive property by mimicking the mussel adhesion principle through the catechol groups of oxidized polyphenols. The hydrogels have strong adhesion to the surfaces of various substrates, such as glass, poly(tetrafluoroethylene) (PTFE), rubbers, wood, and carnelian. As can be seen in Figure 3.12c, the hydrogel can be stably adhered on the fingers to achieve up to 200% recoverable stretching, which indicated the excellent self-adhesiveness of CNC-based hydrogels with skin. Furthermore, the already adhered hydrogel can also be easily peeled off and removed from the skin without any residues or the irritation of skin. Importantly, the stripping lag at the gel–skin interface was crucial to the immediate detachment between CNC-based hydrogels and skin; however, no stripping lag was observed for the pure PAA hydrogels, indicating that the TA@CNC could improve the self-adhesive properties of the hydrogels. Figure 3.12d shows the adhesion strength of the CNC-based hydrogels to the surfaces of different substrates. The aluminium had the highest adhesion strength among them, which may be contributed to the synergy of metal complexation and hydrogen bonding between hydrogel and aluminium. The PTFE also exhibited relative high adhesion strength due to both the hydrophobic interaction and hydrogen bonding. The aforementioned excellent features (mechanical, self-healing, and self-adhesive) of CNC-based hydrogels are potentially attractive for practical applications as wearable strain sensors.

3.5.3 Application of the Nanocellulose-based Hydrogel for Wearable Strain Sensor

The CNC-based hydrogel had a good electric conductivity; it could work as one part of the circuit to lighten the light-emitting diode (LED). Furthermore, the brightness of this LED changed when the hydrogel was starched or compressed, indicating the piezoresistant feature of CNC-based hydrogel (ionic gel). The relative resistance change of the hydrogel under different strains was further tested; the relative resistance change is defined as $\Delta R/R_0 = (R - R_0)/R_0$, where R_0 is the resistance of the hydrogel in releasing state and R is the resistance of the hydrogel after stretching. As can be seen in Figure 3.13a, the relative resistance change is increased linearly by increasing strain. The relative resistance-change curve of the hydrogel was divided into three linear parts and the hydrogel sensor's sensitivity was evaluated by gauge factor (S), which is defined as $S = (\Delta R/R_0)/\varepsilon$, where ε is the applied strain. In the strain range of 0–40%, the gauge factor of the hydrogel sensor was 0.23, while it increased to 4.90 under 65–75% strain. The inset of Figure 3.13a shows the gauge factor changes of hydrogel sensor under different strain; the gauge factor increased from 5.5 to 7.8 by increasing the strain from 100% to 2000%. Besides, the hydrogel sensor also exhibited excellent stability and durability for strain sensing.

The electrical self-healing property of the CNC-based hydrogel was also demonstrated by connecting in series with an LED in the circuit. The LED could be relighted

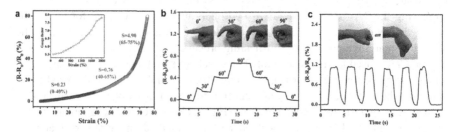

FIGURE 3.13 (a) Relative resistance changes of the hydrogel sensor as a function of the applied strain. (b) Relative resistance changes of the hydrogel-based strain sensor to monitor finger bending with different angles. (c) Relative resistance changes of the sensor for wrist bending.

when two separated hydrogels were recombined after self-healing; the electrical-conductive ability of the hydrogel was almost recovered. This reveals that the hydrogel-based strain sensor had great reliability against physical damage and can be applied in various fields.

Taking advantages of great self-adhesiveness, high sensitivity for strain sensing, excellent electrical stability, and fast self-healing ability, the CNC-based hydrogel was used as wearable strain sensor to detect human motions by attaching it on body skin. As can be seen in Figure 3.13b, the hydrogel-based strain sensor was used to detect the forefinger under bending with different angles. As the bending angles increased from 0° to 90°, the relative resistance changes of the sensor also exhibited monotonical increment and it had the stepwise feature. As expected, the signal intensity was decreased by decreasing the finger bending angles. Importantly, the signal intensity under the same bending angle was consistent in case of both stepwise increased and decreased bending angels. This indicates the great reliability and reproducibility of the hydrogel-based strain sensor, which is vital to the sensor's practical applications. Figure 3.13c illustrates the application of the hydrogel-based sensor to detect elbow's motions, such as repeated bending and releasing. The obtained signals were dramatic and stable, which could demonstrate the bending state of elbow. Besides, the hydrogel-based sensor was able to accurately detect subtle motions, such as pulse and breath. Therefore, it could be used for real-time monitoring of personal healthcare. Furthermore, the obtained signals of strain sensing can be transmitted wirelessly to the smartphones through blue-tooth module, enabling users to analyse the sensing data and interact with the sensor.

3.6 NANOCELLULOSE-BASED CARBON AEROGEL FOR SUPERCAPACITOR

Carbon aerogels with lots of merits, such as low density, large specific surface area, and high electrical conductivity have drawn increasing attentions from researchers in recent years (Bryning et al. 2007, Wu et al. 2013, Bi et al. 2013). As renewable biomass material, cellulose fibres and/or nanofibres have a high carbon content and are the promising precursor for the production of carbon aerogels. However, the conventional cellulose nanofibre-based carbon aerogels have some limitation, such as fragile and easy-destroyed features as well as complicated fabrication processes including costly freeze-drying (Zu et al. 2016).

To prepare the carbon aerogel with outstanding mechanical properties, especially high compressibility and fatigue resistance, the melamine foams (MF) with interconnected networks were used as skeleton. By high-temperature carbonization, the obtained aerogels made of MF and nanocellulose (TOCNF) were converted into carbon aerogels (TMCA). Their mechanical property and electrochemical performances as electrodes for supercapacitor were further demonstrated.

3.6.1 Preparation of the Nanocellulose-based Carbon Aerogel

Figure 3.14a shows the fabrication process of the TOCNF-based carbon aerogel (TMCA). As can be seen, MF and TOCNFs were combined by dip-coating and then drying at ambient pressure. The MF with an interconnected 3D scaffold; worked as

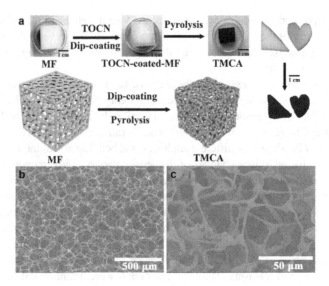

FIGURE 3.14 (a) Schematic illustration and photograph of the nanocellulose-based carbon aerogel. SEM image of (b) melamine foam and (c) nanocellulose-based carbon aerogel.

skeleton to support the nanocellulose materials and avoid the structure shrinkage of aerogel during drying. Subsequently, the obtained hybrid TOCNF/MF aerogel (TMA) was carbonized at 800°C under N_2 atmosphere to prepare the TMCA. As shown by the photographs, the TMCA retained its original shapes after pyrolysis at high temperature. However, the TMCA exhibited obvious shrinkage and its density was decreased from 13.9 to 11.23 mg cm^{-3} after pyrolysis.

Figure 3.14b and c are the SEM images of MF and TMCA, respectively. As observed, the MF has an interconnected network architecture with a smooth triangle fibre shape. After carbonization, the interconnected porous structure of aerogel was maintained, which endowed TMCA with great compressibility. Besides, the carbonized TOCNFs were wrapped around the carbonized MF skeleton, which led to the rougher skeleton fibres in TMCA than that in untreated aerogel. Furthermore, the density and morphology of as-prepared TMCA could be tuned by adjusting the nanocellulose concentration of TOCNF suspensions during the dip-coating process, which would lead to great potential applications for TMCA in wider-range fields.

3.6.2 Characterization of the Nanocellulose-based Carbon Aerogel

The pyrolysis process of TMA was evaluated by thermo gravimetric analysis (TGA). From room temperature to 210°C, the weight loss that ascribed to the water desorption of TOCN was only 9.9%. At 210–325°C, the obvious weight loss was mainly contributed to the dehydration and depolymerization of cellulose. While at temperatures from 325 to 400°C, the decomposition of melamine led to the weight loss of 22.1% for TMA. In the final stage from 400 to 800°C, the weight loss was relatively stable due to the carbonization of residual structures.

X-ray diffraction (XRD) pattern, Raman spectra, and x-ray photoelectron (XPS) were further used to investigate the chemical structures of TMCA. During carbonization, the amorphous carbon was converted into graphitic carbon; the obtained TMCA had high degree of graphitization, which is critical to the electrical conductivity of TMCA. The TMCA had 78.46% carbon, which was from MF (62.19%) and TOCNF (37.81%). Besides, TMCA had various nitrogen functional groups, such as $-NH_2$, pyridinic N, and graphitic N, which would increase the electrical conductivity and even capacitance of TMCA.

The mechanical properties including compressibility and fatigue resistance of TMCA were evaluated by stress–strain measurements. As shown by the photographs in Figure 3.15, the TMCA was cyclically compressed and released with different strains. The stress–strain curves indicate that the TMCA was significantly soft and tough enough against collapse during compression. Besides, the TMCA only exhibited a slight plastic deformation after 500 cycles of compressing and releasing with 50% strain when compared with its original state, which demonstrate the excellent mechanical stability and fatigue-resistance properties of the TMCA. Furthermore, the TMCA's unique structure of interconnected network and wrapped carbon fibres around skeleton was still well maintained after cyclic compression and releasing, which is critical to its stable performances for application in various fields.

The TMCA exhibited great electric property; its conductivity was higher than pure MF carbon aerogel. Besides, the electric conductivity of TMCA was increased by increasing the concentration of TOCNF. Unlike the conventional carbon materials that were always hydrophobic, the TMCA we prepared, was both hydrophilic and oleophilic due to its high nitrogen content. This strong wetting ability can dramatically enhance the infiltration of electrolyte at the electrolyte/TMCA interface, which is beneficial to the TMCA's electrochemical performances.

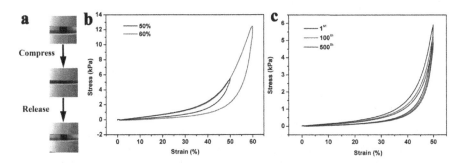

FIGURE 3.15 (a) The compression and releasing process of nanocellulose-based carbon aerogel at 60% strain. (b) The stress–strain curves of nanocellulose-based carbon aerogel at 50% and 60% strain. (c) The stress–strain curves of nanocellulose-based carbon aerogel at the maximum strain of 50% after 1, 100, and 500 cycles.

3.6.3 Application of the Nanocellulose-based Carbon Aerogel for Supercapacitor

The TMCA with high nitrogen content, hydrophilicity, and coherent conductive network can be used as compressible electrodes to prepare high-performance supercapacitors. To investigate the electrochemical performances of TMCA, its cyclic voltammetry (CV) curves, galvanostatic charge-discharge (GCD) curves, and electrochemical impedance spectroscopy (EIS) were measured in a three-electrode configuration. As shown in Figure 3.16a, the curves with different scanning rates exhibited nearly rectangular shape, which indicates the electrical double-layer capacitive features of TMCA. Besides, the N heteroatoms of TMCA might lead to the slight distortion of the CV curves. Based on the GCD curves in Figure 3.16b, the specific and areal capacitance of TMCA were calculated as 92.2 F g^{-1} and 461 mF cm^{-2} at a current density of 1 mA cm^{-2}. Despite relatively high capacitance, the fabrication process of TMCA was facile and cost-effective, revealing the potential commercial applications of TMCA for energy storage.

The electrochemical behaviour of TMCA under different strains was investigated to evaluate its potential application for compressible electrodes of energy-storage devices. Figure 3.16d and e are the CV curves (at a scan rate of 1 mV s^{-1}) and GCD curves (at a current density of 1 mA cm^{-2}) of TMCA under different strains, respectively. The approximately rectangular shapes of CV curves indicate the electric double-layer capacitors (EDLC) behaviour of TMCA under compression with different strains. The GCD curves demonstrated that the capacitance of TMCA was increased by increasing the strain, which may be contributed to the decreased internal

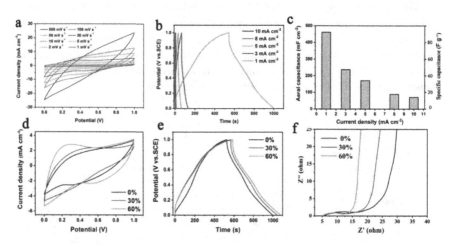

FIGURE 3.16 (a) CV curves of the nanocellulose-based carbon aerogel at the scan rates from 1 mV s^{-1} to 500 mV s^{-1}. (b) GCD curves of the nanocellulose-based carbon aerogel at the current densities from 1 mA cm^{-2} to 10 mA cm^{-2}. (c) Areal capacitance and specific capacitance calculated for the nanocellulose-based carbon aerogel electrode. (d) CV curves of the nanocellulose-based carbon aerogel at a scan rate of 1 mV s^{-1} under different strains. (e) GCD curves of nanocellulose-based carbon aerogel at a current density of 1 mA cm^{-2} under different strains. (f) EIS spectra of nanocellulose-based carbon aerogel under different strains.

resistance of TMCA under compression. The EIS of TMCA at 0.1–100 kHz under different strains was also measured to investigate the electrochemical transfer mechanism of TMCA. As shown in Figure 3.16f, all the Nyquist plots consist of a typical semicircle and a straight line. The equivalent series resistance was decreased from 10.5 to 6.2 by increasing the strain, indicating the improvement of charge-transfer ability at the electrode–electrolyte interfaces and the decline of ion diffusion resistance. This may be ascribed to the enhanced conductivity and the shortened ion diffusion pathway at higher strains.

3.7 CONCLUSIONS

In summary, the fabrication, property, and application of nanocellulose-based functional materials have been briefly introduced in this chapter. The nanocellulose with merits of renewability, high mechanical strength, and low coefficient of thermal expansion, etc. can be used to prepare advanced functional materials for applications in various fields, such as wearable stain sensors, energy-storage devices, and flexible displays. Furthermore, our group's recent works on the various nanocellulose-based functional materials have been discussed in detail. The superhydrophobic, highly transparent, and hazy cellulose nanopaper could improve the solar cells' efficiency and address the issue of dust accumulation. The as-prepared nanocellulose-based composite carbon aerogel with the compressible and conductive property exhibited great potential for applications in compressible electrodes for supercapacitors. Besides, the tough, self-healing, and self-adhesive nanocellulose-based hydrogel was a promising candidate for wearable strain sensors.

REFERENCES

Ahmed, Enas M. 2015. Hydrogel: Preparation, characterization, and applications: A review. *Journal of Advanced Research* 6 (2):105–121.

Bi, Hengchang, Zongyou Yin, Xiehong Cao, et al. 2013. Carbon fiber aerogel made from raw cotton: A novel, efficient and recyclable sorbent for oils and organic solvents. *Advanced Materials* 25 (41):5916–5921.

Bryning, Mateusz B, Daniel E Milkie, Mohammad F Islam, Lawrence A Hough, James M Kikkawa, and Arjun G Yodh. 2007. Carbon nanotube aerogels. *Advanced Materials* 19 (5):661–664.

Cannon, Robert E., and Steven M. Anderson. 1991. Biogenesis of bacterial cellulose. *Critical Reviews in Microbiology* 17 (6):435–447.

Cao, Wen-Tao, Fei-Fei Chen, Ying-Jie Zhu, et al. 2018. Binary strengthening and toughening of MXene/cellulose nanofiber composite paper with nacre-inspired structure and superior electromagnetic interference shielding properties. *ACS Nano* 12 (5):4583–4593.

Chau, Mokit, Kevin J De France, Bernd Kopera, et al. 2016. Composite hydrogels with tunable anisotropic morphologies and mechanical properties. *Chemistry of Materials* 28 (10):3406–3415.

Chen, Chaoji, and Liangbing Hu. 2018. Nanocellulose toward advanced energy storage devices: Structure and electrochemistry. *Accounts of Chemical Research* 51 (12):3154–3165.

Chen, Sheng, Yijia Song, and Feng Xu. 2018a. Highly transparent and hazy cellulose nanopaper simultaneously with a self-cleaning superhydrophobic surface. *ACS Sustainable Chemistry & Engineering* 6 (4):5173–5181.

Chen, Wenshuai, Haipeng Yu, Sang-Young Lee, Tong Wei, Jian Li, and Zhuangjun Fan. 2018b. Nanocellulose: A promising nanomaterial for advanced electrochemical energy storage. *Chemical Society Reviews* 47 (8):2837–2872.

Chen, Yulin, Aaron M Kushner, Gregory A Williams, and Zhibin Guan. 2012. Multiphase design of autonomic self-healing thermoplastic elastomers. *Nature Chemistry* 4 (6):467.

Curvello, Rodrigo, Vikram Singh Raghuwanshi, and Gil Garnier. 2019. Engineering nanocellulose hydrogels for biomedical applications. *Advances in Colloid and Interface Science* 267:47–61.

De France, Kevin J, Todd Hoare, and Emily D Cranston. 2017. Review of hydrogels and aerogels containing nanocellulose. *Chemistry of Materials* 29 (11):4609–4631.

Dong, Hong, James F Snyder, Kristen S Williams, and Jan W Andzelm. 2013. Cation-induced hydrogels of cellulose nanofibrils with tunable moduli. *Biomacromolecules* 14 (9):3338–3345.

Du, Haishun, Wei Liu, Miaomiao Zhang, Chuanling Si, Xinyu Zhang, and Bin Li. 2019. Cellulose nanocrystals and cellulose nanofibrils based hydrogels for biomedical applications. *Carbohydrate Polymers* 209:130–144.

Duan, Bo, Huimin Gao, Meng He, and Lina Zhang. 2014. Hydrophobic modification on surface of chitin sponges for highly effective separation of oil. *ACS Applied Materials & Interfaces* 6 (22):19933–19942.

Fall, Andreas B, Stefan B Lindström, Ola Sundman, Lars Ödberg, and Lars Wågberg. 2011. Colloidal stability of aqueous nanofibrillated cellulose dispersions. *Langmuir* 27 (18):11332–11338.

Fang, Zhiqiang, Gaoyuan Hou, Chaoji Chen, and Liangbing Hu. 2019. Nanocellulose-based films and their emerging applications. *Current Opinion in Solid State and Materials Science* 23 (4):100764.

Fang, Zhiqiang, Hongli Zhu, Yongbo Yuan, et al. 2014. Novel nanostructured paper with ultra-high transparency and ultrahigh haze for solar cells. *Nano Letters* 14 (2):765–773.

Fernandes, Susete N, Pedro L Almeida, Nuno Monge, et al. 2017. Mind the microgap in iridescent cellulose nanocrystal films. *Advanced Materials* 29 (2):1603560.

Fu, Xiaotong, Hairui Ji, Binshou Wang, Wenyuan Zhu, Zhiqiang Pang, and Cuihua Dong. 2020. Preparation of thermally stable and surface-functionalized cellulose nanocrystals by a fully recyclable organic acid and ionic liquid mediated technique under mild conditions. *Cellulose* 27 (3):1289–1299.

Gao, Lichao, and Thomas J McCarthy. 2006. A perfectly hydrophobic surface (θA/θR= 180/180). *Journal of the American Chemical Society* 128 (28):9052–9053.

Gaurav, N, S Sivasankari, GS Kiran, A Ninawe, and J Selvin. 2017. Utilization of bioresources for sustainable biofuels: A review. *Renewable and Sustainable Energy Reviews* 73:205–214.

Gonzalez, Jimena S, Leandro N Ludueña, Alejandra Ponce, and Vera A Alvarez. 2014. Poly (vinyl alcohol)/cellulose nanowhiskers nanocomposite hydrogels for potential wound dressings. *Materials Science and Engineering: C* 34:54–61.

Guidetti, Giulia, Siham Atifi, Silvia Vignolini, and Wadood Y Hamad. 2016. Flexible photonic cellulose nanocrystal films. *Advanced Materials* 28 (45):10042–10047.

Ha, Dongheon, Zhiqiang Fang, Liangbing Hu, and Jeremy N Munday. 2014. Paper-based anti-reflection coatings for photovoltaics. *Advanced Energy Materials* 4 (9):1301804.

Heidarian, Pejman, Abbas Z Kouzani, Akif Kaynak, et al. 2019. Dynamic plant-derived polysaccharide-based hydrogels. *Carbohydrate Polymers*:115743.

Hoare, Todd R, and Daniel S Kohane. 2008. Hydrogels in drug delivery: Progress and challenges. *Polymer* 49 (8):1993–2007.

Hu, Liangbing, Guangyuan Zheng, Jie Yao, et al. 2013. Transparent and conductive paper from nanocellulose fibers. *Energy & Environmental Science* 6 (2):513–518.

Huang, Jia, Hongli Zhu, Yuchen Chen, et al. 2013. Highly transparent and flexible nanopaper transistors. *ACS Nano* 7 (3):2106–2113.

Hubbe, Martin A, Pegah Tayeb, Michael Joyce, et al. 2017. Rheology of nanocellulose-rich aqueous suspensions: A review. *BioResources* 12 (4):9556–9661.

Isogai, Akira, Tsuguyuki Saito, and Hayaka Fukuzumi. 2011. TEMPO-oxidized cellulose nanofibers. *Nanoscale* 3 (1):71–85.

Jia, Chao, Tian Li, Chaoji Chen, et al. 2017. Scalable, anisotropic transparent paper directly from wood for light management in solar cells. *Nano Energy* 36:366–373.

Jiang, Feng, and You-Lo Hsieh. 2016. Self-assembling of TEMPO oxidized cellulose nano-fibrils as affected by protonation of surface carboxyls and drying methods. *ACS Sustainable Chemistry & Engineering* 4 (3):1041–1049.

Khoo, Hwa Seng, and Fan-Gang Tseng. 2008. Engineering the 3D architecture and hydropho-bicity of methyltrichlorosilane nanostructures. *Nanotechnology* 19 (34):345603.

Kistler, Samuel Stephens. 1931. Coherent expanded aerogels and jellies. *Nature* 127 (3211):741–741.

Klemm, Dieter, Friederike Kramer, Sebastian Moritz, et al. 2011. Nanocelluloses: A new family of nature-based materials. *Angewandte Chemie International Edition* 50 (24):5438–5466.

Kotelev, MS, ZV Bobyleva, IA Tiunov, DA Sharipova, and AA Novikov. 2017. Investigation of the influence of conditions of reprecipitation of cellulose on its adsorption properties. *Chemistry and Technology of Fuels and Oils* 53 (5):722–726.

Krogsgaard, Marie, Vicki Nue, and Henrik Birkedal. 2016. Mussel-inspired materials: Self-healing through coordination chemistry. *Chemistry–A European Journal* 22 (3):844–857.

Lavoine, Nathalie, and Lennart Bergström. 2017. Nanocellulose-based foams and aero-gels: Processing, properties, and applications. *Journal of Materials Chemistry A* 5 (31):16105–16117.

Li, Jingjing, Lifang Geng, Gang Wang, Huijuan Chu, and Hongliang Wei. 2017. Self-healable gels for use in wearable devices. *Chemistry of Materials* 29 (21):8932–8952.

Lin, Ning, and Alain Dufresne. 2014. Nanocellulose in biomedicine: Current status and future prospect. *European Polymer Journal* 59:302–325.

Lin, Shin-Ping, Iris Loira Calvar, Jeffrey M Catchmark, Je-Ruei Liu, Ali Demirci, and Kuan-Chen Cheng. 2013. Biosynthesis, production and applications of bacterial cellulose. *Cellulose* 20 (5):2191–2219.

Lin, Shin-Ping, Hsiu-Ni Kung, You-Shan Tsai, Tien-Ni Tseng, Kai-Di Hsu, and Kuan-Chen Cheng. 2017. Novel dextran modified bacterial cellulose hydrogel accelerating cutane-ous wound healing. *Cellulose* 24 (11):4927–4937.

Ma, Lina, Rong Liu, Haijun Niu, Fang Wang, Li Liu, and Yudong Huang. 2016. Freestanding conductive film based on polypyrrole/bacterial cellulose/graphene paper for flexible supercapacitor: Large areal mass exhibits excellent areal capacitance. *Electrochimica Acta* 222:429–437.

McKee, Jason R, Eric A Appel, Jani Seitsonen, Eero Kontturi, Oren A Scherman, and Olli Ikkala. 2014. Healable, stable and stiff hydrogels: Combining conflicting properties using dynamic and selective three-component recognition with reinforcing cellulose nanorods. *Advanced Functional Materials* 24 (18):2706–2713.

Mendoza, Llyza, Warren Batchelor, Rico F Tabor, and Gil Garnier. 2018. Gelation mechanism of cellulose nanofibre gels: A colloids and interfacial perspective. *Journal of Colloid and Interface Science* 509:39–46.

Mohamad, Najwa, Mohd Cairul Iqbal Mohd Amin, Manisha Pandey, Naveed Ahmad, and Nor Fadilah Rajab. 2014. Bacterial cellulose/acrylic acid hydrogel synthesized via electron beam irradiation: Accelerated burn wound healing in an animal model. *Carbohydrate Polymers* 114:312–320.

Nishiyama, Yoshiharu. 2018. Molecular interactions in nanocellulose assembly. *Philosophical Transactions of the Royal Society A: Mathematical, Physical and Engineering Sciences* 376 (2112):20170047.

Nogi, Masaya, Shinichiro Iwamoto, Antonio Norio Nakagaito, and Hiroyuki Yano. 2009. Optically transparent nanofiber paper. *Advanced Materials* 21 (16):1595–1598.

Nogi, Masaya, Makoto Karakawa, Natsuki Komoda, Hitomi Yagyu, and Thi Thi Nge. 2015. Transparent conductive nanofiber paper for foldable solar cells. *Scientific Reports* 5:17254.

Park, Junsu, Jung-Hun Seo, Seung-Won Yeom, et al. 2018. Flexible and transparent organic phototransistors on biodegradable cellulose nanofibrillated fiber substrates. *Advanced Optical Materials* 6 (9):1701140.

Purandare, Sumit, Eliot F Gomez, and Andrew J Steckl. 2014. High brightness phosphorescent organic light emitting diodes on transparent and flexible cellulose films. *Nanotechnology* 25 (9):094012.

Sabo, Ronald, Aleksey Yermakov, Chiu Tai Law, and Rani Elhajjar. 2016. Nanocellulose-enabled electronics, energy harvesting devices, smart materials and sensors: A review. *Journal of Renewable Materials* 4 (5):297–312.

Scherer, George W. 1986. Drying gels: I. General theory. *Journal of Non-Crystalline Solids* 87 (1–2):199–225.

Sehaqui, Houssine, Ngesa Ezekiel Mushi, Seira Morimune, Michaela Salajkova, Takashi Nishino, and Lars A Berglund. 2012. Cellulose nanofiber orientation in nanopaper and nanocomposites by cold drawing. *ACS Applied Materials & Interfaces* 4 (2):1043–1049.

Shao, Changyou, Meng Wang, Lei Meng, et al. 2018. Mussel-inspired cellulose nanocomposite tough hydrogels with synergistic self-healing, adhesive, and strain-sensitive properties. *Chemistry of Materials* 30 (9):3110–3121.

Shi, Zhijun, Yue Zhang, Glyn O Phillips, and Guang Yang. 2014. Utilization of bacterial cellulose in food. *Food Hydrocolloids* 35:539–545.

Sileika, Tadas S, Devin G Barrett, Ran Zhang, King Hang Aaron Lau, and Phillip B Messersmith. 2013. Colorless multifunctional coatings inspired by polyphenols found in tea, chocolate, and wine. *Angewandte Chemie International Edition* 52 (41):10766–10770.

Tang, Hu, Núria Butchosa, and Qi Zhou. 2015. A transparent, hazy, and strong macroscopic ribbon of oriented cellulose nanofibrils bearing poly (ethylene glycol). *Advanced Materials* 27 (12):2070–2076.

Thomas, Bejoy, Midhun C Raj, Jithin Joy, Audrey Moores, Glenna L Drisko, and Clément Sanchez. 2018. Nanocellulose, a versatile green platform: From biosources to materials and their applications. *Chemical Reviews* 118 (24):11575–11625.

Ummartyotin, S, J Juntaro, M Sain, and H Manuspiya. 2012. Development of transparent bacterial cellulose nanocomposite film as substrate for flexible organic light emitting diode (OLED) display. *Industrial Crops and Products* 35 (1):92–97.

Veigel, Stefan, Ulrich Müller, Jozef Keckes, Michael Obersriebnig, and Wolfgang Gindl-Altmutter. 2011. Cellulose nanofibrils as filler for adhesives: Effect on specific fracture energy of solid wood-adhesive bonds. *Cellulose* 18 (5):1227.

Walther, Andreas, Jaakko VI Timonen, Isabel Díez, Antti Laukkanen, and Olli Ikkala. 2011. Multifunctional high-performance biofibers based on wet-extrusion of renewable native cellulose nanofibrils. *Advanced Materials* 23 (26):2924–2928.

Wang, Meng, Yanglei Chen, Yanlin Qin, Tiejun Wang, Jun Yang, and Feng Xu. 2019. Compressible, fatigue resistant, and pressure-sensitive carbon aerogels developed with a facile method for sensors and electrodes. *ACS Sustainable Chemistry & Engineering* 7 (15):12726–12733.

Wang, Sha, Tian Li, Chaoji Chen, et al. 2018. Transparent, anisotropic biofilm with aligned bacterial cellulose nanofibers. *Advanced Functional Materials* 28 (24):1707491.

Wei, Haoran, Katia Rodriguez, Scott Renneckar, and Peter J Vikesland. 2014. Environmental science and engineering applications of nanocellulose-based nanocomposites. *Environmental Science: Nano* 1 (4):302–316.

Wichterle, O, and D Lim. 1960. Hydrophilic gels for biological use. *Nature* 185 (4706):117–118.

Wu, Chao, Xingyi Huang, Xinfeng Wu, Rong Qian, and Pingkai Jiang. 2013. Mechanically flexible and multifunctional polymer-based graphene foams for elastic conductors and oil-water separators. *Advanced Materials* 25 (39):5658–5662.

Xue, Yan, Zihao Mou, and Huining Xiao. 2017. Nanocellulose as a sustainable biomass material: Structure, properties, present status and future prospects in biomedical applications. *Nanoscale* 9 (39):14758–14781.

Yang, Jun, Chun-rui Han, Feng Xu, and Run-cang Sun. 2014a. Simple approach to reinforce hydrogels with cellulose nanocrystals. *Nanoscale* 6 (11):5934–5943.

Yang, Jun, Chun-Rui Han, Xue-Ming Zhang, Feng Xu, and Run-Cang Sun. 2014b. Cellulose nanocrystals mechanical reinforcement in composite hydrogels with multiple cross-links: Correlations between dissipation properties and deformation mechanisms. *Macromolecules* 47 (12):4077–4086.

Yang, Jun, and Feng Xu. 2017. Synergistic reinforcing mechanisms in cellulose nanofibrils composite hydrogels: Interfacial dynamics, energy dissipation, and damage resistance. *Biomacromolecules* 18 (8):2623–2632.

Yao, Kun, Qijun Meng, Vincent Bulone, and Qi Zhou. 2017. Flexible and responsive chiral nematic cellulose nanocrystal/poly (ethylene glycol) composite films with uniform and tunable structural color. *Advanced Materials* 29 (28):1701323.

Zhang, Xinyue, Mingyue Jiang, Na Niu, et al. 2018. Natural-product-derived carbon dots: From natural products to functional materials. *ChemSusChem* 11 (1):11–24.

Zheng, Chunxiao, Yiying Yue, Lu Gan, Xinwu Xu, Changtong Mei, and Jingquan Han. 2019. Highly stretchable and self-healing strain sensors based on nanocellulose-supported graphene dispersed in electro-conductive hydrogels. *Nanomaterials* 9 (7):937.

Zhou, Sukun, Pengpeng Liu, Meng Wang, Hong Zhao, Jun Yang, and Feng Xu. 2016. Sustainable, reusable, and superhydrophobic aerogels from microfibrillated cellulose for highly effective oil/water separation. *ACS Sustainable Chemistry & Engineering* 4 (12):6409–6416.

Zhou, Sukun, Meng Wang, Xiong Chen, and Feng Xu. 2015. Facile template synthesis of microfibrillated cellulose/polypyrrole/silver nanoparticles hybrid aerogels with electrical conductive and pressure responsive properties. *ACS Sustainable Chemistry & Engineering* 3 (12):3346–3354.

Zhou, Sukun, Tingting You, Xueming Zhang, and Feng Xu. 2018. Superhydrophobic cellulose nanofiber-assembled aerogels for highly efficient water-in-oil emulsions separation. *ACS Applied Nano Materials* 1 (5):2095–2103.

Zhu, Hongli, Zhiqiang Fang, Colin Preston, Yuanyuan Li, and Liangbing Hu. 2014. Transparent paper: Fabrications, properties, and device applications. *Energy & Environmental Science* 7 (1):269–287.

Zhu, Mingwei, Yilin Wang, Shuze Zhu, et al. 2017. Anisotropic, transparent films with aligned cellulose nanofibers. *Advanced Materials* 29 (21):1606284.

Zu, Guoqing, Jun Shen, Liping Zou, et al. 2016. Nanocellulose-derived highly porous carbon aerogels for supercapacitors. *Carbon* 99:203–211.

4 Horizons for Future Sustainability
From Trash to Functional Cellulose Fibres

Aleksandra M. Mikhailidi and Nina E. Kotelnikova

CONTENTS

4.1 PAPER WASTES: TERMINOLOGY AND OBJECTIVE DEFINITION

From newspapers to paper wrappings, paper is still everywhere and the most share is ending up in landfills creating a staggering amount of paper waste. There was a time when paper was a rare and precious commodity. Now it fills our planet. It was initially invented as a tool for communication in order to store and transfer information, but today, paper is used more for packaging because the digital age is dictating new information requirements – new format and design (Paper Waste Facts 2020).

The most exact and short definition for the waste paper is as follows: paper discarded as, superfluous, spoiled or unsuitable for use in traditional areas. The term 'waste paper' usually introduces paper and cardboard which are collected from individuals and enterprises, then sorted into grades, and finally, prepared for recycling; while the term 'recovered paper' is more common for the industry.

It is important to use the consistent and correct definitions for proper terminology of the paper wastes. The Paper Processing Coalition (USA) has developed and published an exhaustive list of precise definitions of the terms used for paper wastes and paper recycling which is selectively presented below (Paper Recycling Terminology 2019). This list not only defines the terminology in the branch but also gives some notions on a variety of the waste papers.

Since waste paper is a complicate combination of many constituents, we consider herein the most common components, according to ref. Paper Recycling Terminology (2019). Terms are listed alphabetically.

Contaminant: Any item or material that reduces the quality of paper for recycling or, in large quantities, makes it unrecyclable. Contaminants include metal, foil, glass, plastic, hot melt or pressure-sensitive adhesives, food, hazardous waste, carbon paper, waxed boxes, and synthetic fabrics. Collecting paper commingled with other recyclables during the collection process may increase contaminant levels.

Corrugated cardboard: Layers of paper glued together with a fluted inner layer – the most recycled product in some countries.

Deinking: A process that removes inks, dyes, or other contaminants from collected waste paper (GRN – Recycling Terms Glossary 2007).

Fibre: Small pieces of thread-like material that are woven together to give structure and strength to paper products. Fibre used in papermaking comes primarily from wood and recovered paper; cotton is also used to make certain products.

Hard mixed paper: This *classification* grade of recovered paper typically includes kraft paper, corrugated cardboard, office paper, i.e., all paper with longer fibres, and paperboard packaging.

High grade paper: Usually deinked, these primarily include printed and unprinted white paper collected from converting operations, printing plants, and offices.

Kraft paper: Any paper, usually unbleached, made from sulfate pulp.

Mixed paper: The comingling of various paper grades, such as old mail, paperboard packaging, magazines, copy, and computer paper, egg cartons, etc. for recycling.

Newsprint paper: A low-cost off-white and uncoated paper that is produced from mechanical, chemical, and deinked recycled fibres. Its most evident application is for printing newspapers; however, varieties of newsprint are also used for catalogues, flyers, and some magazines.

Paperboard : A generic term that includes heavy classes of paper. The most common are paperboard packages, which include folding cartons for foods and medicine, set-up boxes for games and jewellery, milk and juice cartons, composite cans for frozen concentrates, and beverage carriers.

Pulp: The fibrous material used to make paper. Pulp is obtained from blending wood, recovered paper, or in some cases, cotton with water.

Recovered paper: Paper and paper by-products that have known recycling potential, and that have been removed or diverted from solid waste, or that have never been discarded as solid waste and are intended for sale, use, reuse, or recycling whether or not such paper require subsequent separation and processing, excluding the virgin content of mill broke.

Recycled fibre (secondary fibre): Fibre derived from recovered paper that is processed into a product or a form usable in the manufacture of a product.

Recycled paper: Paper that is produced entirely from recovered paper.

Recycling: The total system by which recovered materials are collected, separated, processed, and reused or returned to use in the form of a marketable product.

Soft mixed paper: Typically includes magazines, newspapers and paperboard packaging, i.e., paper with shorter fibres.

White office paper: A mix of paper collected for recycling that includes white copy paper and writing paper; white, green-bar, and multi-stripe computer print-out; and white envelopes without plastic windows or labels.

4.2 PAPER RECYCLING

4.2.1 THE ENVIRONMENTAL IMPACT

Nowadays when humanity is facing a global ecological crisis, the concept of sustainable development is the only way to decrease the negative impacts on the environment. While world production is rising more and more, the total amount of wastes is also growing up dramatically. The paper production is a significant sector in the total world industry, and the consumption of paper and board is rapidly growing and varies from one country to another. Statistically, just one person uses approximately 60 kg of paper per year (Bajpai 2014), which gives a total hundred of million tonnes every year. That is why it is crucial to analyse the life cycle of paper and board from cradle (forest) to grave (disposal) and improve the ways of waste treatment that exist now, and find novel ones.

There is an opinion in society that paper is an environment-friendly, biocompatible and recyclable material that has remarkable advantages. However, the first two stages of the paper life cycle, such as processing fibres from raw materials and further turning them into paper, including bleaching and cleaning, cause severe damage to the environment, namely air pollution and water contamination. Also, the first and second stages of the paper life cycle are the most energy-consuming. Hence, it is important to highlight that paper becomes eco-friendly only when secondary fibres, obtained from waste paper and board, are used in its production instead of/or in addition to virgin fibres to decrease the impact of the first two 'dirty' stages of paper life cycle on the environment.

4.2.2 PAPER WASTE – WHAT IS IT GOOD FOR? WHY HUMANITY NEEDS TO RECYCLE PAPER WASTE

In this chapter we will pay attention to the last stage of the paper life cycle – the 'grave' or in the case of sustainable approach the 'new life' – regarding the concept of 3 *Rs* which includes reduction, reuse, and recycling. Paper recycling is the process of extracting the cellulose fibres from the mixture of waste paper and/or board for further *reuse* in different productions, usually in the manufacturing of new paper and board. So, if people collect and *recycle* waste paper, it will *reduce* the amount of forest resources that are used for paper production, and it will help to prevent such serious ecological problems as soil degradation and erosion, habitat damage, and reducing biodiversity. In addition to this, the sustainable approach will *reduce* the number of wastes in landfills.

Reducing wastes is well suited to diminish all three triple bottom-ups, as it is depicted in Figure 4.1.

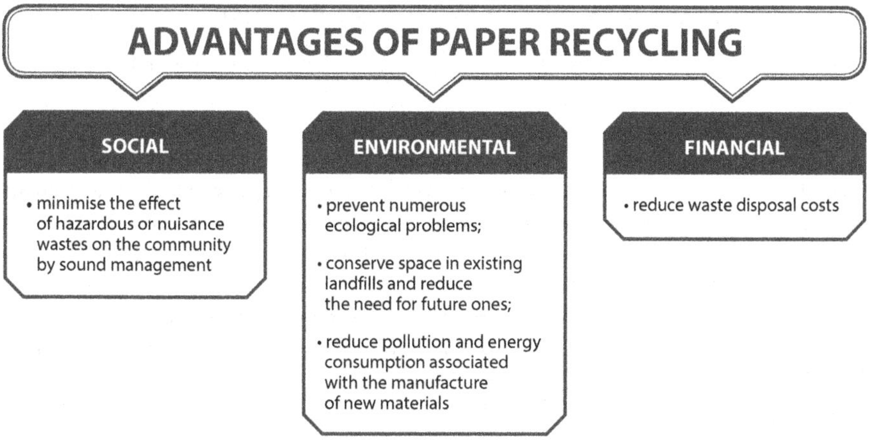

ADVANTAGES OF PAPER RECYCLING

SOCIAL	ENVIRONMENTAL	FINANCIAL
• minimise the effect of hazardous or nuisance wastes on the community by sound management	• prevent numerous ecological problems; • conserve space in existing landfills and reduce the need for future ones; • reduce pollution and energy consumption associated with the manufacture of new materials	• reduce waste disposal costs

FIGURE 4.1 Advantages of paper recycling.

4.2.3 PAPER WASTES AS A VALUABLE SECONDARY RAW MATERIAL

Paper and board products, once they have been collected and processed for further recycling, become a valuable secondary raw material and should no longer be considered as waste (COST Action 2009); however, in this chapter we will use the term 'waste paper' as a synonym for the term 'recovered paper' to define used paper and board collected and processed for further applications.

Recycled fibres play a significant role in paper industry as a substitute for virgin pulps. The paper recycling process is becoming more and more effective for the reuse of secondary fibres due to the fact that ink removal operations develop rapidly (Bajpai 2014). The authorities in collaboration with organizations which are responsible for the monitoring of recovered paper in different countries set new challenges to increase the amount of paper recovery, for instance, in the USA a goal of 70% (see Better Practices Better Planet 2020 initiative of American Forest and Paper Association (AF&PA Sustainability Report 2018)) was established, while in the EU, which is the world leader in paper and board recycling, the same target was 74% (Monitoring Report 2017) by 2020. Developing countries also take part in such programmes. According to Ha Noi 3R Declaration – Sustainable 3R Goals for Asia and the Pacific for 2013–2023 (Ha Noi 3R Declaration 2013), Asian countries set a goal to achieve significant increase in recycling rate of paper as well as other recyclables. For instance, India has created a market for waste paper by incentivizing the production of recycled paper (Achieving the Sustainable Development Goals in South Asia 2017). So, though there are significant regional variations in waste paper collection and utilization, globally the majority of countries prioritize recycling of waste paper.

Obviously, such concerns from authorities and public organizations are caused by a number of good reasons.

4.2.4 Benefits and Downsides of Waste Paper Recycling

Both benefits and downsides of waste paper recycling have been debated multiple times.

Bajpai (2014) discussed the most important advantages of recycling paper.

- Saving resources (wood, water, power): When secondary fibres are used in papermaking, the number of saved trees is about 17 per each tonne. Such additional resources as water and electricity for paper production are also preserved. The reason is that the biggest share of power that is used during the life cycle of paper is consumed at the stage of turning wood into paper. In general, the carbon footprint of paper produced from secondary fibres is up to 50% less compared to that produced from virgin fibres.
- Decreasing of production emissions: The application of secondary fibres in papermaking contributes less to air pollution and water contaminating compared to wood processing. Bleaching step is not always necessary for the utilization of recycled fibres; however, even when it is needed, usually oxygen is used instead of chlorine. This reduces the amount of dioxins that are released into environment as a by-product of the chlorine bleaching process.
- Saving soil from landfills: Nowadays, the landfills' area is enormous with about 20–30% of space for waste paper. Recycling this waste will reduce landfill area. A landfill space of 2.3 m^3 is saved for every tonne of paper used for recycling.
- Positive image for enterprises: Business can promote a positive company image by launching and maintaining a paper recycling programme. A number of corporations are using recycled paper in offices, even though the price of recycled paper is never lower when compared with that of regular paper. This can only be possible as a result of the public attitude to resources and environments.
- Decreasing of transport emissions: Since waste paper is usually collected rather close to recycling facilities, the production of recycled paper reduces transportation footprint and economic expenses. That is why the life cycle of paper becomes more efficient.

However, the recycling of waste paper has not only the benefits but downsides as well. Since recycling process is a production, it also has its own carbon footprint. There are no completely 'clean' technologies with zero waste in the world. Recycling mills produce such residuals as sludge, including ink, adhesives, unusable fibres and other substances removed from the usable fibres. On the other hand, if the paper was not recycled, those materials would still finish the life cycle in landfills or incinerators, while recycling mills have enlarged environment-friendly methods of handling sludge (Bajpai 2014).

4.2.5 Recycling of Waste Paper

The recycling of waste paper has a long history. The first known effort to use waste paper for processing was made in 1567.

4.2.5.1 Background Information on Waste Paper and Cardboard: Composition of Paper and Cardboard

In order to consider what waste paper is recycled for and how it is processed, we present here a basic understanding of the composition and distribution of the main components in waste paper.

Paper and cardboard are one- or multilayer sheets that consist of cellulose fibres with the addition of various chemicals that are used to improve the properties such as opacity, brightness, or glossiness of the sheets. The most common additives, namely limestone (calcium carbonate, $CaCO_3$), clay (kaolin, kaolinite, $Al_2Si_2O_5(OH)_4$), and starch have the largest proportion in the paper content. Their total weight is near 15%, though, the amount depends on the type of paper. Other chemicals, for example, resins, wet strength agents, optical brightening agents, sizing agents, dyestuffs, coatings, retention agents, anti-foaming agents, cleaning agents, and biocides are added to paper sheets in smaller amounts (up to 2 wt.% altogether) for the purpose of obtaining the specific properties of the final product. Finally, there are a few chemicals that may be found in paper in trace quantities, for instance, talc and sodium silicate that are residuals from the deinking procedure.

The structural difference of cardboard from paper is that cellulose fibres are set together in a stiffer way, it is thicker and less foldable. Cardboard has a larger density than paper and it is rarely single layer. As a rule, these are from two to four glued layers that are resistant to deformation and may exhibit barrier properties.

The source of cellulose for paper is usually wood; however, other plants, for example, cotton, rice, seaweed, etc. are also used in the production of paper. The main components of wood are natural polymers cellulose (approximately from 45 to 56 wt.%) and lignin (from 19 to 28 wt.%), at that the proportion of the components depends on the plant species. The function of lignin is holding together cellulose fibres, filling the space between the cell walls, and providing strength. Despite the crucial role in plants, lignin has no essential function in papermaking. Moreover, this component is undesirable since the progressive chemical degradation of lignin rapidly darkens the paper. On the other hand, when the life cycle of paper is short, for example, newsprint paper, or the whiteness is not important, such as corrugating cardboard or inner layers of multilayer sheets, a certain amount of lignin is permissible.

4.2.5.2 Difference between Virgin Fibres and Waste Paper and How to Solve Problems Related to the Complexity of Paper and Paper Recycling

There is a fundamental difference between virgin fibres in paper and fibres obtained from waste paper. Paper can be recycled only a certain number of times (from 3 to 5 times), the only drawback being a progressive shortening of cellulose fibres through processing, which reduces the strength of the resulting paper (Villanueva Krzyzaniak and Eder 2011). In other words, fibres are damaged during the papermaking operation. The fibres collapse during drying on the paper machine and there is hornification at the surface, so that the properties of the fibres are changed, with more fines produced (*Technology of paper recycling* 1995). When paper is recycled, it is turned into lower grade paper; for instance, white office paper becomes newsprint paper, while paperboard reduces to sanitary paper. For the time of paper to paper recycling, the maximum obtainable yield is about 65%. This can give rise to large quantities of waste fibres not suitable for recycling; therefore, this has become a thing of concern to paper manufacturers (Ikeda et al. 2006).

4.2.6 Main Steps in Paper and Board Recycling

Paper recycling includes several stages to convert into secondary fibres, as shown in Figure 4.2. Pre-processing and post-processing stages are described and some useful explanations are clarified below.

4.2.6.1 Collection

The process begins with collection of waste paper and board from individuals and enterprises (so-called private and industrial collection) and then they are delivered to mills. Collecting step is one of the most expensive in paper recycling (Bajpai 2014). One of the main problems here is that usually waste paper is a mixture of many fibre types in different paper and board products, and it has a low value, since mixed papers will not normally provide the properties required for a specific end use, such

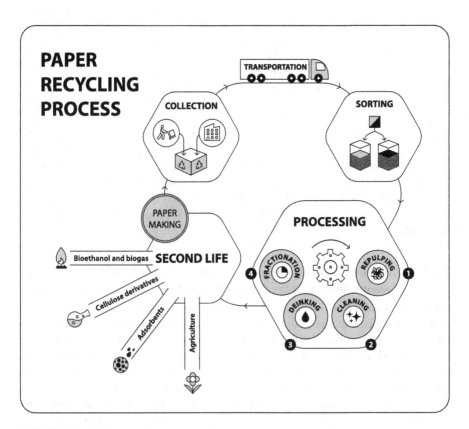

FIGURE 4.2   Main steps in paper recycling.

as necessary strength requirements in packaging, contrast in printing papers, or absorbency in tissue (*Technology of paper recycling* 1995). Hence, mixtures of fibre types can only be used to produce the lowest grades of paper and board.

4.2.6.2 Sorting into Grades

Once waste paper and board have been sorted into grades, then they have a higher value than mixed paper; however, sorting requires time and additional expenses that vary from country to country. There is a huge number of grades, according to classifications CEPI EN 643-2013 in Europe and ISRI PS-2018 in North America, that makes the task of sorting complicated. As a result, the most part of waste paper is processed without sorting into the lowest grades, namely cardboard and tissue.

4.2.6.3 Repulping

When waste paper and board have been collected and sorted into grades (or not sorted), they undergo the processing at recycling facilities. The first step of the processing is repulping. The paper is soaked in large vats, reducing the paper into fibres. Repulping is followed with several additional stages such as the removal of contaminants when washing, screening, or cleaning are used for each type of contaminant. Then selective purification is done by fractionation when the separation of short and long fibres occurs.

4.2.6.4 Deinking, Removal of the Contaminants, and Bleaching

The big share of waste paper is black and white or multicoloured printed paper and packaging board. In order to isolate useful secondary fibres from printed paper, ink usually must be removed during the deinking process, although, paper with a greyish tone without deinking step may be used for some applications. When paper swells during repulping, ink particles separate from the cellulose fibre. Firstly, the pulp is cleaned several times with heat and chemicals. During this procedure, some ink and additives are removed. Then, to withdraw residual ink from the pulp, the flotation process is implemented. Small colloidal particles of contaminants are flocculated and driven to the surface with air bubbles where it can be skimmed away. After the deinking process, the pulp is ready to be manufactured into paper and related products in a similar manner to that by which paper is produced from wood pulp (Bajpai 2014).

The fibres undergo commonly by bleaching to increase brightness. For this purpose hydrogen peroxide (oxidative bleaching) or sodium hydrosulfite (reductive bleaching) are commonly used.

As mentioned earlier, one more problem in the recycling process is that waste paper keeps non-fibrous materials. In general, the rate of formation of such residues is between 5% and 40% depending on the recovered paper grade processed and the paper grade produced (Bajpai 2014). For example, many packaging boards or printing papers contain up to 5% by weight of starch. Printing papers contain about 20–30 wt.% of mineral fillers, while coated papers have added coating fillers so that they contain up to 40 wt.% of mineral fillers and coating binders. As a result, all waste paper has a need for some contaminants' removal (*Technology of paper recycling* 1995). Most non-fibrous materials reduce fibre web strength. As particle size increases, this effect increases, so large particles cause weak spots in the fibre web.

Removal is difficult due to the diversity of materials, organic and inorganic, natural and synthetic, etc. (*Technology of paper recycling* 1995). However, not always all of these non-fibrous materials are considered as contaminants. For example, while in tissue, coating and filler clays are regarded as contaminants, for printing and writing paper they are not due to different demands to these types of paper. Aside from this, the non-fibrous residuals can be used as a raw material in other industries. While some treatments of the remainders include the production of fuels or energy recovery via incineration, other techniques are more eco-friendly due to the real reuse of non-fibrous residuals. For example, Pitroda (2016) showed that so-called hypo sludge which contains useless low-quality paper fibres, as well as calcium oxide, calcium chloride, magnesium oxide, and silicon dioxide can be used in concrete formulations as a supplementary cementitious material. Tofani et al. used recycled fillers (e.g., $CaCO_3$, kaolin) from deinking paper sludge for newspaper production. Deinking paper sludge was incinerated, subsequently bleached, and then reused for papermaking (Tofani et al. 2020). Also, the construction industry is a field for the second life of non-fibrous wastes from paper recycling where they are transformed into building panels or insulation, as well as in such processes as the conversion of plastic and making composite semi-finished products (COST Action 2009).

Thus, besides valuable cellulosic fibres, paper contains non-fibrous impurities, which usually should be removed. Removal impacts the loss of time and chemicals during additional stages of processing, and as a result, residues are generated. However, non-fibrous waste can be used in other manufacturing, which partly makes closed-loop processing and raise the sustainability of paper recycling.

In general, the paper recycling process is a number of several steps, each of them has its own drawbacks; however, all in all, secondary fibres that are extracted during this process are the valuable raw materials for different functionalization.

4.2.7 WHERE RECYCLED PAPER WASTE IS USED?

The paper waste is recyclable into useful products and this is very topical not only to conserve natural resources but also to limit environmental pollution.

4.2.7.1 Papermaking

Waste paper is a low-cost source of fibre for paper and board manufacture, and the use of recycled fibres can greatly reduce the cost of the final paper (Liu et al. 2012). Paper recycling, the final product of which is valuable cellulose fibre, plays an important role in the paper industry. According to (COST Action 2009), in the end of 2000s the global recovered paper utilization rate was about 51%. Traditionally, extracted fibres find the further application in the production of new paper. While for some paper and board grades, for example, corrugating medium and test liner, or newsprint, recycled fibres could be used exclusively, other grades are the mixtures of secondary and virgin fibres. Currently, the percentage of recycled fibre in paper and board of almost all grades is increasing. In the same time, recovered paper is still used mostly as packaging cardboard since the production of these grades requires fewer expenses due to the absence of deinking and bleaching stage. Over 90% of recovered paper in the world is used in grades other than for printing and writing, such as newsprint, tissue, container boards, and other packaging or board products. Approximately 6% of the global recovered paper supply is used in printing and writing grades, and this percentage is forecast to increase only slightly by 2025 (Bajpai 2014). The further development of the usage of recovered fibres for papermaking is limited by the development of the collection system that minimizes the degree of mixing different paper grades, enabling efficient utilization of the best collection sources.

So, in waste paper utilization downgrading usually happens when high-quality paper grades turn into low-quality grades. However, low paper and board grades that still contain a certain quantity of cellulose fibres turn into the end-of-lifecycle products such as sanitary paper or the organic compost, or go to the incinerators.

4.2.7.2 As Adsorbents

Contaminants that are present in the effluent water in dyestuffs, textile, leather, printing, paper and plastic industries are harmful, sometimes cancerogenic, and pose a vital threat to the aquatic ecosystems (Sharma et al. 2011; Crini 2006). Dyes are the most common contaminants which even in a trace amount will decrease the quality of water significantly. Most commercial purifying systems currently use activated carbon as a sorbent to remove dyes in wastewater because of its excellent adsorption ability. However, numerous alternative non-conventional low-cost and renewable adsorbents for dyes, produced from (1) waste materials from agriculture and industry, for example, newspapers (Okada et al. 2003), carbon slurries (Jain et al. 2003), sewage sludge (Otero et al. 2003), (2) fruit waste, for example, orange peel (Rajeshwarisivaraj et al. 2001), (3) plant waste (Sharma et al. 2011), such as corncob, bagasse (Juang et al. 2002), cane pith (Juang et al. 2001), coir pith (Namasivayam et al. 2001), rice husk, straw (Kannan and Sundaram 2001) have been proposed by different research groups. These by-products from the agricultural and wood manufacturing industries are abundant, require little processing, and are effective adsorbents. In addition, these residuals have low economic value and present a disposal problem. Some waste materials can be used as adsorbents without treatments, while others should be processed into activated carbon in order to increase the sorption capacity, selectivity, or other performance (Crini 2006).

A suitable low-cost unconventional adsorbent for dyes should comply with the following requirements:

- efficiency for removal of a broad number of dyes;
- high capacity and rate of adsorption;
- high selectivity for different concentrations;
- tolerance of a wide range of wastewater parameters (Crini 2006).

Activated carbon fits all these requirements perfectly. That is why sometimes it is more effective to treat wastes chemically and physically to obtain activated carbon or another more efficient chemical form. On the other hand, each low-priced adsorbent, produced from wastes without special treatment, has its specific physical and chemical characteristics, for example, porosity, surface area, and physical strength, and as a result, its own intrinsic benefits and drawbacks in wastewater treatment (Crini 2006).

Strange as it may sound, one of the challenges hindering implementation is an excessively large number of adsorbents derived from wastes, which complicates the choice for their industrial use. The efficiency of adsorbents of various origins is rather difficult to compare with each other because of the reasons listed above. Moreover, efficiency strongly depends on the conditions, for instance, pH of the medium plays an important role in the adsorption and pretreatment processes (Sharma et al. 2011). In order to tackle this problem in the most sustainable way, a good solution is to use local waste for the production of adsorbents for the particular needs of the municipality.

The use of waste paper as adsorbent is promising and of high priority because of the contribution to the reduction of costs for waste disposal. Waste paper includes a number of cellulosic materials and is a potential adsorbent due to high cellulose fibre proportion, carbon content, and oxygenated functional groups. Waste paper could be utilized instead of more conventional but expensive adsorbents for such toxic contaminants as metal ions, dyes, phenol, pesticides, and pharmaceuticals, and its modification by means of thermal or chemical processes may help to improve the pollutant adsorption capacity of such waste (Islam et al. 2019). For example, the gel obtained from waste newsprint paper by chemical modification with p-aminobenzoic acid was capable to adsorb gold, palladium, and platinum ions selectively from the number of other coexisting metal ions. In addition, the gel was a regenerative sorbent and the recovery of precious metal ions occurred with a high percentage (up to 95%) (Adhikari et al. 2008). Aerogels obtained from recycled paper waste cellulose fibres after being coated with methyltrimethoxysilane achieved high absorption capacities for crude oil. The adsorption capacities of the aerogels were nearly double those values obtained with polypropylene – the common oil absorbent (Nguyen et al. 2013).

Okada et al. (Okada et al. 2003) produced activated carbons from old newspaper (ONP) and paper prepared from simulated paper sludge. They used two different methods of pretreatment: chemical activation with alkali carbonates and hydroxides and physical activation by steam. The specific surface areas and total pore volume values of physically activated carbons were lower compared with chemically activated carbons due to the smaller size and lower volume of their micropores. On the other hand, they retain the initial fibre shape and the paper sheet morphology after activation.

Adil Sabbar studied waste paper as an adsorbent for phenol which is one of the harmful organic by-products deposited in the environment. The highest phenol removal efficiency was 86% with an adsorption capacity of 5.1 mg g^{-1} at optimization conditions (Adil Sabbar 2019). Kadam et al. processed recycled waste paper into high-yield α-cellulose fibres (αCFs) using chemical treatment with NaOH in the ultrasonic bath (Kadam et al. 2019). Following this, the lyophilized αCFs were modified to a magnetic composite (Fe$_3$O$_4$ nanoparticles) to explore bioadsorption of engineered nanomaterials, namely, CuO, CoO, and ZnO. The biodegradable supermagnetic bioadsorbent obtained from waste paper removed large quantities of engineered nanomaterials from various real-world wastewater samples.

4.2.7.3 Bioconversion of Paper Waste into Bioethanol, Biogas Other Liquid Biofuels Products, and Sugar

In contrast with methods of functionalization of recovered fibres that were listed above, the following applications of waste paper and board do not require extracting fibres, being a part of sustainable development. The materials derived from the renewable bio-resources could be the alternative to petroleum-based synthetic products due to their advantages of relatively low cost, environmental friendliness, easy availability, renewability, and non-toxicity (Sun and Cheng 2002; Mondal 2017). Waste paper is becoming a perspective source to obtain the raw material for the production of liquid biofuels. For example, Brummer et al. subjected offset paper, cardboard, and recycled paper to enzymatic hydrolysis in order to obtain ethanol (Brummer et al. 2014). Chemical pretreatment of material did not bring a significant increase in yields. The highest yields were achieved for cardboard. Kumara et al. obtained 30 ml of bioethanol with a concentration of 86% from 1 kg of waste paper (Kumara et al. 2020). Waste paper combined with food waste could be also used in biogas production (Li et al. 2020).

Wang et al. widely analysed the technologies for the conversion of paper and paper-derived materials to bioethanol (Wang L. et al. 2012a; Wang L. et al. 2012b; Wang et al. 2013). The authors showed that usually applied bioconversion technologies by enzymatic hydrolysis with consequent and sometimes simultaneous saccharification and fermentation yielded up to 80% at solids loading of 5% and 10% (w/w). However, there are two obstacles to commercial use. Firstly, bioethanol affects the growth of microorganisms; secondly, the cost of enzymatic hydrolysis is rather high. Therefore, a few strategies were proposed to decrease the expenses, namely the substrate loading enlargement, the enzyme loading decrease, the residence time reduction and enzyme activity enhancement. Wang et al. used the newspaper, printed office paper, and packaging cardboard as raw materials (Wang L. et al. (2012a). High-solids loading (15 w.%) enzymatic hydrolysis using two enzyme alternatives (Celluclast 1.5 L + Novozyme 188 and Cellic Ctec 1) achieved glucan conversion efficiencies from waste paper of 50% to 76%. It was shown that the bioethanol produced from cardboard resulted in the lowest conversion. The authors estimated the cost and efficiency of bioconversion and the ways to increase the recovery in waste paper recycling.

One of the very popular directions of the utilization of waste paper materials is bioconversion to sugars (Walpot 1986). A number of articles are devoted to this topic due to the fact that this is a way to develop this resource of polysaccharides through the bioconversion into fermentable sugars such as glucose. The catalysis with enzymes is the most usable method. van Wyk et al. (1999) applied saccharification with cellulase from *Penicillium funiculosum* (van Wyk et al. 1999). Filter paper, foolscap paper, newspaper, and office paper (photocopy paper) cut into small pieces were incubated with a standard cellulase solution for 1 h at 55°C. After filtration, resulting sugars in the filtrate were determined by high performance liquid chromatography or by the dinitrosalicylic reagent method. It was found that the foolscap paper was the most susceptible towards the enzyme-catalysed saccharification (23%), while filter paper, newspaper, and office paper had lower saccharifications, 19%, 13%, and 5%, respectively. To make the process economically viable, it is imperative to increase the extent of bioconversion; various pretreatment methods of the waste paper samples such as mechanical or physical pre-processing of the samples and the use of the cellulase from other microorganisms would be beneficial.

One of the first papers on the enzymatic treatment of waste paper was devoted to the production of gluconic acid from waste office automation paper using an enzymatic hydrolysate in a culture of *Aspergillus niger* (EH) (Ikeda et al. 2006). In repeated batch cultures using flasks, the saccharified solution medium did not reveal any inhibitory effects on gluconic acid production compared to glucose medium (GM). The median gluconic acid yields were 92% (EH) and 80% (GM). In repeated batch cultures in a turbine blade reactor, the gluconic acid yielded about 60% (EH) and 67% (GM) with 80–100 g L^{-1} of gluconic acid. When air was supplied by pure oxygen the production rate increased to four times.

4.2.7.4 Agriculture and Farming

For the reason of rapid urbanization and industrialization per capita, land availability is decreasing and challenging the soil-based agricultural practices. Soilless plant culture is a method of growing plants without the use of soil as a rooting medium which includes inorganic, organic and synthetic substrates, solid media culture, and aeroponics. Waste paper along with proportionate amount of soil and household wastage can find the application as a suitable medium for plantation (Kakati 2020).

4.2.7.5 Other Areas of Application

In general, only 5–8% of the total amount of recovered paper is utilized outside the paper industry (COST Action 2009). Areas of application may include the following: building panels, roofing and insulation materials, as well as moulded products, for example, egg boxes and cup holders. Another field of usage is 'from wastes to energy' approach that we will not discuss here since this is not recycling. Eventually, one of the debatable approaches for utilizing paper and board is to compost them. On the one hand, composting is a sustainable technique that returns carbon, nitrogen, etc. to soil and rounds out the cycle of those elements in nature. On the other hand, humanity produces more than 25% of food wastes from the total amount of wastes, accounted for millions of tones, which are quite enough if they are composted properly (COST Action 2009). So, people do not need additional raw materials for composting while recovered cellulose fibres can be used for other purposes. We will discuss a number of the most popular directions below.

4.3 RECENT NOVEL PATHWAYS IN WASTE RECYCLING

Poor recycling techniques, especially in developing countries, generate high levels of environmental pollution that affects both the ecosystems and the people living within or near the main recycling areas. The recovery of materials from urban waste has become more progressive with wastes having a great potential of material resources. In recent decades, some novel pathways in waste recycling have been developed (Khanna et al. 2014).

4.3.1 AGROWASTES

Agricultural residues management is considered to be a vital strategy in order to accomplish resource conservation and to maintain the quality of the environment (Mostafa et al. 2018). Recovering of useful matters from agrowastes became an important sector in waste recycling, and considerable literature pertaining to the isolation of cellulose from agro-wastes is available. Novel and unconventional techniques have been described in many publications (Khanna et al. 2014; Jiang and Hsieh 2015; Mariño et al. 2015; Macedo de Melo 2018; Tanga et al. 2020). While reviewing worldwide production and extraction of cellulose from agro-wastes and the present status of converting them into value-added products for food and pharmaceutical applications, Sundarraj and Ranganathan noted that agro-industrial waste disposal is a serious issue of concern in developing countries (Sundarraj and Ranganathan 2018). Some examples are considered below.

4.3.1.1 Fruit and Vegetable Wastes

One of the types of agricultural waste suitable for obtaining useful products seems to be waste biomass.

A sustainable approach for the production of microfibrillated cellulose (MFC) and high-value chemicals from depectinated orange peel residue (OPR) in the context of a zero waste orange peel biorefinery has been presented in reference Macedo de Melo (2018). The OPR has been subjected to an energy-efficient acid-free hydrothermal microwave treatment at temperature from 120 to 220°C. A solid fraction, containing MFC and hydrolysate, was rich in pectin, lignin microparticles, sugars, soluble organic acids, and furans. MFC characteristics strongly depended on the temperature during the treatment. MFC produced at 120°C could easily form hydrogel and had improved rheological performance. This valorization of ORD showed its prospects as a valuable source for the manufacturing of bio-based materials with numerous potential applications.

Mariño et al. addressed some of the extraction routes for obtaining nanocellulose from citrus waste via a three-step physico-chemical and enzymatic procedure (Mariño et al. 2015). The prepared nanofibres had 55% crystallinity, an average diameter of 10 nm, and a length of 458 nm.

Tanga et al. prepared a new type of carboxylated cellulose nanofibre (CNF) with super absorption/flocculation ability by the direct thermo-oxidizing of pomelo peel with H_2O_2 (Tanga et al. 2020). Carboxylated CNF had a high aspect ratio and density of carboxyl groups which made possible grafting functional polyetherimide on the surface of CNF. The product showed reusability and excellent adsorption performance to pollutants, such as malachite green and Cu(II) ions. The simple preparation process and executive properties made the new product a viable candidate as environmental remediation material.

Jiang and Hsieh describe pure cellulose isolation from tomato peels by either acidified sodium chlorite or chlorine-free alkaline peroxide routes at 10.2–13.1% yields (Jiang and Hsieh 2015). After hydrolysis of cellulose with sulfuric acid (64% H_2SO_4), negatively charged and flat spindle-shaped cellulose nanocrystals (CNCs) were obtained at a 15.7% yield. Highly crystalline (80.8%) cellulose Iβ fibrous mass containing mostly sub-micron fibres (ϕ = 260 nm) and few interconnected nanofibres (ϕ = 38 nm), with 21.7 m^2 g^{-1} specific surface and 0.049 m^3 g^{-1} pore volume was assembled from dilute aqueous suspensions. More uniform nanofibres with average 42 nm width and specific surface area 101.8 m^2 g^{-1}, mesoporosity and pore volume 0.4 m^3 g^{-1} were prepared from CNCs in 1:1 v/v tert-butanol/water mixture.

4.3.1.2 Crops, Cereals, Weeds, Grass, Algae Wastes, Bast, and Staple Fibres

The agricultural waste of cereals, weeds, grasses including rice straw and husk, wheat straw and stems, hemp stalks, soya hulls are also sources of microcrystalline and/or nanofibrillated cellulose. Alemdar and Sain isolated cellulose nanofibres from the agricultural residues, wheat straw, and soya hulls by the chemical and mechanical treatments (Alemdar and Sain, 2008). The wheat straw nanofibres had diameters in the range of 10–80 nm and lengths of a few thousand nanometres. The soya hull nanofibres had diameter 20–120 nm and shorter lengths than the wheat straw nanofibres. The nanofibres revealed an improved crystallinity, according to X-Ray diffraction results, and the thermal properties due to the results obtained by thermogravimetric analysis (TGA). The properties of the nanofibres exhibited their usability in biocomposite applications and in reinforced polymer manufacturing. The wheat straw as a raw material has been also applied, as described in the ref. Espinosa et al. (2019). The straw was subjected to soda pulping followed by treatment in a defibrillator to obtain an aqueous suspension of semi-chemical pulp. The suspension was supplied to enzyme treatment followed by either ultra-fine grinding or homogenization. The screw extruder was also used. The most energy-efficient (less costly) process was homogenization. Combined with the enzymatic treatment, this required 10 times less energy to obtain the lignocellulose nanofibres with the yield up to 55%, degree of polymerization ranging 400–600, and rather high crystallinity. The authors concluded that the processes studied to prepare lignocellulose nanofibres provided the energy efficient, i.e., less costly methods compared to the mechanical refining currently used.

Virtanen et al. studied physico-chemical characteristics of potential new cellulose sources (rice husk, hemp stalk) and microcrystalline cellulose (MCC) manufactured from them (Virtanen et al., 2012). The rice husk was ground and chemically purified and the hemp stalks were cut into small pieces and boiled with NaOH and water. MCC was prepared by acid hydrolysis from these samples with the suitable yield 46.4 w% (rice husk) and 75.7 w% (hemp stalk). The extracted samples had high values of specific surface, $310\ m^2g^{-1}$ and $270\ m^2g^{-1}$, respectively. These properties were valuable for MCC production and promising for biomedical application. Jiang et al. have investigated the isolation of cellulose from rice straw, which was steam-exploded, *via* delignification with recyclable mixed solvent system and bleaching with alkaline hydrogen peroxide (Jiang et al. 2014). The bleached cellulose had the moderate degree of polymerization (549) and contained no detectable acid-insoluble lignin.

A detailed study on the suitability of rice straw, wheat straw, corn stalks, and dhaincha waste for the extraction of cellulose microfibrils was conducted by Nuruddin and co-workers (Nuruddin et al. 2011). CNF were prepared by sulfuric acid hydrolysis from these raw materials, which had previously been submitted to a fractionation process by formic acid/peroxyformic acid/peroxide. Cellulose, amounting to 37–43%, lignin (16–20%), hemicelluloses (20–33%), and other compounds were extracted. All CNF samples presented the triclinic structure and showed good thermal stability.

Recently, El Achaby et al. have shown that high-quality cellulose nanocrystals have been successfully extracted from red algae waste as raw material (El Achaby et al. 2018). The ability of CNC to reinforce polymers has been studied. Algae waste was chemically treated via alkali, bleaching, and acid hydrolysis. Needle-like shaped

CNC had diameters and lengths ranging from 5.2 to 9.1 nm, and from 285.4 to 315.7 nm, respectively, and crystallinity ranging from 81 to 87%. CNC were tried as nanofillers for the production of PVA-based nanocomposite films with improved thermal and tensile properties, as well as optical transparency.

Mostafa et al. reported an efficient method for the production of the cellulose acetate biofibre with yields of 81% and 54% from flax fibres and cotton linters, respectively (Mostafa et al. 2018). The fibres were first washed, bleached, dried, and acetylated with a mixture of acetic anhydride, glacial acetic, and sulfuric acids and cooled. The following stages included dilution of the mixture with the concentrated acetic and sulfuric acids and left it to rest for 15 h. The resulting viscous fluid was centrifuged and plasticized with polyethylene glycol. For shaping, the product was diluted with acetone before forming or moulding. These newly produced cellulose acetates were biodegradable and were not affected by acid or salt treatment. Moreover, they were affected by acids to a lesser extent than polypropylene and polystyrene. Therefore, these products can be applied in both the food industry and medicine.

4.3.2 Recycling of Paper Waste for Conversion into Cellulose, Cellulose Derivatives, Cellulose Composites, Nanocellulose, and Nanofibrillated Cellulose

Cellulose is the major constituent of paper, paperboard, card stock, and the main ingredient of textiles made from cotton, linen, and other plant fibres (Conte 2016).

Studies of the transformation of paper waste into cellulose, cellulose derivatives, cellulose composites, nanocellulose, and nanofibrillated cellulose occupy a crucial and special place among the directions of its processing. Waste paper is particularly attractive as feedstock for isolation of cellulose fibres due to high content of cellulose and to the fact that cellulose fibres from paper waste exhibit high flexibility and good mechanical properties (Conte 2016).

Generally, the isolation of the cellulosic components of the municipal solid wastes is obtained through the transformation under heat and pressure of the whole cellulose biomass into a fairly uniform material that is separated from most of the metals, plastics, textiles, and glass by vibratory or trommel screening. Some of these contaminants are removed from the wet or dry material by stoner processing (Conte 2016).

4.3.2.1 Cellulose and Cellulose Derivatives Produced from Waste Paper

The extracted cellulose from waste paper has been often used as a component with valuable properties in different composite systems, despite the fact that sophisticated equipment and reagents are required in some cases. Zhang et al. recycled old newspaper (ONP) as a typical waste paper to prepare natural fibre composites with capping agent maleic anhydride grafted high-density polyethylene (MAPE) (Zhang W. et al. 2019). The dried ONP strips were treated with methyltrichlorosilane for hydrophobization at 60°C, then washed to pH neutral, and dried at 120°C. The treated strips mixed with MAPE pellets in high-speed mixer at 100°C were subjected to further extrusion–pelletization processes to obtain the pelletized composites. ONP/MAPE composites prepared with fibres modified with 4% (v/w) MTCS showed the best mechanical properties and satisfactory water-resistance properties.

Guo et al. studied the production of recycled cellulose fibres from waste paper (newsprint fibres and kraft fibres) using ultrasonic wave processing (Guo et al. 2015). It was shown that the ultrasonic cavitation effect was feasible for the preparation of the secondary fibres. The fibres exhibited high values of water absorption and surface area. Due to these properties, the recycled cellulose fibres after processing fulfilled several technical indexes; therefore, they could be considered as a filling material for used in cement-based materials.

Cellulosic fibres can be prepared from different sources of recycled waste papers (newspapers, magazines, and cardboard). Hospodarova et al. showed that unbleached recycled cellulosic fibres were obtained right after repulping in the raw state without further treatment (Hospodarova et al. 2018). These fibres were grey coloured and contained 80% of cellulose. In the amount of 0.5% of the weight of other components (fillers and binders), they were used as an additive in cement composites together with other components. The cellulose fibres positively affected the physical properties of fibre/cement composites.

Cellulosic waste materials, such as waste printing paper and cardboard boxes, can be successfully converted into nanostructured SiO_2 ceramics and carbon/SiO_2 nanocomposites by submersing cellulosic materials into silica sol, followed by calcination at 550°C in air and nitrogen, respectively (Pang et al. 2011). Waste paper was cut into smaller pieces and underwent the initial maceration process to disperse the cellulosic fibres. After that, the macerated waste paper fibres were ground into powdery form using a blender. Once residual lignin had been removed, conversion of cellulosic samples into nanostructured ceramics was performed by handling in silica sol. To obtain the nanostructured SiO_2 ceramics with defined and specific microstructure, the samples were heated at 550°C and then cooled. This method provided a cost-effective synthesis approach for the preparation of nanostructured ceramics and nanocomposites.

One more application for recycled cellulose fibres was suggested in the paper of Ma and co-workers (Ma et al. 2016). High-strength fibres have been prepared using cellulosic waste, A4 copy paper sheets, cardboard (fluting board mill), and the ionic liquid 1,5-diazabicyclo[4.3.0]non-5-ene-1-ium acetate as a solvent. Cellulosic waste materials have been dissolved, and solutions with visco-elastic properties, suitable for dry-jet wet fibre spinning, have been obtained. The resulting samples were cost-efficient fibres with high tensile strength and Young's modulus. The advantage of the process was that carbohydrates were preserved almost entirely. Bleaching was possible but not necessary as the resulting colour could serve as a natural dye in the garment production.

Liu et al. studied the possibility to recycle mixed office waste paper, which is an available and inexpensive source of high-quality bleached chemical fibre with improved drainability (Liu et al. 2012). Firstly, the waste paper was deinked and the pulp suspensions were prepared by disintegrating to separate the fibres completely and in the same time without considerable changing of their structure. Then, the pulp suspensions were warmed and enzyme endoglucanase was added to the suspensions. The optimal pretreatment conditions were improved on the basis of the simulated model of enzymatic reaction. The obtained lignocellulose fibres had a high surface area and showed a positive impact on the drainability of cardboard. Enzymatic hydrolysis of the waste paper fractions with different mesh (from 80 to 180) was also supplied using enzymatic treatments with commercial cellulase 'Celluclast' or cellobiase enzyme 'Novozyme' (Li et al. 2015). Waste paper pulp was treated with enzymes for 96 h and after adding of sodium acetate buffer the solid was extracted and dried. The BET specific area of the obtained pulps was the highest for the mixed fraction (80–180 mesh); however, the presence of lignin and ash in the fraction with 180 mesh inhibited the effect of hydrolysis.

Hasan and Sauodi extracted pure cellulose amounting 17.4%, 20%, and 18.2%, respectively, from agricultural and industrial waste sources: rice husk, waste office paper, and sugar cane *via* fast and simple technique (Hasan and Sauodi 2014). Cellulose from waste paper exhibited crystallinity 47.7% and FTIR confirmed that it was of a high purity.

The cellulose extracted from waste paper can be derivatized to its esters and other compounds. Ünlü obtained carboxymethylcellulose (CMC) with commercial grade from the recycled newspaper (Ünlü 2013). Cellulose was recovered from newspapers under oxidative alkaline conditions and the recovery was determined as 75–90% (w/w) of starting material. Degree of substitution of CMC was between 0.3 and 0.7% and 84–94% CMC content.

All of the above in this chapter is only a part of the existing data on the conversion of waste paper to cellulose, its derivatives, and composites. However, this shows that the methods of obtaining these materials are very diverse and differ both in complexity of implementation and in their cost. In addition, they have not been exhausted and new advanced technologies are likely to be developed in the future to produce valuable cellulose-containing materials.

4.3.2.2 Nanocellulose from Wastes and Its Functionalization

Nanocellulose is known since 1949 when the first cellulose nanocrystals were produced (Nechyporchuk et al. 2016); however, cellulose has gained importance as a nanostructured material only during last decades.

Besides biodegradability, renewability, and biocompatibility, the nanodimensional cellulosic fibres exhibit promising properties such as high strength and elastic modulus, high surface area, high aspect ratio, chemical functionality, dimensional stability, low density, moisture absorption properties, and thermal stability (Yano and Nakahara 2004; De Mesquita et al. 2010; Sarno and Cirillo 2019).

Investigations on nanocellulose are the most cited of all publications on cellulose study. There are hundreds of research groups around the world studying the production, properties, and applications of nanocellulose. Dozens of monographs (Dufresne 2013; *Biopolymer Nanocomposites* 2013; *Polysaccharide-Based Nanocrystals* 2014; *Nanocellulose Polymer Nanocomposites* 2015; Hamad 2017; *Handbook of Nanocellulose and Cellulose Nanocomposites* 2017; *Nanocellulose and Sustainability* 2018; *Nanocellulose and Nanohydrogel Matrices* 2017; *Sustainable Nanocellulose and Nanohydrogels from Natural Sources* 2020), hundreds of reviews, and a huge number of articles provide a comprehensive overview of this undoubtedly interesting and already widely used cellulose material. Contemporaneous status in nanocellulose study has been summarized in several recent reviews (Habibi et al. 2010; Klemm et al. 2011; Abdul Khalil et al. 2014;

George and Sabapathi 2015; Moon et al. 2016; Nechyporchuk et al. 2016; Trache et al. 2017; Satyanarayana et al. 2017; Bejoy et al. 2018; Xie et al. 2018; Klemm et al. 2018; Trache et al. 2020; Miyashiro et al. 2020).

Within a few recent years the number of the studies on nanocellulose has increased enormously. In the last 3 years alone, over 1000 articles have been published. We give herein a short overview on the preparation and main properties of nanocellulose. The references are representative ones and do not comprise a comprehensive list.

The term 'nanocellulose' includes all cellulose-based nanomaterials with at least one of its dimensions less/equal to 100 nm (Abdul Khalil et al. 2014). There are several types of nanocellulose: nanocrystalline cellulose, nanofibrillated cellulose, and bacterial nanocellulose. (Kettunen 2013; Phanthong et al. 2018). Nanocellulose of all the types except bacterial can be extracted from biomass-based polysaccharides by chemical and mechanical methods or a combination thereof. The structure of cellulose fibril and the methods of CNC and CNF production are shown schematically in Figure 4.3.

Nanocrystalline cellulose, also called cellulose nanocrystals (CNCs), is usually extracted from cellulose fibres by strong acid hydrolysis which degrades and removes the noncrystalline (or amorphous) domains, while the highly crystalline residue remains and may be converted into a stable suspension by subsequent vigorous mechanical shearing action (Visakh and Thomas 2010). CNC have short rod-like whisker shape, that is why fibres are often called nanowhiskers with 2–35 nm in diameter and 100–500 nm in length (Eichhorn et al. 2010; Nechyporchuk et al. 2016; Phanthong et al. 2018). It is worth noticing that the dimensions of CNC vary depending on the source of cellulose and the particular extraction method (Kettunen 2013). The main distinctive features of CNC as well as CNF are presented in Table 4.1.

Nanofibrillated cellulose or cellulose nanofibrils (CNF) (also called microfibrillated cellulose) are the long, flexible, and entangled native cellulose nanofibres which can be extracted from cellulose fibrils by milder treatments. The fibrils consist of alternating crystalline and amorphous domains (Abdul Khalil et al. 2014). In general, CNF is produced by mechanical defibrillation after pretreatments. A combination of high mechanical shearing forces and mild enzymatic hydrolysis can also be used to prepare CNF. CNF consist of nanofibril aggregates, with lateral dimensions ranging between 10 and 30 nm or more and from 500 nm to several μm in length (Eichhorn et al. 2010; Kettunen 2013; Abdul Khalil et al. 2014; Nechyporchuk et al. 2016). Different methods of production provide CNF with different geometric dimensions.

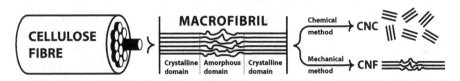

FIGURE 4.3 Schematic method of obtaining CNC and CNF from cellulose fibres.

TABLE 4.1

Comparison between Nanocrystalline and Nanofibrillated Cellulose

Actions and Outcome	Cellulose Nanocrystals (CNCs)	Cellulose Nanofibrills (CNF)	
Method of extraction	Chemical method (acid hydrolysis) + mechanical treatment (e.g., sonication)	Biological or/and chemical pretreatment: enzymatic hydrolysis, carboxylation, carboxymethylation, quaternization, sulfonation, solvent-assisted pretreatment	+ Mechanical methods: homogenization, grinding, extrusion, blending, ultrasonication, cryocrushing, steam explosion, ball milling, aqueous counter collision, mechanical disintegration; grinding
What happens	Amorphous regions of cellulose are removed by a chemical treatment	Networks of nanofibrils with crystalline and amorphous regions are produced using shearing forces	
Behaviour in water	Form suspension	Form gels	
Shape	Short-rod or whiskers	Thread/hair	
Diameter	2–35 nm	5–50 and sometimes even to 100 nm	
Length	100–500 nm	500–2000 nm	

Nanocellulose also includes bacterial cellulose which is synthesized with a bottom-up method from glucose by different families of bacteria (Klemm et al. 2011; Abdul Khalil et al. 2014), but this one will not be considered in this context as we pay our attention on powder and nanocellulose produced from waste paper and cardboard.

Nanocellulose due to its geometry and excellent properties is used in the biomedical field as stimuli-responsive, drug carriage, food additives; in energy and environment fields as transistors and batteries, selective membranes, and oil adsorbents; as surface modifiers in nanocomposite materials, in production of transparent displays, and, eventually, in making paper with special functions or 'smart paper' (Okahisa et al. 2009; Heath and Thielemans 2010; Wang and Roman 2011; Kettunen 2013; Donius et al. 2014; Danial et al. 2015; Phanthong et al. 2018).

For example, both CNC and CNF are used as reinforcing agents to improve the tensile modulus and strength of the nanocomposite films with other polymers, such as poly(vinyl alcohol), natural rubber, poly(styrene-co-butyl acrylate) latex, polylactic acid, polyethylene glycol, etc. (Abdul Khalil et al. 2012; Duran et al. 2012). Another interesting field of application is the controlled drug delivery, for instance, water-soluble antibiotics and anticancer drugs (Duran et al. 2012).

Nanocellulose paper is transparent, optically clear, and easily bendable; therefore, it is preferable for printed electronics, for example, for solar cells, flexible organic light-emitting diode displays, and electric circuits instead of the conventional papers (Okahisa et al. 2009; Hu et al. 2013). Modified nanofibrillar cellulose affects the morphology of paper sheets positively, for instance, when carboxymethylated cellulose nanofibrils were added to paper sheets, the sheets exhibited more uniform formation (Park et al. 2018).

4.3.2.3 Methods of Production of Nanocellulose and Nanofibrillated Cellulose

Since nanocellulose is a material of great interest nowadays and is intensely studied, a number of techniques of extraction were introduced. All the methods lead to different types of nanocellulosic materials, depending on the origin of cellulose pulp and its pretreatment (Abdul Khalil et al. 2012). The chemical, biological, and mechanical treatments are implemented in different combinations. We describe briefly some of the methods below.

4.3.2.3.1 Chemical and Biological Treatments

In order to obtain CNC, acid hydrolysis of the cellulose pulp in concentrated sulfuric acid is commonly applied (Table 4.1). However, hydrochloric, phosphoric, and hydrobromic acids are also used to extract CNC. The acid contributes not only to the isolation of CNC but also to the dispersing of them due to the esterification (Phanthong et al. 2018). After neutralization of the suspension, it may be subjected to mechanical treatment to improve the separation of nanocrystals. In most cases, sonication is implemented for this purpose (Kettunen 2013).

Enzymatic hydrolysis is a frequently used biological technique to obtain CNF (Table 4.1). Wood pulp is slowly degraded in mild conditions with one or more enzymes, for example, endoglucanases, which attack only amorphous regions of fibre; after that it may be subjected to homogenization or another mechanical treatment. Enzymatic hydrolysis is usually used in a combination with other methods in order to accelerate the process.

Cellulose nanofibrils could be prepared by oxidation processes, for example, by using TEMPO (2,2,6,6-tetramethylpiperidine-1-oxyl radical) or its derivatives as a catalyst and hypochlorite as an oxidant in different media. The method may consist of one or several steps in which lignocellulose is treated by oxidizing agents to remove both lignin and hemicelluloses and to obtain oxycellulose simultaneously (Duran et al. 2012). TEMPO-oxidized cellulose nanofibres mostly have uniform width (3–4 nm) with a high aspect ratio (Phanthong et al. 2018).

Carboxymethylation is one more chemical pretreatment for obtaining CNF. The purpose is to introduce anionic charges on the fibril surface to increase the electrostatic repulsion between the fibrils, and thus facilitate the fibrillation. Etherification is carried out with monochloroacetic acid and NaOH in an alcohol media. After carboxymethylation, the fibres are homogenized using one of the mechanical methods (Kettunen 2013).

4.3.2.3.2 Mechanical Treatments

The mechanical high shear force techniques are used to cleavage the cellulose fibres in longitudinal axis to extract CNF. The main disadvantage of these methods is the high level of power consumption. High pressure homogenization is one of the conventional methods when impact and shear forces in fluid are generated to cleavage cellulose microfibrils into nanometre size in diameter (Phanthong et al. 2018).

The treatment of lignocellulosic biomass with short-time high pressure steam, followed by decompression (explosion) leads to fiberization or 'mulching' of fibres. As a result, CNF is obtained. The pressure increases and drops resulting in a substantial break down of the lignocellulosic structure, hydrolysis of the hemicelluloses fraction, depolymerization of the lignin components, and defibrillization (Abdul Khalil et al. 2012).

In a number of studies, a combination of chemical and mechanical methods is applied. One of the examples was described in the paper Phanthong et al. (2018). High-intensity ultrasonication process implements the hydrodynamic forces of the ultrasound. After the removal of lignin and hemicelluloses by chemical methods, the samples are rinsed in water until the residues get neutralized, and then the ultrasonic treatment is carried out. The mechanical oscillating power is produced resulting in the formation, expansion, and implosion of microscopic gas bubbles when the liquid molecules absorb ultrasonic energy.

One of the popular methods, especially before the 2000s, is CNF production by ball milling. When the centrifugal force creates the shear forces between balls and the surface of a dish, the cellulose fibrils are split into the smaller size in diameter reaching nanometre sizes (Phanthong et al. 2018).

4.3.2.4 Direct Acid Hydrolysis of Waste Paper

In recent years, the viability of the direct production of nanocellulose from recycled paper was proved by numerous reports. The first approach to extract nanocellulose is acid hydrolysis. This method requires such pretreatment as mercerization and bleaching. The aim of the alkali treatments is to ensure the hydrolysis of hemicelluloses and removal of undesirable amorphous type polymer components, while the bleaching treatment is primarily aimed at removing lignin (Cherian et al. 2010; Li et al. 2009; Ndazi et al. 2007; Danial et al. 2015).

4.3.2.4.1 CNC and CNF from Waste Paper

The new types of fibre materials such as functionalized fibres and nanofibres produced from cellulose provide the possibility to improve the recycling process of the wastes. The preparation of these materials in a convenient and environment-friendly way from abundant waste materials is still a challenge (Sarno and Cirillo 2019). Recently, a large number of articles and some reviews on the extraction of the valuable fibres and conversion of waste paper into the functional materials have been issued.

Kumar et al. published an extensive and critical review summarizing the preparation of CNC from waste paper and their physico-chemical properties (Kumar et al. 2020). The authors underlined that although the properties of CNC from waste paper are comparable with the nanocellulose obtained from lignocellulosic raw materials, yet waste paper is an underutilized source for nanocellulose preparation due to its ordinary fate of recycling, dumping, and incineration. They analysed characteristics of CNC fibres from various types of waste paper and cardboard prepared with yields varying from 1.5% to 64%. The quality of nanocellulose strongly depends on type of waste paper used as cellulose source and preparation methods. The diameters of these nanocelluloses were reported to be in the range of 2–100 nm and crystallinity ranging around 54–95%. Thermal degradation of waste paper nanocellulose varied from 187°C to 371°C.

Danial et al. extracted cellulose nanofibres by subjecting waste paper (ONP) to alkali and bleaching treatments followed by hydrolysis in sulfuric acid (Danial et al. 2015). The resulting rod-like particles of CNC had the dimensions ranging from 3 to 10 nm (an average value 4 nm) in diameter and from 100 to 300 (an average 170 nm) in length. The authors compared these dimensions to those of CNC extracted from different other sources of cellulose. The diameter was similar to the values for CNC obtained from bacterial, ramie, and sisal fibres. However, the average diameter and length of the CNC from newspaper were lower than those reported for CNC extracted from wood pulps. The aspect ratio of the CNC was higher compared to CNC industrially produced from Avicel.

Another group of researchers has used the same procedure to extract CNC from two types of recycled paper: ONP with ink and 100% recycled newsprint paper without ink (Campano et al. 2017). The direct acid hydrolysis in sulfuric acid without any pretreatment or with alkali pretreatment and subsequent bleaching was carried out. In the case of unpretreated samples, purity and crystallinity of CNC were lower than when alkali pretreatment and bleaching were applied. Obviously, the better quality the raw material had the better purity the obtained CNC exhibited. The crystallinity indexes (CI) of all obtained CNC were higher compared to the initial materials and were similar to CI for CNC obtained from pure microcrystalline cellulose (CI = 95.5%). It was also shown that the hydrolysis yield for unpretreated samples was almost similar, approximately 60%, while for the pretreated sample the total hydrolysis yield was only 35%. This decrease was explained by removal of impurities as by the loss of cellulose and other compounds during pretreatments. Regarding the sizes of CNC, Campano et al. obtained similar values to those reported by Mohamed et al. (2015) and by Danial et al. (2015) which confirmed that this method provides the reproducible results.

In the same paper, financial efficiency and some particular qualities of the production of CNC from different kinds of waste paper were compared. Briefly, it can be summarized as follows:

1. CNC obtained from ONP by the direct acid hydrolysis had relatively low quality due to the higher levels of inks and lignin; however, the production was cost-effective and the yield was high;
2. CNC obtained from the newsprint paper (100% recycled fibres, no inks) by the direct acid hydrolysis had higher quality but lower yield, as opposed to newspapers; however, the price increased up to 5 times;
3. CNC obtained from the newsprint paper (100% recycled fibres, no inks) using alkali pretreatment and bleaching before the hydrolysis had perfect quality; however, the total yield was low (35 wt.%) and the price was high, so this method seems to be impractical (Campano et al. 2017).

In the examples listed above, the multi-stage processing was applied, namely swelling in the 5 wt.% NaOH followed by bleaching in 2 wt.% NaClO, subsequent milling, and eventually, heating in 60 wt.% H_2SO_4. It is worth noticing that chlorine-based bleaching is not environment friendly. Conducting the isolation process under milder conditions, one can obtain powder celluloses with partly agglomerated particles. This has been shown in the publication (Mikhailidi et al. 2019) when old newspapers and multicolour cardboard were used as raw sources of cellulose fibres. Mercerization was carried out in NaOH with lower concentration of 0.5–4 wt.%, bleaching was performed with a 'greener' compared to NaClO reagent H_2O_2 with concentration ranging from 3 to 9%. The hydrolysis was conducted in nitric acid with 'middle' concentration from 9 to18 wt.%. The total yield of the powder samples was about 48 wt.%. Hence, using mild procedure it is possible to increase the total yield of the process from 35% to 48%; however, in this case, the size of cellulose particles was more in micro- rather than in nanoscale. The powder cellulose obtained (Mikhailidi et al. 2019) from both sources had the structure of cellulose I and were of a high-grade purity. The cross-section dimension of the majority of fibres obtained from cardboard varied in micron dimensions ranging from 1 to 20 µm. Although the micron-sized cellulose particles did not possess all the variety of fascinating properties of nanocelluloses, they could be used as adsorbents for dyes or as scaffolds for intercalation of different additives, such as metal nanoparticles.

Bousfield presented the overview on the conversion of waste paper to cellulose nanoscale high-value products and described the benefits of using paper in a variety of applications, including nanofibre production (Bousfield 2013). The semi-industrial grinding plant at the University of Main (USA) produces CNF from waste paper. CNF in the slurry form used in paperboard coatings replaces polymer-based binders and stiffens the boards. As a green adhesive, CNF fully replaces also urea-formaldehyde resin and provides low formaldehyde products with lower formaldehyde emission. The author confirmed that the materials with CNF have the potential to be used in a number of applications such as paints, coatings, medical devices, and construction materials. The last but not the least option is the use of these materials to displace plastics in food packaging or other single-use applications.

Nanocellulose production from recycled paper sludge was performed by ozonation pretreatment, followed by maleic acid hydrolysis (Peretz et al. 2019). The total yield of CNC was 0.8 wt.%, but it was improved with higher ozone doses. Maleic acid was recovered, thus contributing to the green-industry concept, as it leaves no chemical traces and can be applied at ambient conditions.

De Souza et al. compared compositional (FTIR), thermal (TGA), morphological (SEM), and dimensional [dynamic light scattering (DLS) and atomic force microscopy (AFM) analyses] properties of CNS obtained through chemical and mechanical isolation processes from primary residues of the paper industry (De Souza et al. 2020). For the isolation by chemical method, the acid hydrolysis with sulfuric acid was implemented and by mechanical method, the ball milling was applied. Both processes gave purified cellulose but both obtained samples had different morphology and displayed a tendency to agglomerate. The average CNC sizes that varied with the method were 246 nm and 281 nm for CNC-chemical and CNC-mechanical samples, respectively.

The same authors obtained cellulose nanostructures (CNS) from industrial paper wastes by acid hydrolysis and two chemical pretreatments consisting of an alkaline (CNS-I) or an acid (CNS-II) treatment (De Souza et al. 2019). A reduction in the non-cellulosic compounds in pretreated samples with different levels of components removal and lower lignin values for alkali treatment was observed. However, CNSs presented lower thermal stability than the pristine wastes as revealed with TGA. The DLS and AFM analyses revealed that spherical nanoparticles had the dimensions of approximately 195 nm and 350 nm for CNS-I and for CNS-II, correspondingly. It was concluded that the conversion of paper wastes into CNS was an excellent opportunity to recover the paper residues and aggregate value to them.

Nanocellulose was also successfully obtained for the first time from waste cellulose from used diapers (Sarno and Cirillo 2019). Since cellulose of diapers is already purified from non-cellulosic (hemicelluloses, lignin) material, the first step was bleaching using sodium chlorite ($NaClO_2$). Bleached cellulose was treated with concentrated sulfuric acid, heated at 40°C and then washed, adjusted to a neutral pH and then centrifuged. After drying at 80°C, a nanocrystalline powder with a good level of purity and high crystallinity was gained.

In the study (Van-Pham et al. 2020), CNC from waste newspapers contained inks that were extracted by means of alkali treatment and bleaching followed by acid hydrolysis. The remarkable removal of surface contaminants and the reduction in fibre diameter during pretreatments were observed and the nano-dimension of the extracted CNC was revealed with the average diameter of 12.3 ± 2.8 nm. CNC exhibited a high whiteness index of 80%, and high transparency of about 80% of the light at 600 nm calculated for a 0.02 mm thick nanocellulose film. FTIR indicated that lignin, hemicelluloses, and other colouring agents were removed. TGA analysis showed that the product had a typical maximum thermal degradation at 300°C. The authors claim that the implemented method of the production of CNC from waste newspaper had the shortest processing time ever reported for acid hydrolysis with conventional alkali and bleaching pretreatments.

Tang et al. prepared cationic cellulose nanofibres having three different contents of positively charged quaternary ammonium groups from waste pulp residues ('fibre sludge') with cellulose and hemicellulose contents of 95% and 4.75%, respectively, according to a water-based modification method involving etherification of the pulp with glycidyltrimethylammonium chloride followed by mechanical disintegration (Tang et al. 2015). The cationic CNF had a maximum cationic charge content of 1.2 mmol·g^{-1} and positive ξ-potential at various pH values. Sorption of negatively charged contaminants (fluoride, nitrate, phosphate, and sulfate ions) and their selectivity onto cationic CNF were evaluated. Maximum sorption of ~0.6 mmol·g^{-1} of these ions by CNF was achieved and selectivity adsorption studies showed that cationic CNF are more selective towards multivalent ions (PO_4^{3-} and SO_4^{2-}) than monovalent ions (F^- and NO_3^-). In addition, it was demonstrated that cationic CNF can be manufactured into permeable membranes capable of dynamic nitrate adsorption by utilizing a simple papermaking process.

Nagarajan et al. isolated CNC and CNF from used disposal paper cups through citric acid hydrolysis (Nagarajan et al. 2020). The effect of acid concentration on microstructure and yield of CNC were investigated. The optimized yield of CNC was 55 wt.% and the particles exhibited rod-like structures with a width of 13.7 ± 0.6 nm. TGA analysis showed that CNC and CNF were thermally stable until 308°C and 305°C. Authors considered that obtained CNC has the potential to replace the artificial nanoparticles.

Tang et al. reported on the preparation of CNC from old corrugated container pulp fibres using a combined chemical—mechanical process involving hydrolysis in phosphoric acid with the concentration of 60 wt.%, enzymatic hydrolysis using cellulase and sonication (Tang et al. 2015). CNC was extracted with a yield of about 24 wt.%. Enzymatic hydrolysis imparted the obtained CNC improved dispersion, increased crystallinity, and thermal stability.

Van-Pham et al esterified the cellulose substrate, for example, pure wood pulp filter paper, with a mixture of 17.5M acetic and 18.4M sulfuric acid with the aid of ultrasonication (Van-Pham et al. 2020). The effects of esterification and ultrasonication time and temperature (68–75°C) on the yield and degree of substitution (DS) of CNC were estimated. The yield and DS of the sample obtained without pretreatment, with values of 48.16% and 0.22, respectively, were not as big as for the sample obtained by ultrasonication and heating, with a yield of 85.38% and a DS of 0.46. Characterization revealed that hydroxyl groups of cellulose were successfully esterified and rod-shaped CNC were formed (width 10–100 nm) by a simple and convenient method.

Analysis of the cited publications showed the following:

- There is a great scientific and experimental interest in recycling waste paper as a cellulose-containing raw material for the use of valuable products. In this regard, the search for low-cost, convenient, and environment-friendly methods to obtain these products is intensively developing;
- Nanocellulose and nanofibrillated cellulose are among the hot favourite cellulose-based materials and the development of the methods for their production occupies a significant sector in this search;
- The most popular methodology of the waste paper recycling to produce CNC and CNF fibres includes several stages, for example, alkaline pretreatment and bleaching followed by mechanical or/and acid treatments. Most of the reagents used for the isolation of CNC and CNF are applied in high concentrations; they are toxic and the treatments are performed at harsh conditions;
- In order to process CNC and CNF quality and to minimize environmental impact, enzyme hydrolysis technique and ionic liquids are applied. However, so far these methods were financially costly.

4.3.2.4.2 Cellulose-based Aerogels Obtained from Waste Paper

The application of waste paper and cardboard, which are one of the largest municipal waste sources for production of biodegradable and environment-friendly hydrogels and aerogels, first became known in early 2000. In the report of the National University of Singapore, December 2015, the needs and opportunities for aerogel production from waste paper have been formulated (*Cellulose aerogels from recycled waste* 2015). It was stated that, for example, the thermal and acoustic insulation sector is one of the most interested in the use of waste paper. Herein, we shall give a short overview of the publications related to that and other techniques to synthesize the aerogels derived from waste paper and cardboard.

A series of papers published by a research group led by H. M. Duong showed prospective ways for transforming waste paper into aerogels (Feng et al. 2014;

Nguyen et al. 2014; Thai et al. 2019; Zhen et al. 2019; Duong et al. 2020). Feng et al. fabricated green biodegradable cellulose aerogels from paper waste (recycled cellulose) *via* a cost-effective and scalable method by applying sodium hydroxide/urea and a sample repetitive thawing—freezing (Feng et al. 2014). The aerogels yielded densities changing from 0.03–0.1 g cm^{-3}, the absorption capacities varied from 9 to 20 times of their own weight, the thermal conductivity ranged from 0.029 to 0.032 W mK^{-1}. The authors confirmed that the developed method revealed excellent properties of the recycled cellulose aerogels: macroporous structure, extremely light weight (0.04 g cm^{-3}), good mechanical properties, flexibility, high absorbance ability, self-cleaning, and low thermal conductivity (Nguyen et al. 2014). Coating the aerogels with methyltrimethoxysilane improved their hydrophobicity without affecting its absorbency. The material had a high potential to be used in the diaper production as it is biodegradable in contrast with most super absorbent polymers. Mechanically, the aerogel was flexible making a wide range of applications possible.

In the production of aerogels, cross-linking of the pretreated waste paper is often used to enhance the properties of the resulting aerogels. Zhen et al. suggested a new way to enhance the heat insulation performance of water canteen bottles using eco-friendly super-hydrophobic recycled cellulose aerogels made from paper waste (Zhen et al. 2019). The aerogels were prepared via ultrasonic treatment of recycled cellulose sample followed by adding to obtained suspension of the Kymene cross-linker solution and sonication. The mixture poured into the mould was freeze-dried and after that was heated in the oven at 120°C. The tested samples of a 1.5 cm thickness of 1.0 wt.% aerogel wrapped around the water canteen bottle provided excellent thermal insulation performance, while not adding significant weight to the bottle. This result was better than that provided by available commercial. The authors also noted that the cost-effective method making the recycled cellulose aerogel by using Kymene cross-linker is more advanced than using sodium hydroxide and urea.

Jin et al. fabricated cellulose-based aerogels from waste paper without any pretreatment using 1-allyl-3-methyimidazolium chloride ([Amim][Cl]) as a solvent by regeneration and freeze-drying technique (Jin et al. 2015). The aerogels were rendered both hydrophobic and oleophilic after treatment with a trimethylchlorosilane *via* a simple thermal chemical vapour deposition process. The silane-coated aerogels exhibited good absorption performance for oils and organic solvents, for example, chloroform, making them diversified absorbents for potential applications including sewage purification.

In 2018, Long et al. published the detailed review on the research progress towards the technologies for the preparation of the cellulose aerogels, their properties, and applications (Long et al. 2018). The authors summarized these specificities of a number of the aerogels including natural cellulose aerogels and, as relative to them, aerogels derived from waste paper. Application of ionic liquids as solvents for the pretreatment of waste paper before regeneration and formation of the aerogels is very popular compared with other methods. Fan et al. prepared waste newspaper aerogels using pretreatment of the newspapers with a series of chemical extraction, in order to remove ink and glue (Fan et al. 2017). Then the samples were treated consequently with solution of NaOH and the acidified sodium chlorite (NaClO$_2$) solution following the dispersing in the ionic liquid [Amim][Cl] and centrifugation. After the suspension was dialysed against water and rinsed, the extracted cellulose was lyophilized. The resulting aerogels demonstrated prominent adsorption for oils and organic solvents and excellent filtration capacity comparable to that of HEPA membrane for lampblack.

Chen et al. developed a facile one-step refining route, which combines mild maceration of never-dried waste pulp sludge and a mechanical shearing process, to prepare microfibrillated cellulose with a high storage modulus (Chen et al. 2016). The maceration was conducted with the mixture of glacial acetic acid and hydrogen peroxide. Two mechanical processes, i.e., disc refining and ultrasonication plus homogenization were applied to the cellulose after maceration and resulted in MFC with a highly tangled fibril network. The resultant cellulosic suspensions exhibited a gel-like and shear-thinning behaviour with storage moduli ranging from 200 to 4000 Pa. Specific surface areas and water retention values of MFC were accordingly increased with the enhancement of shear force, while the storage moduli were not consistently increased. The strong MFC gels can be used as reinforcing fillers for polymer composites, template for further functionalization, and immobilization.

A series of publications deals with the fabrication of the lightweight, hydrophobic, and porous carbon aerogels from cellulose-based waste newspaper as raw material (Duong et al. 2020). Carbon aerogels are receiving great attention in academia and industry; however, the preparation of high-performance carbon aerogels is challenging (Li et al. 2017). Compressible and conductive carbon aerogels (3C aerogels) were prepared from waste paper (office paper, newspaper and cardboard) using a general oxidation-oven drying-carbonization method. The 3C aerogels were lightweight (23.6 mg cm^{-3}) and showed good electrical conductivity (0.051 S cm^{-1}), high compressibility, super hydrophobicity/super oleophilicity, and high oil absorption capacity ($33–70$ g g^{-1}). The 3C aerogels can also be used for the removal of free oils from water *via* different approaches.

Carbon microbelt aerogels were also prepared from waste paper scraps (Bi et al. 2014). They were mixed with water and hydrochloric acid (10%), and left for 12 h. The solid was rinsed with water and after centrifugation was dried at 60°C. The dried sample was vigorously agitated and the uniform mixture was freeze-dried to form aerogels. The aerogels were subjected to pyrolysis at 850°C for 2 h and the resulting low-density carbon aerogels were obtained. The waste paper-produced carbon aerogel can absorb a wide range of organic solvents and oils with a maximum sorption capacity up to 188 times the weight of the pristine sample. The aerogel can be recycled and repeatedly used *via* a simple method of distillation or squeezing. The method was simple and cost-effective for various industrial applications, such as barrier separation, water purification, as three-dimensional (3D) electrode material for energy storage devices, as well as building blocks for functional composite materials.

Han et al. preliminary soaked and freeze-dried the pristine newspaper with the composition of 50.1% cellulose, 16.8% hemicelluloses, 18.1% lignin, and 6.8% ash (Han et al. 2016). The post-pyrolysis of the pretreated sample was performed first at 400°C and then at 900°C. The obtained carbon aerogel exhibited a low density of 18.5 mg cm^{-3} and excellent hydrophobicity with a water contact angle of 132° and selective absorption for organic reagents. The green and facile fabrication process, excellent hydrophobicity, and oleophilicity of these carbon aerogels used as an absorbent material have great potential in the application of organic pollutant solvents' absorption and protection of the environment.

Guo et al. produced recycled cellulose fibres from waste paper (newsprint fibres and kraft fibres) *via* ultrasonic wave processing (Guo et al. 2015). The ultrasonic cavitation effected on feasibility of the preparation of the secondary fibres. The ultrasonic treatment lasted 10 min and revealed that water absorption of both samples increased significantly and reached the highest values of 12.5 g g^{-1} and 11.2 g g^{-1}, respectively, i.e., about two times higher than that of untreated fibres. The rise of the treatment time to 20 min contributed to the increase in the length-to-diameter ratio and in the specific surface area. The recycled cellulose fibres fulfilled several technical indexes; therefore, they can be considered as a filling material for use in cement-based materials.

Waste paper flakes were added to wood particle mixtures with various ratios (75/25, 50/50, 25/75, and 0/100 wt./wt.) to prepare the laboratory-made single-layer particleboards (Nourbakhsh and Ashori 2010). The effects of press temperature and content of maleic anhydride on the properties of the boards were estimated and it was shown that evaluated properties that satisfied requirements of European Norms standards were modulus of rupture, modulus of elasticity, internal bonding strength, and thickness swelling. The results revealed that waste paper could be a potential substitute material for particleboard manufacture.

Other types of waste paper can also be successfully used for the various applications, for example, for oil sorption. Waste napkin paper was valorized to low-density (27.2 mg cm^{-3}) cellulose aerogels for oil sorption material (Sanguanwong et al. 2020). Two simple methods with different gel coagulators, ethanol, and sulfuric acid were used for the preparation of aerogel. The alkaline treatment of the raw material and the pre-freezing temperature in the lyophilization process were applied. It was found that the water and oil sorption capacities of the aerogels were not significantly affected by alkaline treatment, while they could be adjusted by changing the pre-freezing temperature. Although the produced aerogels were initially amphiphilic, hydrophobic surfaces were obtained by vapour deposition of methyltrimethoxysilane (MTMS), and these materials possessed high sorption capacities, up to 32.24 cm^3 g^{-1} (28.56 g g^{-1}) for pump oil and 26.77 cm^3 g^{-1} (39.59 g g^{-1}) for chloroform. This was comparable to aerogels prepared from fresh cellulosic materials *via* the sol–gel method, as their sorption capacities varied in the range of 14–45 g g^{-1}.

The waste tissue paper was used as a raw material to derive 3D carbon aerogel (CA) with a 3D hierarchical architecture integrated with cerium dioxide (CeO$_2$) nanotubes decorated nitrogen-doped reduced graphene oxide nanosheets (NRGO) as a competent anode (Senthilkumar et al. 2020). The direct growth of NRGO and CeO$_2$ over CA in the form of freestanding and binder-free composite NRGO/CeO$_2$ (1:2)/ CA was performed. The 3D architecture of CA with open porous structure provided easy access for bacteria, thus increased the bacterial colonies per unit volume and yielded excellent extracellular electron transfer efficiency. The electrostatic interaction between the NRGO and bacteria cells improved the bacterial adhesion and biofilm formation. This study thus faced access advanced avenues to converting waste matters of tissue paper, human urine, and wastewater into profitable constituents for the development of efficient and durable power producing systems.

A biomass-derived banana peel/waste paper hybrid aerogel with hierarchical porous structure has been fabricated via combination of freezing-cast, freeze-drying, and pyrolysis approach (Yue et al. 2018). The microsheets derived from banana peel were immobilized on the surface of the waste paper fibres. The resulting aerogels exhibited a compression strain of 75%, high-hydrophobicity of 149.3° under air superoleophilicity of 0°, and high porosity. The aerogels can absorb free oils in water with high oil sorption capacity of 35–115 times its own weight. They can also effectively separate various surfactant-stabilized water-in-oil emulsions driven solely by gravity, with the high separation efficiency above 99.6%. The obtained banana peel/waste paper aerogel can be a potential candidate for application in oil adsorption and water-in-oil emulsions separation.

Summarizing the results considered, the following conclusions were made:

- Technology for recycling waste paper into aerogels is very diverse and in most cases, it is adapted to the subsequent application of aerogels.
- The very popular technology for the recycling of the waste paper uses modified alkaline/urea and freeze-drying/supercritical drying process to achieve a cost-effective synthesis technique (*Cellulose aerogels from recycled waste* 2015).
- The enzymatic treatments often used on laboratory scale for the bioconversion of the waste paper are quite effective and scalable, however, economically inefficient and low-productive.
- The waste paper derived aerogels could be modified to alter the absorption properties of the aerogel and optimized for high absorbance applications for polar liquids (e.g., water) or non-polar liquids (e.g., oil).
- The cellulose aerogels have intrinsically very low thermal conductivity properties so they could also be optimized for use of thermal or acoustic insulation by coating its porous structure with hydrophobic materials.
- The composites with valuable properties can be produced from waste paper when combined with a wide variety of additives.

4.3.2.4.3 Some Successful Examples for Obtaining Hydro- and Aerogels from Waste Paper

In the previous part, we gave some examples illustrating the recycling of the aerogels from the solutions of waste paper (Jin et al. 2015; Fan et al. 2017; Long et al. 2018). The very popular technologies for the conversion of the waste paper to the aerogels applied the ionic liquids and alkaline/urea as solvents for the dissolving of waste paper before subsequent regeneration following eco-friendly freeze-drying or modified thawing–freeze drying methods. However, most of the known procedures require preliminary treatments (mercerization, bleaching), significant time, and energy expenditure and are economically expensive. In many cases, an additional activation of the pristine waste samples was also required (Wang Z. et al. 2012). In our opinion, the methods of direct dissolution of waste paper and subsequent regeneration or self-regeneration from the solutions are more attractive. Herein, we present some successful examples that confirm the lower cost and effectiveness of these methods and, consequently, their future viability.

Jeong et al. reported on the method to dissolve cellulose waste papers (Whatman filter paper, grade 1) completely using the ionic liquid [Amim][Cl] accompanied by under microwave irradiation (Jeong et al. 2018). Subsequent cellulose regeneration by an additional antisolvent, absolute ethanol, was performed to obtain 'cellulose material', so-called carbon dots (CDs). The samples were spherical-shaped and, in fact, this material consisted of spherical-shaped aerogel particles. The CDs that exhibited low cytotoxicity can probably act as a promising fluorescent probe for bioimaging. Due to the high price of ionic liquid, recovery and its reutilization was important in the process of regeneration.

The simple, low-cost, and prospective method for the preparation of hydrogels from newspaper (NP) and cardboard (CB) wastes by direct dissolving the wastes in the solution of solvent system DMAc/LiCl and subsequent regeneration from the solution was informed by Mikhailidi and Kotelnikova (2019). The procedure was a low number-step process that included thermal defibration of the wastes in boiling water and dissolving in DMAc/LiCl at 30–40°C. The solutions were stored for several days, during which spontaneous gelation occurred. The rest solvent was rinsed with water that led to replacement of the solvent to water. The self-assembly of the cellulose chains in the solutions revealed the formation of the super-swollen hydrogels which exhibited a high capacity to retain large amounts of water: up to 3900 wt.% for NP and 1750 wt.% for CB hydrogels. The structure modification of cellulose I of the pristine wastes transformed to a partial ordered cellulose modification II, and, in general, to the overall changes in the cellulose structure. The results revealed that the processing allowed us to utilize the waste paper and to prepare prospective materials.

4.4 CONCLUSIONS

Utilization of waste paper as a renewable resource of lignocellulosic constituents has the opportunity to promote a cleaner environment and to prepare prospective and helpful materials. This chapter summarizes the information on the main highlights of waste paper recycling and some novel and prospective approaches for conversion of the waste paper to valuable functional cellulose products. The key attention is paid to the novel methods of transforming the waste paper into cellulose, cellulose derivatives, cellulose composites, CNCs, and cellulose nanofibrills (CNF). The examples of obtaining CNC and CNF by direct processing of waste paper as well as the properties, characteristics, and functionalization of the products are considered. To transform the waste paper to valuable and functional cellulose materials, this is commonly subjected to alkaline pretreatment and bleaching followed by mechanical or/and acid treatments. Aerogels and hydrogels for various applications could be produced from waste paper by modification in alkaline/urea and freeze-drying/supercritical drying process or enzymatic treatments; however, these techniques are cost unprofitable. One of the prospective and low-cost way is a direct dissolution of waste paper in the solvent systems and subsequent regeneration from the solutions in the shape of hydrogels/aerogels, spherical particles, films, etc. The waste paper derived aerogels could be modified to alter the absorption properties of the aerogels and optimized for high absorbance applications for polar liquids, for example, water or non-polar liquids (oils). The composites with valuable properties can be also synthesized from the waste paper when combined with a wide variety of additives.

DECLARATION OF COMPETING INTEREST

The authors declared that there is no conflict of interest.

REFERENCES

Abdul Khalil, A., H. Bhat, and A. F. Yusra. 2012. Green composites from sustainable cellulose nanofibrils: A review. *Carbohydr. Polym.* 87:963-979. doi:10.1016/j.carbpol.2011.08.078

Abdul Khalil, H. P. S., Y. Davoudpour, Md. Nazrul Islam, et al. 2014. Production and modification of nanofibrillated cellulose using various mechanical processes: A review. *Carbohydr. Polym.* 99:649-665. doi:10.1016/j.carbpol.2013.08.069

Achieving the Sustainable Development Goals in South Asia, 2017. *Key policy priorities and implementation challenges.* United Nations Economic and Social Commission for Asia and the Pacific. https://www.unescap.org/publications/achieving-sustainable-development-goals-south-asia-key-policy-priorities-and.

Adhikari, Ch. R., D. Parajuli, K. Inoue, K. Ohto, H. Kawakita, and H. Haradaa. 2008. Recovery of precious metals by using chemically modified waste paper. *New J. Chem.* 32:1634-1641. doi:10.1039/B802946F

Adil Sabbar, H. 2019. Adsorption of phenol from aqueous solution using paper waste. *Iraqi J. Chem. Petrol. Eng.* 20(1):23-29. doi:10.31699/IJCPE.2019.1.4

AF&PA Sustainability Report, 2018. Advancing U.S. Paper and wood products industry sustainability performance. American Forest and Paper Association. https://sustainability.afandpa.org/

Alemdar, A., and M. Sain. 2008. Isolation and characterization of nanofibers from agricultural residues – wheat straw and soy hulls. *Bioresour. Technol.* 99: 1664-1671. doi:10.1016/j.biortech.2007.04.029

Bajpai, P. 2014. *Recycling and deinking of recovered paper.* Waltham: Elsevier Inc. doi:10.1016/C2013-0-00556-7

Bejoy, Th., M. C. Raj, B. Athira, et al. 2018. Nanocellulose, a versatile green platform: From biosources to materials and their applications. *Chem. Rev.* 118(24):11575–11625. doi:10.1021/acs.chemrev.7b00627

Bi, H., X. Huang, X. Wu, et al. 2014. Carbon microbelt aerogel prepared by waste paper: An efficient and recyclable sorbent for oils and organic solvents. *Small* 10(17):3544–3550. doi:10.1002/smll.201303413

Biopolymer nanocomposites: Processing, properties, and application, ed. A. Dufresne, S. Thomas, and L. A. Pothan. 2013. Hoboken: John Wiley & Sons, Inc.

Bousfield, D. 2013. The conversion of waste paper to cellulose nanofibres and to high value products. Presentation P-886. The University of Maine, USA. https://www.aiche.org/system/files/aiche-proceedings/conferences/300296/papers/-886/P-886.pdf

Brummer, V., T. Juren, V. Hlavacek, et al. 2014. Enzymatic hydrolysis of pretreated waste paper – Source of raw material for production of liquid biofuels. *Bioresour. Technol.* 152: 543–547. doi:10.1016/j.biortech.2013.11.030

Campano, C., R. Miranda, N. Merayo, C. Negro, A. Blanco. 2017. Direct production of cellulose nanocrystals from old newspapers and recycled newsprints. *Carbohydr. Polym.* 173:489–496. doi:10.1016/j.carbpol.2017.05.073

Cellulose aerogels from recycled waste. 2015. *Climate Technology Centre & Network. National University of Singapore.* https://www.ctc-n.org/products/cellulose-aerogels -recycled-waste

Chen, N., J. Y. Zhu, and Zh. Tong. 2016. Fabrication of microfibrillated cellulose gel from waste pulp sludge via mild maceration combined with mechanical shearing. *Cellulose* 23(4):2573–2583. doi:10.1007/s10570-016-0959-1

Cherian, B. M., A. L. Leão, S. F. De Souza, S. Thomas, L. A. Pothan, and M. Kottaisamy. 2010. Isolation of nanocellulose from pineapple leaf fibres by steam explosion. *Carbohydr. Polym.* 81(3):720–725. doi:10.1016/j.carbpol.2010.03.046

Conte, R. 2016. Cellulose recovery from waste. *IST* Autumn 2016. https://www.academia. edu/32433517

COST Action E48, 2009. The final report: 'The limits of paper recycling'. https://www.cost. eu/cost-action/the-limits-of-paper-recycling/#tabslName:overview

Crini, G. 2006. Non-conventional low-cost adsorbents for dye removal: A review. *Bioresour. Technol.* 97(9):1061–1085. doi:10.1016/j.biortech.2005.05.001

Danial, W. H., Z. A. Majid, M. N. M. Muhid, S. Triwahyono, M. B. Bakar, and Z. Ramli. 2015. The reuse of wastepaper for the extraction of cellulose nanocrystals. *Carbohydr. Polym.* 118:165–169. doi:10.1016/j.carbpol.2014.10.072

De Mesquita, J. P., C. L. Donnici, and F. V. Pereira. 2010. Biobased nanocomposites from layer-by-layer assembly of cellulose nanowhiskers with chitosan. *Biomacromolecules* 11(2):473–480. doi:10.1021/bm9011985.

De Souza, A. G., R. F. S. Barbosa, and D. S. Rosa. 2020. Nanocellulose from industrial and agricultural waste for further use in PLA composites. *J. Polym. Environ.* 28:1851–1868. doi:10.1007/s10924-020-01731-w

De Souza, A. G., D. Belcior, R. Fabiany, S. Kano, D. S. Rosa. 2019. Valorization of industrial paper waste by isolating cellulose nanostructures with different pretreatment methods. *Resour. Conserv. Recycl.* 143:133–142. doi:10.1016/j.resconrec.2018.12.031

Donius, A. E., A. Liu, L. A. Berglund, and U.G.K. Wegst. 2014. Superior mechanical per-formance of highly porous, anisotropic nanocellulose–montmorillonite aerogels pre-pared by freeze casting. *J. Mech. Behav. Biomed. Mater.* 37:88–99. doi:10.1016/j. jmbbm.2014.05.012

Dufresne, A. 2013. *Nanocellulose: From nature to high performance tailored materials.* Berlin: De Gruyter. doi:10.1515/9783110254600

Duong, H. M., D. Kh. Le, Q. B. Thai, et al. 2020. Advanced thermal properties of carbon-based aerogels in *Thermal behaviour and applications of carbon-based nanomaterials*, ed. D. V. Papavassiliou, H. M. Duong and F. Gong. Amsterdam: Elsevier. doi:10.1016/ b978-0-12-817682-5.00009-x

Duran, N., A. B. Seabra, and A. P. Lemes. 2012. Review of cellulose nanocrystals patents: Preparation, composites and general applications. *Recent Patents on Nanotechnology* 6(1). doi:10.2174/187221012798109255

Eichhorn, S. J., A. Dufresne, and M. Aranguren. 2010. Review: Current international research into cellulose nanofibres and nanocomposites. *J. Mater. Sci.* 45(1):1–33. doi:10.1007/ s10853-009-3874-0

El Achaby, M., Z. Kassab, A. Aboulkas, C. Gaillard, and A. Barakat. 2018. Reuse of red algae waste for the production of cellulose nanocrystals and its application in polymer nano-composites. *Int. J. Biol. Macromol.* 106:681–691. doi:10.1016/j.ijbiomac.2017.08.067

Espinosa, E., F. Rol, J. Bras, and A. Rodriquez. 2019. Production of lignocellulose nano-fibers from wheat straw by different fibrillation methods. Comparison of its viabil-ity in cardboard recycling process. *J. Clean. Prod.* 239:118083. doi:10.1016/j. jclepro.2019.118083

Fan, P., Y. Yuan, J. Ren, et al. 2017. Facile and green fabrication of cellulosed based aerogels for lampblack filtration from waste newspaper. *Carbohydr. Polym.* 162:108–114. doi:10.1016/j.carbpol.2017.01.015

Feng, J. D., S. T. Nguyen, and H. M. Duong. 2014. Recycled paper cellulose aerogel synthesis and water absorption properties. *Adv. Mater. Res.* 936:938–941. doi:10.4028/www.scientific.net/amr.936.938

George, J., and S. N. Sabapathi. 2015. Cellulose nanocrystals: Synthesis, functional properties, and applications. *Nanotechnol. Sci. Appl.* 8:45–54. doi:10.2147/NSA.S64386

GRN – Recycling Terms Glossary, 2007. http://www.encyclo.co.uk/local/20096

Guo, X., Zh. Jiang, H. Li, W. Li. 2015. Production of recycled cellulose fibers from waste paper via ultrasonic wave processing. *J. Appl. Polym. Sci.* 132(19):41962. doi:10.1002/app.41962

Ha Noi 3R Declaration, 2013. Sustainable 3R goals for Asia and the Pacific for 2013–2023. http: https://www.env.go.jp/recycle/3r/en/declaration/hanoi-declaration.html

Habibi, Y., L. A. Lucia, and O. J. Rojas. 2010. Cellulose nanocrystals: Chemistry, self-assembly, and applications. *Chem. Rev.* 110:3479–3500. doi:10.1021/cr900339w

Hamad, W. Y. 2017. *Cellulose nanocrystals: Properties, production, and applications.* Chichester: John Wiley & Sons Ltd. doi:10.1002/9781118675601

Han, S., Q. Sun, H. Zheng, J. Li, and C. Jin. 2016. Green and facile fabrication of carbon aerogels from cellulose-based waste newspaper for solving organic pollution. *Carbohydr. Polym.* 136:95–100. doi:10.1016/j.carbpol.2015.09.024

Handbook of nanocellulose and cellulose nanocomposites, ed. H. Kargarzadeh, I. Ahmad, Sabu Th.., and A. Dufresne. 2017. Weinheim: Wiley-VCH Verlag GmbH & Co. KGaA. doi:10.1002/9783527689972

Hasan, H.R., and M. H. Sauodi. 2014. Novel method for extraction of cellulose from agricultural and industrial wastes. *Chem. Technol. An Indian J. (CTAIJ)* 9(4):14–153.

Heath, L., and W. Thielemans. 2010. Cellulose nanowhisker aerogels. *Green Chem.* 12(8), 1448–1453. doi:10.1039/c0gc00035c

Hospodarova, V., N. Stevulova, J. Briancin, and K. Kostelanska. 2018. Investigation of waste paper cellulosic fibers utilization into cement based building materials. *Buildings* 8(43):1–12. doi:10.3390/buildings8030043.

Hu, L., G. Zheng, J. Yao, et al. 2013. Transparent and conductive paper from nanocellulose fibers. *Energy Environ. Sci.* 6(2):513–518. doi:10.1039/C2EE23635D

Huang, J., P. R. Chang, N. Lin, and A. Dufresne (ed.) 2014. *Polysaccharide-based nanocrystals: Chemistry and applications.* Weinheim: Wiley-VCH Verlag GmbH & Co. KGaA. doi:10.1002/9783527689378

Ikeda, Y., E.Y. Park, and N. Okuda. 2006. Bioconversion of waste paper to gluconic acid in a turbine blade reactor by the filamentous fungus: *Aspergillus niger. Bioresour. Technol.* 97:1030–1035. doi:10.1016/j.biortech.2005.04.040

Islam, M.A., D. W. Morton, B. B. Johnson, and M. J. Angove. 2019. Application of pulp and paper industry wastes in environmental pollutants removal. *Bangladesh J. Sci. Ind. Res.* 54:76. doi:10.3329/bjsir.v54i1.40731

Jain, A. K., V. K. Gupta, A. Bhatnagar, and Suhas. 2003. Utilization of industrial waste products as adsorbents for the removal of dyes. *J. Hazard. Mater.* B101:31–42. doi:10.1016/S0304-3894(03)00146-8

Jawaid, M. and F. Mohammad. (eds.) 2017. *Nanocellulose and nanohydrogel matrices. Biotechnological and biomedical applications.* Weinheim: Wiley-VCH Verlag GmbH & Co. KGaAdoi:. 10.1002/9783527803835

Jiang, F. and Y.-L. Hsieh. 2015. Cellulose nanocrystal isolation from tomato peels and assembled nanofibers. *Carbohydr. Polym.* 122:60–68. doi:10.1016/j.carbpol.2014.12.064

Jeong, Y., K. Moon, S. Jeong, W.-G. Koh, and K. Lee. 2018. Converting waste papers to fluorescent carbon dots in the recycling process without loss of ionic liquids and bioimaging applications. *ACS Sustain. Chem. Eng.* 6:4510.4515. doi:10.1021/acssuschemeng.8b00353

Jiang, M., Z. Zuowan, M. Chunyu, Z. Jun, Z. Mengmeng, and Z. Shibu. 2014. Isolation of cellulose from steam-exploded rice straw with aniline catalyzing dimethyl formamide aqueous solution. *Renew. Energy* 63:324–329. doi:10.1016/j.renene.2013.09.016

Jin, C., S. Han, J. Li, and Q. Sun. 2015. Fabrication of cellulose-based aerogels from waste newspaper without any pretreatment and their use for adsorbents. *Carbohydr. Polym* 123(5): 150–156. doi:10.1016/j.carbpol.2015.01.056

Juang, R. S., R. L. Tseng, and F. C. Wu. 2001. Role of microporosity of activated carbons on their adsorption abilities for phenols and dyes. *Adsorption* 7(1):65–72. doi:10.1023/a:1011225001324

Juang, R. S., F. C. Wu, and R. L. Tseng. 2002. Characterization and use of activated carbons prepared from bagasses for liquid-phase adsorption. *Colloid Surf. A: Physicochem. Eng. Aspect* 201(1–3):191–199. doi:10.1016/s0927-7757(01)01004-4

Kadam, A. A., S. Lone, S. Shinde, et al. 2019. Treatment of hazardous engineered nanomaterials by supermagnetized α-cellulose fibers of renewable paper-waste origin. *ACS Sustain. Chem. Eng.* 7:5764–5775.

Kakati, D. 2020. Growing plant on waste paper. *Indian science congress-2020*, Bangalore. https://www.researchgate.net/publication/338411508

Kannan, N., and M. M. Sundaram. 2001. Kinetics and mechanism of removal of methylene blue by adsorption on various carbons – a comparative study. *Dyes Pigments* 51(1):25–40. doi:10.1016/s0143-7208(01)00056-0

Kettunen, M. 2013. Cellulose nanofibrils as a functional material. PhD Diss., Aalto University.

Khanna, R., R. Cayumil, P. S. Mukherjeeb, and V. Sahajwalla. 2014. A novel recycling approach for transforming waste printed circuit boards into a material resource. *Procedia Environ. Sci.* 21:42–54. doi:10.1016/j.proenv.2014.09.006

Klemm, D., E. D. Cranston, D. Fischer, et al. 2018. Nanocellulose as a natural source for groundbreaking applications in materials science: Today's state. *Materialstoday* 21(7):720–748. doi:10.1016/j.mattod.2018.02.001

Klemm, D., F. Kramer, S. Moritz, et al. 2011. Nanocelluloses: A new family of nature-based aterials. *Angew. Chem. Int. Ed.* 50(24):5438–5466. doi:10.1002/anie.201001273

Kumar, V., P. Pathak, and N. K. Bhardwaj. 2020. Waste paper: An underutilized but promising source for nanocellulose mining. *J. Waste Manag.* 102:281–303. doi:10.1016/j.wasman.2019.10.041

Kumara, S., Y. Hartantio, and R. S. Hartati. 2020. Analysis of bioethanol processing from paper waste as generator fuel. doi:10.31219/osf.io/b3z94

Lee, K.-Y. (ed.). 2018. *Nanocellulose and sustainability: Production, properties, applications, and case studies.* Boca Raton: CRC Press/Tailor & Francis Group.

Li, L., Zh. Kong, Y. Qin, et al. 2020. Temperature-phased anaerobic co-digestion of food waste and paper waste with and without recirculation: Biogas production and microbial structure. *Sci. Total Environ.* 724:138–168. doi:10.1016/j.scitotenv.2020.138168

Li, L., B. Li, H. Sunm and J. Zhang. 2017. Compressible and conductive carbon aerogels from waste paper with exceptional performance for oil/water separation. *J. Mater. Chem. A* 5(2):14858–14864. doi:10.1039/c7ta03511j

Li, R., J. Fei, Y. Cai, Y. Li, J. Feng, and J. Yao. 2009. Cellulose whiskers extracted from mulberry: A novel biomass production. *Carbohydr. Polym.* 76(1):94–99. doi:10.1016/j.carbpol.2008.09.034

Li, X., H. Yu, D. Sun, J. Jiang, and L. Zhu. 2015. Comparative study of enzymatic hydrolysis properties of pulp fractions from waste paper. *BioResources* 10(2):3818–3830. doi:10.15376/biores.10.3.3818-3830

Liu, J., H. Hu, J. Xu, and Y. Wen. 2012. Optimizing enzymatic pretreatment of recycled fiber for improving its draining ability using response surface methodology. *BioResources* 7(2):2121–2140. doi:10.15376/biores.7.2.2121-2140

Long, L.-Y., Y.-X. Weng, and Y.-Zh. Wang. 2018. Cellulose aerogels: Synthesis, applications, and prospects. *Polymers* 10(6)623–628. doi:10.3390/polym10060623

Ma, Y., M. Hummel, M. Määttänen, A. Särkilahti, A. Harlin, and H. Sixta. 2016. Upcycling of waste paper and cardboard to textiles. *Green Chem.* 18:858. doi:10.1039/c5gc01679g

Macedo de Melo E. 2018, Microfibrillated cellulose and high-value chemicals from orange peel residues. PhD Diss., Univ. of York.

Mariño, M., L. Lopes da Silva, N. Durán, and L. Tasic. 2015. Tasic Enhanced materials from nature: Nanocellulose from citrus waste. *Molecules* 20:5908–5923. doi:10.3390/molecules 20045908

McKinney, R.W. J. (ed.) 1995. *Technology of paper recycling*. London: Blackie Academic and Professional/Chapman & Hall.

Mikhailidi, A.M., and N.E. Kotelnikova. 2019. New pathway to prepare nanocrystalline materials from paper wastes. *6th International Polysaccharide Conference EPNOE*, Aveiro. P. 298. https://epnoe2019.sciencesconf.org/resource/page/id/10

Mikhailidi, A. M., Sh. Karim Saurov, V. I. Markin, and N. E. Kotelnikova. 2019. Functional materials from paper wastes I. From waste newsprint paper and cardboard to high-grade cellulose fibers. *Russ. J. Bioorg. Chem.* 45(7):888–894. doi:10.1134/S1068162019070069

Mohammad, F., H. Al-Lohedan, and M. Jawaid (eds.) 2020. *Sustainable nanocellulose and nanohydrogels from natural sources*. Amsterdam: Elsevier.

Mohamed, M. A., W. N. W. Salleh, J. Jaafar, S. Asri, and A. F. Ismail. 2015. Physicochemical properties of 'green' nanocrystalline cellulose isolated from recycled newspaper. *RSC Adv.* 5(38):29842–29849. doi:10.1039/c4ra17020b

Mondal, S. 2017. Preparation, properties and applications of nanocellulosic materials. *Carbohydr. Polym.* 163:301. doi:10.1016/j.carbpol.2016.12.050

Monitoring Report, 2017. European declaration on paper recycling 2016–2020. European Paper Recycling Council (EPRC). https://www.paperforrecycling.eu/download/900/

Moon, R. J., G. T. Schueneman, and J. Simonsen. 2016. Overview of cellulose nanomaterials, their capabilities and applications. *JOM* 68(9):2383–2394. doi:10.1007/s11837-016-2018-7

Mostafa, N. A., A. A. Farag, H. M. Abo-dief, and A. M. Tayeb. 2018. Production of biodegradable plastic from agricultural wastes. *Arab. J. Chem.* 11:546–553. doi:10.1016/j.arabjc.2015.04.008

Miyashiro, D., R. Hamano, and K. Umemura. 2020. A Review of Applications Using Mixed Materials of Cellulose, Nanocellulose and Carbon Nanotubes. *Nanomaterials.* 10(2):186. doi:10.3390/nano10020186

Nagarajan, K. J., A. N. Balaji, S. Thanga Kasi Rajan, and N.R. Ramanujam. 2020. Preparation of bio-eco based cellulose nanomaterials from used disposal paper cups through citric acid hydrolysis. *Carbohydr. Polym.* 235:115997. doi:10.1016/j.carbpol.2020.115997

Namasivayam, C., M. Dinesh Kumar, K. Selvi, R. Begum Ashrffunissa, T. Vanathi, and R. T. Yamuna. 2001. 'Waste' coir pith – a potential biomass for the treatment of dyeing wastewaters. *Biomass Bioenergy* 21(6):477–483. doi:10.1016/s0961-9534(01)00052-6

Ndazi, B. S., C. Nyahumwa, and J. Tesha. 2007. Chemical and thermal stability of rice husks against alkali treatment. *Bioresources* 3(4):1267–1277. doi:10.1016/j.compositesa.2006.07.004

Nechyporchuk, O., M. Belgacem, and J. Bras. 2016. Production of cellulose nanofibrils: A review of recent advances. *Ind. Crops Prod.* 93:2–25. doi:10.1016/j.indcrop.2016.02.016

Nguyen, S. T., J. Feng, N. T. Le, et al. 2013. Cellulose aerogel from paper waste for crude oil spill cleaning. *Ind. Eng. Chem. Res.* 52:18386–18391. https://dx.doi.org/10.1021/ie4032567

Nguyen, S. T., J. Feng, Sh. Kai Ng, J. P. W. Wong, V. B. C. Tan, and H. M. Duong. 2014. Advanced thermal insulation and absorption properties of recycled cellulose aerogels. *Colloid Surface A* 445:128–134. doi:10.1016/j.colsurfa.2014.01.015

Nourbakhsh, A., and A. Ashori. 2010. Particleboard made from waste paper treated with maleic anhydride. *Waste Manag. Res.* 28(1):51–55. doi:10.1177/0734242X09336463

Nuruddin, N., A Chowdhury, S.A. Haque et al. 2011. Extraction and characterization of cellulose microfibrils from agricultural wastes in an integrated biorefinery initiative. *Cell. Chem. Technol.* 45(5–6):347–354.

Okada, K., N. Yamamoto, Y. Kameshima, and A. Yasumori. 2003. Adsorption properties of activated carbon from waste newspaper prepared by chemical and physical activation. *J. Colloid Int. Sci.* 262:194–199.

Okahisa, Y., A. Yoshida, S. Miyaguchi, and H. Yano. 2009. Optically transparent wood–cellulose nanocomposite as a base substrate for flexible organic light-emitting diode displays. *Compos. Sci. Technol.* 69(11–12):1958–1961. doi:10.1016/j.compscitech.2009.04.017

Otero, M., F. Rozada, L. F. Calvo, A. I. Garcia, and A. Moran. 2003. Kinetic and equilibrium modelling of the methylene blue removal from solution by adsorbent materials produced from sewage sludges. *Biochem. Eng. J.* 15:59–68.

Pang, S. C., S. F. Chin, and V. Yih. 2011. Conversion of cellulosic waste materials into nanostructured ceramics and nanocomposites. *Adv. Mat. Lett.* 2(2):118–124. doi:10.5185/amlett.2011.1203

Paper Recycling Terminology. The paper recycling coalition, 2019. https://www.paperrecyclingcoalition.com/faqs/paper-recycling-terminology/

Paper Waste Facts. TheWorldCounts. 2020. https://www.theworldcounts.com/stories/Paper-Waste-Facts

Park, S. Y., H. Park, S. Jonghyun, and J. H. Huyn 2018. Youn effect of morphological properties of pulp fibers on the sheet uniformity in a local closed place. *J. Korea TAPPI* 50(2):36–43. doi:10.7584/JKTAPPI.2018.04.50.2.36

Peretz, R., E. Sterenzon, Y. Gerchman, V. V. Kumar, T. Luxbacher, and H. Mamane. 2019. Nanocellulose production from recycled paper mill sludge using ozonation pretreatment followed by recyclable maleic acid hydrolysis. *Carbohydr. Polym.* 216:343–351. doi:10.1016/j.carbpol.2019.04.003

Phanthong, P., P. Reubroycharoen, X. Hao, G. Xu, A. Abudula, and G. Guan. 2018. Nanocellulose: Extraction and application. *Carbon Resour. Convers.* 1(1):32–43. doi:10.1016/j.crcon.2018.05.004

Pitroda, J.R. 2016. Using paper waste effectively. *Construction World* 18(5):68–69.

Rajeshwarisivaraj, S., C. Namasivayam, and K. Kadirvelu. 2001. Orange peel as an adsorbent in the removal of acid violet 17 (acid dye) from aqueous solutions. *Waste Manage.* 21(1):105–110. doi:10.1016/s0956-053x(00)00076-3

Sanguanwong, A., P. Pavasant, T. Jarunglumlert, K. Nakagawa, A. Flood, and Ch. Prommuak. 2020. Hydrophobic cellulose aerogel from waste napkin paper for oil sorption applications. *Nord. Pulp Pap. Res. J.* 35(1):137–147. doi:10.1515/npprj-2018-0075

Sarno, M., and C. Cirillo. 2019. Production of nanocellulose from waste cellulose. *Chem. Eng. Trans.* 73:103–108. doi:10.3303/CET1973018

Satyanarayana, K. G., A. Rangan, V. S. Prasad, and W. L. E. Magalhães. 2017. Preparation, characterization, and applications of nanomaterials (cellulose, lignin, and silica) from renewable (lignocellulosic) resources. In: *Handbook of composites from renewable materials. Nanocomposites: Science and fundamental*, V. 7, ed. V. K. Thakur, M. K. Thakur, and M. R. Kessler, 1–66. Beverly: Scrivener Publishing, and Hoboken: John Wiley & Sons .

Senthilkumar, N., Md. A. Aziz, M. Pannipara, et al. 2020. Waste paper derived three-dimensional carbon aerogel integrated with ceria/nitrogen-doped reduced graphene oxide as freestanding anode for high performance and durable microbial fuel cells. *Bioproc. Biosyst. Eng.* 43:97–109. doi:10.1007/s00449-019-02208-4

Sharma, P., H. Kaur, M. Sharma et al. 2011. A review on applicability of naturally available adsorbents for the removal of hazardous dyes from aqueous waste. *Environ. Monit. Assess.* 183:151–195. doi:10.1007/s10661-011-1914-0

Sun, Y., and J. Cheng. 2002. Hydrolysis of lignocellulosic materials for ethanol production: A review. *Bioresource Technol.* 83(1):1–11. doi:10.1016/S0960-8524(01)00212-7

Sundarraj, A. A., and T. V. Ranganathan. 2018. A review on cellulose and its utilization from agro-industrial waste. *Drug Invent. Today* 10(1):89–94.

Tang, Y., X. Shen, J. Zhang, D. Guo, F. Kong, and N. Zhang. 2015. Extraction of cellulose nano-crystals from old corrugated container fiber using phosphoric acid and enzymatic hydrolysis followed by sonication. *Carbohydr. Polym.* 125: 360–366, doi:10.1016/j.carbpol.2015.02.063

Tanga, F., H. Yua, and Y. Li. 2020. Green acid-free hydrolysis of wasted pomelo peel to produce carboxylated cellulose nanofibers with super absorption/flocculation ability for environmental remediation materials. *Chem. Eng. J.* 395:125070. doi:10.1016/j.cej.2020.125070

Thai, Q. B., D. K. Le, Th. Ph. Luu, N. H. D. Nguyen, and H. M. Duong. 2019. Aerogels from wastes and their applications. *Mater. Sci.* 5(3):555663.

Thakur, V. K. (ed). 2015. *Nanocellulose polymer nanocomposites. Fundamentals and applications.* Hoboken: John Wiley & Sons and Salem: Scrivener Publishing LLC. doi:10.1002/9781118872246

Tofani, G., I. Cornet, and S. Tavenier. 2020. Alternative filler recovery from paper waste stream. *Waste Biomass Valor.* doi:10.1007/S12649-020-01011-7

Trache, D., M. H. Hussin , M. K. M. Haafiz, V. K. Thakur. 2017. Recent progress in cellulose nanocrystals: Sources and production. *Nanoscale* 9(5):1763–1786. doi:10.1039/c6nr09494e

Trache, D., A. F. Tarchoun, M. Derradji, et al. 2020. Nanocellulose: From fundamentals to advanced applications. *Front. Chem.* 8:392. doi:10.3389/fchem.2020.00392

Ünlü, C. H. 2013. Carboxymethylcellulose from recycled newspaper in aqueous medium. *Carbohydr. Polym.* 97(1):159–164. doi:10.1016/j.carbpol.2013.04.039

Van-Pham D.-Th., Th. Y. Nh Pham, M. Ch Tran, N. Nguyen, and Q. Tran-Cong-Miyata. 2020. Extraction of thermally stable cellulose nanocrystals in short processing time from waste newspaper by conventional acid hydrolysis. *Mater. Res. Express* 7(6):065004. doi:10.1088/2053-1591/ab9668

Villanueva Krzyzaniak, A., and P. Eder. 2011. End-of-waste criteria for waste paper: Technical proposals. EUR 24789 EN – 2011. European Commission. Joint Research Centre. Institute for Prospective Technological Studies. doi:10.2791/57814

Virtanen, T., K. Svedström, S. Andersson, et al. 2012. A physico-chemical characterisation of new raw materials for microcrystalline cellulose manufacturing. *Cellulose* 19:219–235. doi:10.1007/s10570-011-9636-6.

Visakh, P. M., S. Thomas. 2010. Preparation of bionanomaterials and their polymer nanocomposites from waste and biomass. *Waste Biomass Valor.* 1(1):121–134. doi:10.1007/s12649-010-9009-7

Walpot, J. I. 1986. Enzymatic hydrolysis of waste paper. *Conser. Recycl.* 9(1):127–136. doi:10.1016/0361-3658(86)90139-6

Wang, H., and M. Roman. 2011. Formation and properties of chitosan–cellulose nanocrystal polyelectrolyte–macro ion complexes for drug delivery applications. *Biomacromolecules* 12(5):1585–1593. doi:10.1021/bm101584c

Wang, L., M. Sharifzadeh, R. Templer, and R. J. Murphy. 2012a. Technology performance and economic feasibility of bioethanol production from various waste papers. *Energy Environ. Sci.* 5 (2):5717–5730. doi:10.1039/c2ee02935a

Wang, L., M. Sharifzadeh, R. Templer, and R. J. Murphy. 2013. Bioethanol production from various waste papers: Economic feasibility and sensitivity analysis. *Appl. Energy* 111:1172–1182. doi:10.1016/j.apenergy.2012.08.048

Wang, L., R. Templer, and R. J. Murphy. 2012b. High-solids loading enzymatic hydrolysis of waste papers for biofuel production. *Appl. Energy* 99:23–31. doi:10.1016/j.apenergy.2012.03.045

Wang, Z., S. Liu, Y. Matsumoto, and S. Kuga. 2012. Cellulose gel and aerogel from LiCl/DMSO solution. *Cellulose* 19(2):393–399. doi:10.1007/s10570-012-9651-2

van Wyk, J. P. H., M. A. Mogale, and K. S. Moroka. 1999. Bioconversion of waste paper materials to sugars: An application illustrating the environmental benefit of enzymes. *Biochem. Educ.* 27(4):227–228. doi:10.1016/s0307-4412(99)00053-9

Xie, H., H. Du, X. Yang, and Ch. Si. 2018. Recent strategies in preparation of cellulose nanocrystals and cellulose nanofibrils derived from raw cellulose materials. *Int. J. Polym. Sci.* 2018:1–25. doi:10.1155/2018/7923068

Yano, H. and H. S. Nakahara. 2004. Bio-composites produced from plant microfiber bundles with a nanometer unit web-like network. *J. Mater. Sci.* 39(5):1635–1638. doi:10.1023/B:JMSC.0000016162.43897.0a

Yue, X., T. Zhang, D. Yang, F. Qiu, and Zh. Li. 2018. Hybrid aerogels derived from banana peel and waste paper for efficient oil absorption and emulsion separation. *J. Clean. Prod.* 199: 411–419. doi:10.1016/j.jclepro.2018.07.181

Zhang, W., J. Gu, D. Tu, L. Guan, and Ch. Hu. 2019. Efficient hydrophobic modification of old newspaper and its application in paper fiber reinforced composites. *Polymers* 11:842. doi:10.3390/polym11050842

Zhen, L. W., Q. B. Thai, Th. X. Nguyen, et al. 2019. Recycled cellulose aerogels from paper waste for a heat insulation design of canteen bottles. *Fluids* 4(3):174. doi:10.3390/fluids4030174

5 Cellulose Valorization for the Development of Bio-based Functional Materials via Topochemical Engineering

Lokesh Kesavan, Liji Sobhana, and Pedro Fardim

CONTENTS

5.1 INTRODUCTION

Bio-based natural resources are of significant scientific importance in this era due to their rich existence, renewability, easy accessibility, low cost, etc. Cellulose, the naturally occurring biopolymer abundant in plants and woods has been a potential starting material for valorization as their supply from the forest products industry and agricultural wastes is huge and inexpensive. Cellulose is a polymer substance, with the repeating units of monosaccharide containing secondary and tertiary –OH groups. These –OH groups are highly prone to undergo any modification using chemical reagents or by physical methods. Topochemical engineering is one strategy through which directed assembly of new guest molecules on substrate or disassembly of that host substrate units can be performed to produce multifunctional materials. This book chapter aims to disseminate the knowledge around topochemical engineering of cellulose-based functional materials with the emphasis on valorization techniques, material compositions, and applications, for the readers of sustainable biomaterials chemistry.

We are living in the age of rapid industrialization and modernization of our lives. But these developments are highly dependent on natural resources. There are two types of resources: (1) Fossil deposits and (2) Forests. Fossil resources are non-renewable hence unsustainable, whereas forest products are highly abundant and renewable. These resources have turned to be potential replacement to non-renewable fossil resources as they are depleting fast. Wood Pulp is a lignocellulosic material containing cellulose, hemicelluloses, and lignin. The annual production of pulp exceeds 160 million tonnes. This pulp feedstock is hugely acquired from forest products industries to get converted into bulk chemicals and other technical materials. Bioresources offer sustainable chemical or material production with environment compatibility and recyclability. Thus, the sustainable development goals fixed by United Nations can be achieved via bio-based chemistry.

The raw material, pulp fibres, from wood and annual plants is usually obtained by mechanical, chemo-mechanical, or chemical treatment and then bleached prior to further modifications. Mechanical and chemo-mechanical treatments lead to higher amounts of lignin and hemicelluloses left unremoved from pulp, whereas pulp treated by chemical reagents yields high-quality cellulose fibres without contaminants. Kraft pulping and sulfite pulping are the techniques majorly used in chemical pulping. These techniques combined with other bleaching methods like Elemental Chlorine Free (ECF) or Total Chlorine Free (TCF) sequences produce cellulose fibres for high-end applications. Pulps treated with Kraft process and ECF bleaching offer variety of applications like paper, tissues, absorbents, and packaging materials. On the other hand, the process in which pulps are processed until pre-hydrolysis step of Kraft process or by sulfite followed by TCF bleaching delivers pulps with high soluble cellulose fibres as this combined process eliminates lignin and hemicelluloses very efficiently. This dissolved cellulose can be chemically modified into cellulose derivates further. Thus, different methods of pulping and bleaching techniques have great impact on the quality of the resultant pulp, which offers variety of pulp grades suitable for different purposes depending on residual amounts of lignin and hemicelluloses present in the material. All pulps possess anionic groups (AGs) that are active sites for interactions with other reagents. The amount of AGs on fibres depends on the macromolecular properties, amount of hemicelluloses of wood raw material, and dissolution/reaction of biopolymers during pulping and bleaching. Cationic chemicals, polyelectrolytes are usually applied to interact with AGs in the production of fibre-based materials (1).

Cellulose is a biopolymer containing repeating units of anhydroglucose. Hence, it is a polysaccharide. These units contain 1° hydroxyl groups at C6 position and 2° hydroxyl groups at C2, C3 positions on its carbon Skelton. These hydroxyl functional groups are chemically reactive; hence they are sites of opportunity to modify cellulose, for example, cellulose nitrates, esters, and ethers. Oxidation is an important chemical reaction that can be carried out on cellulose. During oxidation the primary –OH groups at position C6 converted into –COOH and secondary –OH groups at positions C2, C3 into –CHO and –COOH (2, 3). Thus cellulose can be modified from neutral to anionic or cationic, hydrophilic to hydrophobic character, with increased mechanical strength and chemical reactivity. Chemical functionalizations can be tuned in cellulose to result desired hydrophilicity/hydrophobicity, crystallinity, stereo-regularity, multi-chirality, thermal and mechanical stability, biocompatibility, and sustainability.

For a long time, cellulose has been viewed as just a pulp and paper source material. But due to the new inventions that led to the potential applications of cellulose fibres in thin films, textiles and personal care products, the scenario in pulp market changed very fast. These are the motivations to explore chemical and physical properties of cellulose for value addition and finally the manufacturing of sustainable novel applied materials. This chapter focuses on topochemical engineering approach on cellulose that can be done either by directed assembly or disassembly via the design of intermolecular interactions in a topological space. The challenge is to disturb the intra/intermolecular hydrogen bonding in cellulose fibres to separate them as independent cellulose molecules with right employment of solvents, reagents, process of extrusion, and drying. Post disassembly, these new morphologies are functionalized or assembled with new guest molecules possessing desired properties. These modifications unveil new surfaces and interfaces with targeted functionalities suitable for a variety of technical material applications. The product materials include well-defined objects such as tissues, sponges, textile fibres, film or sheets, and spherical beads having desired specifications.

5.2 TOPOCHEMICAL ENGINEERING ON CELLULOSE BY CHEMICAL MOLECULES

A topochemical reaction is chemical reaction in which the nature and properties of products are administered by the three-dimensional topological environment of molecules or atoms (4). Photochemical reactions in organic crystals (5) are examples of topochemical reactions that have been widely studied by synthetic organic chemists. Langmuir was the pioneer who identified the kinetic consequences of reaction zones in a topological space (6), i.e. the rate of a topochemical reaction is proportional only to the amount of substrate material present in the reaction zone, but not to the amount of unreacted substrate. The science of topochemistry is fascinating and various inter-pretations on topochemical reactions have been suggested (7). Reactions in three-dimensional, confined spaces are crucial in cell metabolism, biosynthesis of DNA, RNA, proteins, and other numerous biological interactions as well as chemical reac-tions. Thus, topochemistry plays a vital role in structure–activity relationship of natural and man-made systems.

Topochemical engineering is a method of directed assembly of new functional groups on hosting substrate material or directed disassembly of the host itself with the help of chemical reagents. The directed assembly involves electrostatic interac-tions, hydrophobic interactions, hydrogen bonding, or solid-state cross-linking reac-tions of components to create new functional spheres and interfaces. The directed disassembly of a component from a multicomponent system is based on controlled cleavage of intra- or intermolecular bonds that enhances separation and selected frac-tionation of the targeted component.

Topochemical disassembly or assembly in cellulose fibres involves degradation of fibre primary wall layers to enhance cellulose dissolution in water-based solvents, removal of lignin to enhance purification of cellulose, reacting cellulose fibres with polyelectrolytes followed by shaping of functional cellulose into different forms like beads, sponges, etc. Disassembly sometimes leads to complete destruction or depo-lymerization of cellulose into its monomers (monosaccharide sugar molecules like glucose) via enzymatic hydrolysis.

5.3 TOPOCHEMICAL ACTIVATION OF PULP FIBRES BEFORE BLEACHING

Pulps obtained from mechanical treatments are often brown muddy in appearance. Hence, they are bleached later on to improve their brightness and colour into pure white. But in a new attempt Iamazaki and co-workers have tried activators on pulp fibres before bleaching and found promising results. They used weak acid esters and a weak base such as lactose octaacetate (LOA, colloidal particles, and low water solubility), sucrose octaacetate (SOA) (partially soluble in water), and tetraacetyle-thylenediamine (TAED, water soluble). Their motivation stems from TAED being used as first ever activator in hydrogen peroxide bleaching for enhanced whitening performances. Also, TAED was utilized in mechanical and chemical pulps (8–16) as well later on. Alright, how do these activators work?

These activators generate active oxygen species from hydrogen peroxide (the bleaching agent) via peroxyacid intermediates. They are peroxyacid anions that are formed in situ during the hydrolysis of amides or esters. These anions bringing in new carboxylic and amino groups on the fibre surfaces at different depths yielding enhanced mechanical strength for the pulp (17). Iamazaki's work showed that LOA was very effective for topochemical activation of pulp fibres even in low concentra-tion. The whiteness increased from 19% (of H_2O_2) to 24% after activation and the surface coverage of lignin was minimal (in Table 5.1 and Figure 5.1 below). Next to LOA, activations by SOA and TAED showed improved brightness and whiteness. Though the activators are highly water soluble, their influence of surface specificity on fibres was negligible.

TABLE 5.1

Brightness and Whiteness Parameters on Bleached Thermomechanical Pulp (Tmp) with Different Surface Activators

Sample	pH		Residual (kg ton⁻¹)		Brightness % ISO	Whiteness %	Yellowness %
	Initial	Final	Peroxide	Alkali			
Unbleached (reference)	4.5	4.5	–	–	58.0 ± 0.1	1.3 ± 0.1	27.9 ± 0.1
H_2O_2	11.4	9.0	1.42	0.01	65.6 ± 0.1	19.1 ± 0.2	22.8 ± 0.1
TAED	11.3	8.2	2.41	0.02	66.9 ± 0.2	22.6 ± 0.2	21.6 ± 0.1
SOA	11.5	8.2	2.58	0.02	67.1 ± 0.1	23.1 ± 0.2	21.4 ± 0.1
LOA	10.7	7.4	1.14	0.01	68.2 ± 0.2	24.4 ± 0.3	21.3 ± 0.1

Source: Reprinted with permission from Springer Nature (Ref. 17:Iamazaki, E.T.; Orblin, E.; Fardim, P. Topochemical activation of pulp fibres, *Cellulose* 2013, 20, 2615–2624)

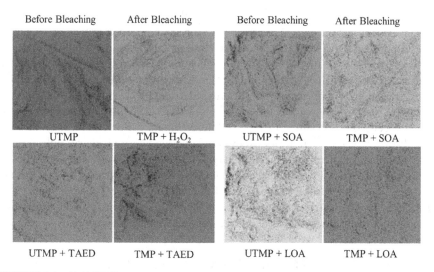

FIGURE 5.1 ToF-SIMS positive mode image of lignin distribution before and after bleaching of thermomechanical pulp (TMP) with and without activators. The size of the images is 100 μm * 100 μm. *Reprinted with permission from Springer Nature (Ref. 17: Iamazaki, E.T.; Orblin, E.; Fardim, P. Topochemical activation of pulp fibres, Cellulose 2013, 20, 2615–2624)*

Thus, the activation of cellulose surface in the pulp fibres prior to peroxide bleaching by the employment of precursors like peracids of mono and disaccharides is a proven method to enhance brightness and whiteness in mechanical pulps as this achieve reduced coverage of lignin on fibres. The disadvantage with this method could be the toxic wastes produced in the process, such as unutilized TAED and H_2O_2.

5.4 PHOTOACTIVE FIBRE INTERFACES FROM CELLULOSE DERIVATIVES

As discussed earlier, the cellulosic –OH groups can be made into its derivatives carrying multiple functions groups. These can be achieved via chemical or enzymatic treatments. By doing so, we may get a new type of fibres but that might affect the inherent properties of original fibres such as tensile strength and other mechanical characteristics (18–23). Chemical methods often come with shortcomings like toxic solvent, radical initiators, high temperature, and uncontrolled self-assembly of guest molecules on the host fibre surface etc. These drawbacks were solved by Vega et al. in their seminal work in which they treated pulp fibres with novel water-soluble cellulose derivative to create a charge induced reactive interface on the fibres at room temperature. In their work, cellulose was made to react with (3-carboxypropyl) trimethylammonium chloride ester of 6-deoxyazidocellulose to yield a derivative first. This derivative was the output of charge directed self-assembly on the bleached Kraft pulp fibres (from pine wood) in aqueous environment at 25°C. This charge-mediated self-assembly was facilitated by the newly formed ionic groups in the cellulose derivative as they were readily soluble in H_2O.

Now let's look into the mechanism of this self-assembly. As mentioned earlier in the introduction (Section 5.1), there are AGs on the pulp fibres which act as active sites. In this case, the AGs on the bleached Kraft pulp had an electrostatic interaction with azidocellulose cation. To show an application, Vega and co-workers used this (3-carboxypropyl) trimethylammonium chloride ester of 6-deoxyazidocellulose (N_3-cell$^+$) treated pulp fibres in copper (I)-catalysed azide-alkyne Huisgen cycloaddition (CuAAc) reaction (Figure 5.2a) to prove its reactivity. The final products were photoactive cellulose and amino (–NH_2) cellulose (Figure 5.2b,c). Thus, they showed that the azide groups were forming covalent bond with desired alkyne molecules in the presence of transition metal catalyst (e.g. Cu^{1+}). They also synthesized Azidocellulose derivative of bleached Kraft pulp fibres in organic solvent medium and they found the surface morphology of the pulp fibres didn't change much (24). These kinds of derivatives of pulp fibres can be employed in the sustainable production of photo-luminescent tapes and labels for authentication purposes.

5.5 CELLULOSE FIBRE SURFACES MODIFIED WITH HEMICELLULOSE DERIVATIVE

Hemicelluloses are other components that exist with cellulose and lignin in wood. During advanced pulping process like Kraft pulping, hemicelluloses and lignin are removed as much as possible. So that bright and white pulp can be produced and converted into paper in downstream processes. However, hemicelluloses and lignin also need attention for value addition as they are the huge waste collected from pulp industries. Can we use hemicelluloses to modify pure cellulose fibre surfaces? Yes, it can be done. In 1950s, hemicelluloses was first attempted to modify cellulose fibre surfaces via physisorption (adsorption) (25–28). But the main disadvantage with hemicelluloses was that it was not ready to form homogeneous composite with cellulose, after its laborious separation process. Further research introduced treatment

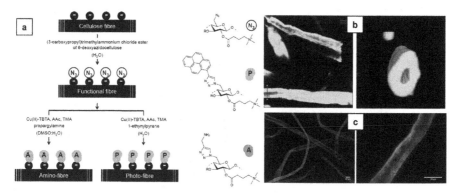

FIGURE 5.2 (a) Synthesis scheme for the preparation of cellulose fibres decorated with photoactive molecules (photo-fibres) and amino functional groups (amino-fibres) prepared from cellulose fibres decorated with azide functions (reactive fibres). Cu(II)-TBTA, Cu(II)-tris[(1-benzyl-1H-1,2,3-triazol-4-yl)methyl]amine complex solution; AAc, ascorbic acid; TMA, triethylammonium acetate buffer. (b) A section (xy plane) of scanned xyz-volume of photo-fibres obtained with TPM (left). A 3D rendering of a cross-section of an individual fibre at the position marked as "ROI2" on the left figure (right). The figures illustrate the dense labelling of the fibres and the preserved 3D shape during the activation and labelling. (c) Image of a photo-fibre observed with Olympus BX60 epi-fluorescence microscope at 10×-(left) and 40×-magnifications (right). Excitation filter, 330–385 nm; dichroic mirror, 400 nm; barrier filter, >420 nm. *Reprinted with permission from Elsevier (Ref. 24: Vega, B.; Wondraczek, H.; Bretschneider, L.; Näreoja, T.; Fardim, P.; Heinze, T. Preparation of reactive fibre interfaces using multifunctional cellulose derivatives. Carbohydr. Polym. 2015, 132, 261–273)*

of cellulose fibre surfaces by hemicellulose derivative for charged fibre surface applications. Also, this made cellulose fibres more reactive and led to chemical conversions.

Now, let's see an example! In an attempt of making hemicellulose derivative, Vega and co-workers extracted xylan (a group of hemicelluloses) from birch wood using pressurized hot water in an autoclave. According to their finding, Xylan can be derivatized into carboxymethyl xylan (CMX), xylan sulfate (XS), and xylan-4-[N,N,N-trimethylammonium] butyrate chloride (XTMAB) through chemical reactions (Figure 5.3a). These charged new derivatives were utilized as polyelectrolytes (PE) and made to absorb pine Kraft pulp.

What characterization techniques were used to analyse this kind of materials? (1) Wet chemical titrations were applied to determine the adsorbed PEs on the fibre surfaces quantitatively. (2) XPS technique was used to identify surface anionic groups. (3) Sorption isotherm measurement was used to understand the adsorption. (4) Time-of-Flight Secondary Ion Mass Spectrometry (ToF-SIMS) was used for imaging the morphology of fibres.

They found that cellulose fibre surface had more affinity towards cationic xylan derivative (XTMAB) than the anionic derivatives CMX and XS. The surface AGs' value calculated using XPS technique showed that the anionic groups present on the pulp surface was stoichiometrically equal to cationic groups that can be adsorbed

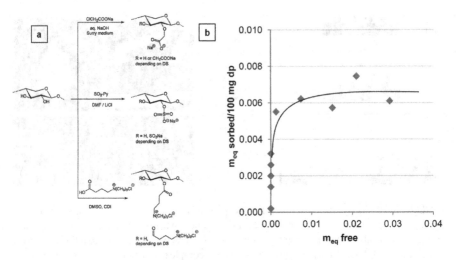

FIGURE 5.3 (a) Scheme paths for the synthesis of ionic xylan derivatives illustrated using an idealized xylan. (b) Sorption isotherm showing the amount of xylan-4-[N,N,N-trimethyl-ammonium]butyrate chloride (XTMAB) per 100 mg of bleached pine Kraft pulp versus "meq free" representing the amount of XTMAB still in solution after sorption. *Reprinted with permission from Elsevier (Ref. 29: Vega, B.; Petzold-Welcke, K.; Fardim, P.; Heinze, T. Studies on the fibre surfaces modified with xylan polyelectrolytes, Carbohydr. Polym. 2012, 89, 768–776)*

onto the fibres surface. Thus the sorption of XTMAB on the pulp fibre surfaces was primarily due to electrostatic forces. The adsorption of XTMAB followed Langmuir model of sorption isotherm (Figure 5.3b). This showed charge interactions (chemisorption) overruled hydrogen bond interactions (physisorption). Further, the distribution of XTMAB on fibre surfaces was found to be uniform as evidenced from ToF-SIMS. This piece of research proves XTMAB can be an effective fibre-modifying agent (29). As Xylan is renewable, XTMAB can be used as sustainable biopolyelectrolyte in pulp and papermaking processes to introduce charged surfaces and enhance mechanical strength.

5.6 CELLULOSE FIBRES FUNCTIONALIZED BY BIOPOLYELECTROLYTE COMPLEXES

Biopolymer-derived polyelectrolyte complexes (PEC) are another class of molecules to functionalize cellulose on pulp fibres via charge-directed assembly (Figure 5.4a). There are many reported methods in which polysaccharide-based polyelectrolyte multilayer moieties were physisorbed or chemisorbed on pulp to improve paper qualities (30–37). But here is one example (38) which uses bio-derived PEC as surface-modifying agents. Those complexes were obtained from the reaction of never dried bleached pine pulp (Kraft) with cellulose (3-carboxypropyl) trimethyl ammonium chloride ester (CN⁺), carboxymethyl xylan (CMX⁻), and xylan sulfate (XS⁻) (Figure 5.4b). The complexes formed were CN⁺, [CN⁺ CMX⁻], and [CN⁺ XS⁻]. Anionic polysaccharide-based

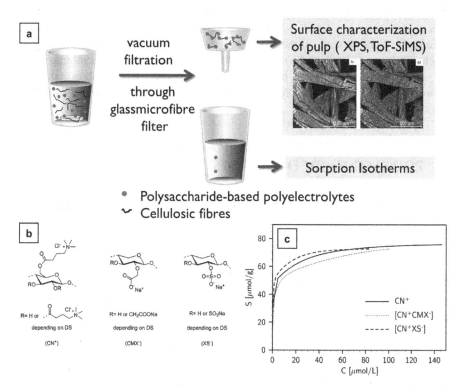

FIGURE 5.4 (a) Graphical representation of the functionalizing cellulose with PEC, (b) Cellulose (3-carboxypropyl) trimethylammonium chloride ester (CN+), carboxymethyl xylan (CMX−), and xylan sulfate (XS−). (c) Sorption isotherms obtained from polyelectrolyte titrations of [CN+CMX−] (system composed of carboxymethyl xylan (CMX−) and cellulose (3-carboxypropyl) trimethylammonium chloride ester (CN+)), [CN+XS−] (system composed by xylan sulfate (XS−)), and CN+, S, amount of adsorbed charge [μmol/g]; C, equilibrium concentration in the solution [μmol/L]. *Reprinted with permission from ACS (Ref. 38: Vega, B.; Wondraczek, H.; Zarth, C.S.P.; Heikkila, E.; Fardim, P.; Heinze, T. Charge-Directed Fiber Surface Modification by Molecular Assemblies of Functional Polysaccharides. Langmuir 2013, 29 (44), 13388–13395)*

polyelectrolytes like xylan sulfates (XS−) and carboxymethyl xylans (CMX−) are of importance for their biological activities (e.g., prevention of blood coagulation), whereas cellulose derivatives containing quaternary ammonium groups (cationic polyelectrolytes) show bacteriostatic behaviour.

The adsorption of CN+ and PECs ([CN+ CMX−], [CN+ XS−]) on the pulp fibres was determined by electrochemical titrations. These complexes got naturally attracted towards cellulose fibres through electrostatic force and yielded modified pulp. PECs were found to be chemically stable throughout the reaction and even after the adsorption. PECs (neutral) treated fibres were having twice the quantity of adsorption compared to CN+ treated fibres. This was due to higher charge density on CN+ which caused less adsorption. The adsorption was purely electrostatic as the amount of net charge developed on the fibre surface was same irrespective of the modifying complex applied (Figure 5.4c). ToF-SIMS revealed the uniform distribution of PECs on fibre surface. The presence of these complexes on modified fibres may bring superior mechanical properties than the reference fibres (38). This type of assembling novel polysaccharide derivative like polyelectrolyte complexes on pulp fibres can offer improved sizing in paper, ion exchange resins, charge transfer reagents, or even biosensor in detection of virus.

5.7 PHOTO-CONTROLLED FIBRE-TO-FIBRE INTERACTIONS USING POLYSACCHARIDE DERIVATIVES

It is a well-known fact that photo radiation can make or break bonds in chemical molecules. Grigoray et al. have carried out photo-induced cross-linking between cellulose derivatives to produce novel photo-responsive pulp fibres. This is another example for charge-directed self-assembly of multifunctional polysaccharide derivatives. Bleached eucalyptus Kraft pulp surface was chemically modified by introducing functional groups like (3-carboxypropyl) trimethylammonium chloride ester and 2-[(4-methyl-2-oxo-2H-chromen-7-yl) oxy] acetate (coumarin) in aqueous environment. Further, these photoactive cationic cellulose derivatives (PCCD) were allowed to undergo ($2\pi+2\pi$) cycloaddition under UV light (320–390 nm). This photo-addition caused a fast photo-cross-linking of the covalent bonds originating from the photoactive groups of chemically modified pulp fibres, in addition to hydrogen bonds and hydrophobic interactions. Thus a creation of new fibre-to-fibre bonds was achieved (Figure 5.5a, b). Upon PCCD functionalization on fibres, the flexibility of the fibres increased from 83×10^9 to 130×10^9 $N^{-1}m^{-2}$. This increase was 58% with respect to unmodified reference fibres (bleached eucalyptus Kraft pulp).

Self-assembly of polysaccharide derivatives (photoactive cationic cellulose) on pulp fibres using external stimuli like UV radiation can lead to change in its mechanical properties such as tensile strength and stiffness (39, 40). Dynamic mechanical analysis (DMA) and Z-directional tensile testing have shown an increase in strength of fibre network from 81 to 84% and unidirectional stiffness of fibres by 60% (40). Thus, mechanical properties of classical pulp fibres can be tailored to improve the performance for smart bio-based materials.

5.8 FLUORESCENT CELLULOSE FIBRES

Fluorescence is a form of luminescence caused by the emission of light by a chemical substance which has already absorbed light or electromagnetic radiation. It has many practical applications like chemical sensors, biosensors, medicine, geology, etc. Grigoray and co-workers have reported a method to introduce fluorescence properties to unrefined bleached Kraft pulp fibres of eucalyptus. They synthesized multifunctional cellulose derivatives (MCD) like N-(3-propanoic acid)- and N-(4-butanoic acid)-1,8-naphthalimide esters of cellulose with (3-carboxypropyl) trimethylammonium chloride cationic moieties from microcrystalline cellulose (Figure 5.6A). Then the fluorescent pulp fibres were prepared by mixing water-soluble fluorescent MCDs with pulp fibres at room temperature in water. The mechanism of supramloecular

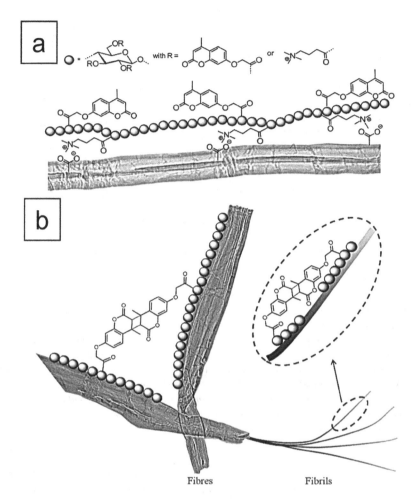

FIGURE 5.5 (a) Schematic drawing of the structure of the photoactive cationic cellulose derivative and its interaction with pulp fibers; (b) visualization of the $2\pi + 2\pi$ cycloaddition reaction creating covalent bonds between pulp fibers and along the fibrils of the fiber. *Reprinted with permission from John Wiley and sons (Ref. 40: Grigoray, O.; Wondraczek, H.; Daus. S.; Kühnöl, K.; Latifi, S.K.; Saketi, P.; Fardim, P.; Kallio, P.; Heinz, T. Photocontrol of Mechanical Properties of Pulp Fibers and Fiber-to-Fiber Bonds via Self-Assembled Polysaccharide Derivatives. Macromol. Mater. Eng. 2014, 300(3), 277-282)*

functionalization (adsorption) of fluorescent multifunctional cellulose derivative (FMCD) on the pulp fibres is ion-exchange (41). The cationic cellulose derivatives quickly interact with negatively charged cellulose fibres driven by the entropy gain from release of counter ions.

The adsorption was also dependent on the length of the aliphatic chain connecting the naphthalimide group to cellulose skeleton. Longer the hydrophobic aliphatic chain, weaker the adsorption. They have demonstrated that propanoic naphthalimide had better adsorption on pulp fibres than butanoic counterpart. The fluorescent

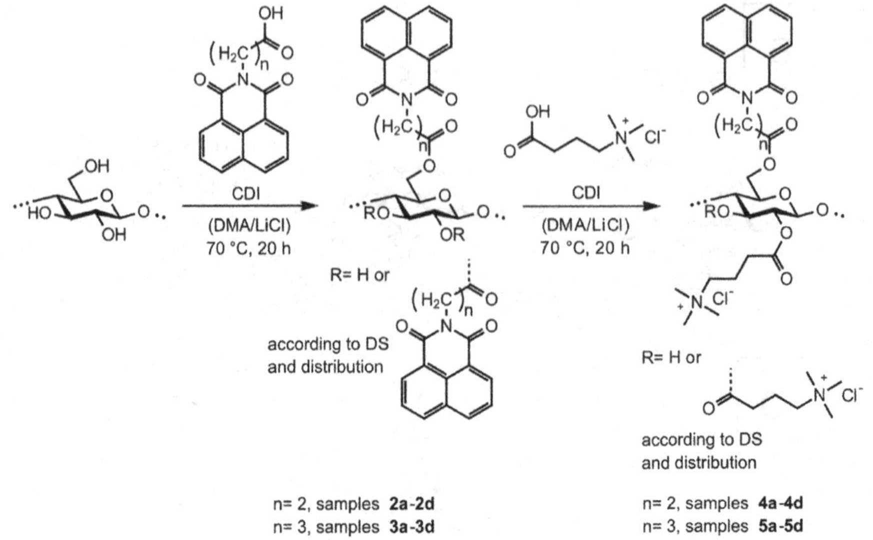

FIGURE 5.6A Synthesis scheme of *N*-(3-propanoic acid)-1,8-naphthalimide and *N*-(4-butanoic acid)-1,8-naphthalimide esters of cellulose and the corresponding mixed naphthalimide (3-carboxypropyl)trimethylammonium chloride esters of cellulose via in situ activation of *N*-(3-propanoic acid)-1,8-naphthalimide, *N*-(4-butanoic acid)-1,8-naphthalimide, and (3-carboxypropyl) trimethylammonium chloride (4) with *N*,*N*-carbonyldiimidazole (CDI) in *N*,*N*-dimethylacetamide/LiCl (DMA/LiCl). *Reprinted with permission from ACS (Ref. 42: Grigoray, O.; Wondraczek, H.; Pfeifer, A.; Fardim, P.; Heinze, T. Fluorescent multifunctional polysaccharides for sustainable supramolecular functionalization of fibers in water, ACS Sustainable Chem. Eng. 2017, 5 (2), 1794–1803)*

properties of the modified fibres were studied using spectrofluorometer with the irradiation of wavelength of UV absorbance maximum. The emission band at 393–398 nm was observed and it was characteristic to naphthalimides. Though the butanoic naphthalimide moiety had less affinity towards fibres, they exhibited superior performance in fluorescence. The microscopic images of modified fibres showed morphology and microstructural elements in both UV and white lighting environments (Figure 5.6Ba,b). This clearly showed that the modified fibres absorb UV light and emit. This fluorescent property was converted into application by making hand sheets of those modified pulp fibres. These hand sheets were further cut into small pieces like labels and then irradiated under UV light. Figure 5.6.B(c) shows the fluorescence illumination of labels made up of propanoic and butanoic naphthalimide moieties modified fibres.

This kind of fluorescent behaviour of the MCD-modified fibres can be utilized in sustainable smart packaging or fluorescent labelling as an authenticity indicator (42). These fluorescent MCD containing invisible security fibres are superior to fibres modified with fluorescent whitening agents (FWA) as they suffer from low affinity to fibre surface. Hence, FWA-modified fibres need fixing agents/salts to maintain surface anchorage (43, 44).

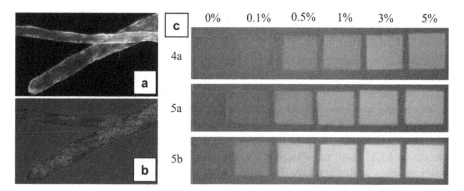

FIGURE 5.6B Visualization of fluorescent pulp fibers by epi-fluorescence microscope under exposure of UV light (**a**) and white light (**b**). The fibers were modified with 5b and the dosage was 2% (w/w). N-(4-Butanoic acid)-1,8-naphthalimide-(3-carboxypropyl)trimethylammonium chloride ester of cellulose (5b, DS_{photo} 0.22, DS_{cat} 0.33). (**c**) Picture of fiber hand-sheets under black light illumination. The quadrates and the background made of FMCDs treated and reference fibers, respectively. N-(3-propanoic acid)-1,8-naphthalimide-(3-carboxypropyl)trimethylammonium chloride ester of cellulose (4a, DS_{photo} 0.07, DS_{cat} 0.31) and N-(4-butanoic acid)-1,8-naphthalimide-(3-carboxypropyl)trimethylammonium chloride esters of cellulose (5a, DS_{photo} 0.11, DS_{cat}0.32; 5b, DS_{photo} 0.22, DS_{cat} 0.33). *Reprinted with permission from ACS (Ref. 42: Grigoray, O.; Wondraczek, H.; Pfeifer, A.; Fardim, P.; Heinze, T. Fluorescent multifunctional polysaccharides for sustainable supramolecular functionalization of fibers in water, ACS Sustainable Chem. Eng. 2017, 5 (2), 1794–1803)*

5.9 TOPOCHEMICAL ENGINEERING OF CELLULOSE FIBRES USING ORGANIC–INORGANIC HYBRIDS

Organic–inorganic hybrid materials are usually a composite of carbon containing compound and metal hydroxides. The carbon precursors can be long aliphatic chain compounds having alcohol, acidic, or ester groups, whereas metal compounds can be layered double hydroxides synthesized from metal nitrates and precipitating hydroxide and carbonate agents. Layered double hydroxide (LDH) is an inorganic clay material written in an empirical way: $[M_{1-x}^{II}M_x^{III}(OH)_2]^{x+} [A_{x/n}^{n-}.yH_2O]^{x-}$, where M^{II} and M^{III} denote the divalent (Ca^{2+}, Mn^{2+}, Ni^{2+}, Mg^{2+}, Fe^{2+}, Co^{2+}, Cu^{2+}, Zn^{2+} etc.) and trivalent metal cations (Ni^{3+}, Cu^{3+}, Ti^{3+}, Cr^{3+}, Fe^{3+}, Al^{3+} etc.). A^{n-} is the intercalated anion (Cl^-, Br^-, CO_3^{2-}, SO_4^{2-}, NO_3^- etc.) located in the cationic inter layer hydrated galleries (45–47). The metal cationic layers are interconnected through anionic layers by the compensation of electric charge. The hybridization of LDH with carbon moieties is mainly driven by the charged sites in the mineral structures as well as on the carbon chain. These hybrids are formed through weak electrostatic interactions arising from hydrogen bonding or polarization of covalent bonds. Pulp fibres can also be composed with those hybrids (48) to form novel composites with improved properties like hydrophobicity, tensile strength, etc. This produces a new kind of sandwich material.

5.10 CELLULOSE-LAYERED DOUBLE HYDROXIDES COMPOSITE FOR PULP UPGRADING

Haartman *et al.* have reported an LDH catalysed oxygen delignification process in eucalyptus Kraft pulp and peroxide bleaching in thermomechanical pulp (TMP). LDH was found to be improving the quality parameters of pulps. They modified a commercially available Mg-Al LDH material by heat treatment (525°C, 4 h) and then dispersed this LDH in anion modifier solutions (benzoate, terephthalate, etc.) in separate experiments. These LDHs were used to modify pulp fibre surfaces as metal cations in LDH have natural affinity to anionic cellulose groups. The presence of LDH on fibres can be confirmed by XPS and water-contact angle analyses.

These LDH particles served as binding sites for applied bleaching and brightening agents in the pulp upgrading process. The results showed that LDH has increased the ISO brightness up to 10% and decreased the kappa number to 2 units in Kraft pulp oxygen delignification process. LDH has also managed to selectively remove the undesirable lignin which contributes brown colour to the pulp. LDH with carbonate anions replaced by terephthalate anions made the hydrogen peroxide (H_2O_2) bleaching process economical as the consumption of expensive and toxic H_2O_2 was brought down considerably. LDH enhanced the opacity of TMP to 3 units. Further, the dosage of optical brightening agent (OBA) can be increased during pulp process as LDH-modified pulp fibres showed high retention of OBA. Thus the waste stream from pulp bleaching plants, contaminating water bodies can be avoided (49). Therefore, LDH-cellulose fibres composite pulp can be utilized in the sustainable production of paper and cardboard which requires high ISO brightness of pulp feedstock.

5.11 LDH BRIDGED STEARIC ACID–CELLULOSE COMPOSITE

Layered double hydroxides (LDH) can also be used to compose functional organic compounds with cellulose by utilizing LDH's interlayer hydroxyl groups. Cellulose, the natural polymer containing –OH groups is known to have hydrophilicity, due to hydrogen bonds it forms with water. This makes cellulose products hydrophilic and liphophobic in nature. There will be always research interest to change a naturally occurring property associated with particular material. Pulp and papermaking industries always wanted to bring hydrophobic property to cellulose, so that modified cellulose can be used in water repellent coating, packaging, thin films applications. Liji Sobhana and co-workers have found this scope in cellulose materials chemistry and attempted to make cellulose into hydrophobic. They used an external reagent (stearic acid) on cellulose, not directly but through LDH linker material. They carefully noted down the bond making sites in cellulose, stearic acid, and LDH and designed a synthesis to produce LDH sandwiched stearic acid–cellulose composite. Stearic acid is a fatty acid that is highly renewable which contains long aliphatic carbon chain and –COOH group (50). These carboxyl groups have tendency to form hydrogen bonding or electrostatic attraction with LDH's interlayer –OH groups and metal cations. Similarly, cellulosic –OH groups also have affinity towards LDH's –OH groups. Therefore, they used LDH as bridging entity between stearic acid (SA) and cellulose (CEL). There are many reported methods where hydrophobic moieties

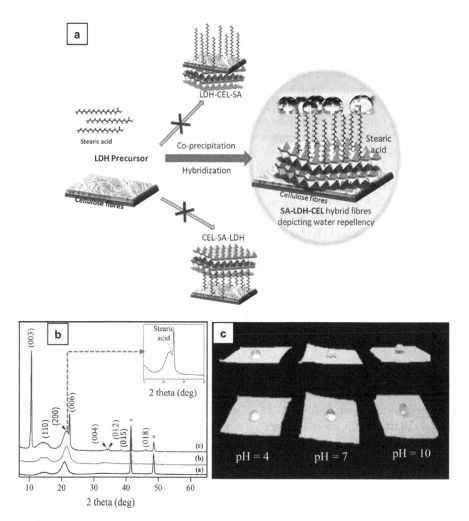

FIGURE 5.7 (a) Schematic representation showing the formation of hybrid fibres (b) XRD patterns the pristine and modified fibres.*Peaks from the copper sample holder (to be ignored). (c) Influence of pH on water contact angle as demonstrated with SA-LDH-CEL hybrid fibres. *Reprinted with permission from Elsevier (Ref. 51: Liji Sobhana, S. S.; Zhang, Xue.; Kesavan, L.; Liias, P.; Fardim, P. Layered double hydroxide interfaced stearic acid - Cellulose fibres: A new class of super-hydrophobic hybrid materials. Colloids Surf A Physicochem. Eng. Asp. 2017, 522, 416-424)*

are directly applied on cellulose to make it hydrophobic, but that can be only done with regenerated cellulose only. Whereas LDH-mediated SA-LDH-CEL composite can be produced even from wet pulp fibres (Figure 5.7a). Here LDH-CEL hybrid fibres were formed first, then stearic acid was introduced on this hybrid material.

The characterization experiments like XRD revealed that the hydrophobic moiety, stearic acid was anchored only on the cationic brucite layers of LDH and not on the interlayer galleries of LDH containing carbonate anion (Figure 5.7b).

This suggested that stearic acid and cellulose self-assembled around LDH on either side, forming SA-LDH-CEL hybrid composite material. The self polarized cellulosic – OH and Stearic –COOH groups were electrostatically attracted towards the positively charged brucite layers of LDH particles and interlayer –OH groups via electrostatic interaction. This led to the non-polar aliphatic carbon chain of stearic acid facing out from the surface of LDH, which ultimately accounted for hydrophobicity in pulp fibres. Now, how to check the hydrophobic behavior of this composite material? They used a simple water contact angle (CA) measurement study to prove the synthesized materials were hydrophobic. This sandwich material exhibited water contact angle of 150° (directly proportional to water-repellence), which is a desirable property in water-proof materials (51). Also, water solutions of different pH were drop casted on SA-LDH-CEL material and contact angle values were measured (Figure 5.7c)

The introduction of LDH between SA and CEL made the process highly efficient as the material showed high order of hydrophobicity for even minimal loading of SA. Also this process reduced the operation time from 5 days to just 1 day by maintaining the same properties reported elsewhere (52). This synthesis strategy can be extended to incorporate other long chain fatty acids such as palmitic, arachidic, lignoceric acids (having number of carbons 16, 20, 24 respectively) in pulp fibres. These composite materials can be used in water-proof packaging materials, films, and oil adsorbents.

5.12 *IN SITU* HYBRIDIZATION OF CELLULOSE FIBRES WITH LDH

Layered double hydroxides (LDH) are often synthesized by co-precipitation of metal nitrate precursors in the presence of carbonate and hydroxide base source. The product LDH will be subjected to aging and crystallization before it is filtered from the preparation mixture and dried. What if we prepare LDH in the presence of biomass? How would it affect LDH formation and biomass properties? Lange *et al.* have reported few methods for this (53, 54). In one of their similar work, they carried out an *in-situ* synthesis of Mg-Al LDH in the presence of pine pulp at pH ≥ 9. Here the pH plays a significant role that the properties of LDH crystals altered according to pH. The size of the particles (from nm to μm) is the one parameter which undergo variation based on the strength of the alkali condition applied. They found that high and low super saturated sodium carbonate solution (*hss* and *lss*) medium, yielded 100–200 nm and 70 nm respectively. They also used urea as a carbonate source upon decomposition at 90°C and produced LDH particles of size (2–5 μm). Thus the nature of the alkali condition plays a role in controlling particle size. It was found that LDH formation was achieved not only on the outer surface of pulp fibres, but inside the cellulose fibre walls as well by urea. The alkali condition will always depolymerize cellulose and it was observed that urea hydrolysis caused more depolymerization than sodium carbonate in the order of *Urea>lss>hss*. They also determined fibre charge by methylene blue (MB) and metanil yellow (MY) adsorption measurements after LDH formation on the cellulose. This revealed the fact that low alkali presence retains most of the original charge on cellulose with maximum anionic groups MY (10 μmol g^{-1}).

Therefore, this kind of LDH modified cellulose fibres can be utilized in charge directed adsorption applications (55).

5.13 CELLULOSE AS TEMPLATE FOR 3D INORGANIC CLAY NANOARCHITECTURES

In materials chemistry, template molecules are often used to bring desirable properties to the end material. These templates will be destructed after synthesis by heat treatments or by solvent washing. Geraud and co-workers have used polystyrene beads as templates by 'inverse opal method' for the synthesis of MgAl-LDH (56). Cellulose, a source of carbon skeleton can also be employed as template to synthesize porous inorganic sold materials with enhanced surface area. Thus we can replace conventionally used templates such as micelles, micro emulsions, and polymers (57) by renewable cellulose. Liji Sobhana *et al.* has reported a method in which they used microcrystalline cellulose (MCC) as a template material for synthesizing high surface area LDH material. MCC is a purified form of cellulose which is inert, stable and easily dispersible in the preparation medium was perceived to be an ideal candidate for the building three dimensional LDH structures (58). MCC could very well stop unorganized LDH nanoparticle growth as the –OH groups keeps check on LDH development in three dimensional space via hydrogen bonding (59). Also cellulose template can be removed under mild conditions (300°C) (60) after synthesis.

In a typical template-mediated synthesis, MCC was made to swell in alkaline conditions and then the swollen MCC was added drop wise to LDH metal precursor solution ($M^{2+}:M^{3+}=3:1$). The mixture was subjected to microwave hydrothermal treatment under autogenic pressure. Later the LDH solid grown around the MCC template was washed in water and freeze-dried to improve surface area. Usually, carbon containing template materials are burned at high temperature, so that the template can etched out as CO_2 gas. In this case as well, the MCC template was removed by calcinations and the final product exhibited the network structure of LDH (Figure 5.8A). As expected the pore volume and specific surface area were increased from 0.083 to 0.417 cm^3 g^{-1} and 36 to 152 m^2 g^{-1} respectively when compared with template free LDH synthesis. It is perfectly reasonable that pore volume and surface area would increase when the internal space occupying template is burnt and removed. Characterization technique like XRD proved that the synthesized LDH didn't undergo much structural change and clarified that there was no chemical bonding between MCC and metal cations during the synthesis. Hence the interaction between MCC and LDH was only a weak Van der Waals force. Scanning electron microscopy revealed the morphology and dimension of the synthesized LDH with smaller particles and fibril nature (Figure 5.8B). As we all know, smaller the particle, larger the surface area. This high surface area property was wisely utilized by them to study adsorption properties of chemicals. They used Orange II dye (methyl orange), a widely used coloring agent and obviously a potential water pollutant, as model molecule.. The adsorption of orange II on LDH was found to be directly proportional to its surface area (61). Therefore apart from the regular use of cellulose in pulp and papermaking, its role as template polymer in material synthesis is worth to explore.

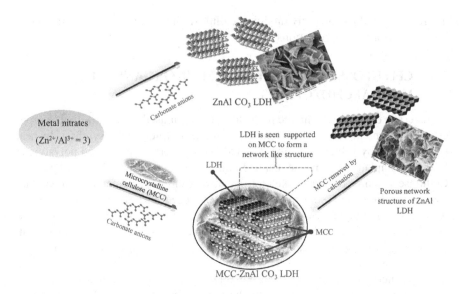

FIGURE 5.8A Schematic synthesis of LDH networks using cellulose biotemplates. *Reprinted with permission from Elsevier (Ref. 61: Liji Sobhana, S. S.; Bogati, D. R.; Reza, M.; Gustafsson, J.; Fardim, P. Cellulose biotemplates for layered double hydroxides networks, Micropor. Mesopor. Mat, 2016, 225, 66–73)*

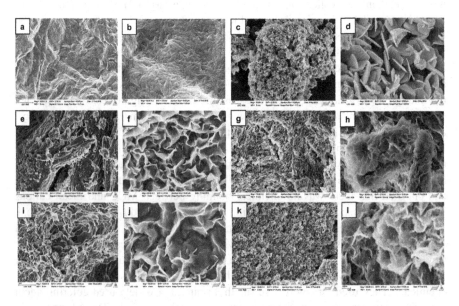

FIGURE 5.8B SEM images of as-synthesized LDH particles (a, b) MCC (c, d) LDH (e, f) LDH-0.5% MCC, (g, h) LDH-1.5% MCC, (i, j) LDH-0.5% MCC removed, and (k, l) LDH-1.5% MCC removed. *Reprinted with permission from Elsevier (Ref. 61: Liji Sobhana, S. S.; Bogati, D. R.; Reza, M.; Gustafsson, J.; Fardim, P. Cellulose biotemplates for layered double hydroxides networks, Micropor. Mesopor. Mat, 2016, 225, 66–73)*

5.14 TOPOCHEMICAL ENGINEERING FOR BIOMASS DECONSTRUCTION

Lignocellulosic biomass obtained from forests is the most abundant natural material. We have seen so far how cellulose fibres can be modified or functionalized using chemical or bimolecular reagents. These methods have always proven to be increasing the desired qualities in pulp fibres and offering new technological solutions. Now, we will explore on how cellulose can be converted altogether into a new molecule for various downstream applications. Cellulose, the polysaccharide biopolymer can be deconstructed or depolymerized into its monomers (mono and disaccharides) through acid/base or enzymatic hydrolysis. Enzymatic hydrolysis is preferred over acid- or base-mediated hydrolysis, because of its eco-friendly process. Enzyme-based methods do not produce much chemical wastes; do not contaminate soil and water. The rate of enzyme catalysed hydrolysis is dependent on the co-constituents present with cellulose in the wood biomass. They are mainly hemicelluloses and lignin which act as protective layers around cellulose fibre network and inhibits the enzyme binding on cellulose directly (62, 63). Often, pressurized hot water treatment, strong base, or ionic liquids are used for the partial removal of lignin. Therefore the removal of those co-constituents is the limiting step which has to be overcome so that hydrolysis of cellulose into monosaccharide sugar, glucose (64) can be achieved.

5.15 TOPOCHEMICAL PRETREATMENT OF WOOD BIOMASS FOR SUGAR PRODUCTION

Mou and co-workers have shown how woody biomass can be dissolved and enzymatically hydrolysed to produce monosugar, Glucose. In their method, milled birch wood and pine wood were dissolved by ionic liquids, hydrothermal, and hydrotropic methods for lignin removal. They used ionic liquids like 1-ethyl-3-methylimidazolium acetate (EmimAC), 1-butyl-3-methylimidazolium chloride (BmimCl). Further, the dissolved cellulose was hydrolysed by enzymes like cellulose and β-glycosidase (65). Their finding suggested that just pressurized hot water was not sufficient enough to remove lignin. Similarly, ionic liquids did not remove lignin, though it increased the surface area of cellulose fibres via swelling. The high surface area could help enzymes to access cellulose network. What they found best was the hydrotropic treatment using sodium xylene sulfonate (SXS)-water solution. This method was proven to be efficient in removing bound lignin around cellulose and then eventually paved a way for enzymes to attack cellulose for hydrolysis. The hydrotropic-enzymatic method yielded glucose of up to 80%. Thus topochemical disassembly of lignin from cellulose fibre surfaces led to cellulose depolymerization. This could be applicable in regenerating sugars from lignin-free pulp in biofuel industries, and hence proves cellulose as sustainable raw material for sugars.

5.16 TOPOCHEMICAL DISASSEMBLY LIGNIN BY HYDROTROPES IN PULP

Lignin is the second most abundant polymer on earth and it is the source of aromatic compounds next to fossil sources. Therefore, when lignin is deconstructed like cellulose, it will yield many organic compounds of commercial interest. Hydrotropic lignin is a raw material for the production of many chemicals and phenol substitutes in phenol-formaldehyde resin development. Gabov et al. extracted lignin from birch wood chips by conventional and modified hydrotropic methods. The leftover lignin in the pulp after extraction was 4.5% and 1.7%, respectively, for conventional and modified methods. This amounts to degrees of delignification 91% and 97% correspondingly. The difference between these two methods was the addition of formic acid and hydrogen peroxide to hydrotropic salt sodium xylene sulfonate in modified method. Their extraction yielded 16% of lignin (dry wood basis) with very low amount of other contents for both the methods. Another important point to note is the hydrotropic methods caused structural change in lignin leaving more phenolic hydroxyl groups in lignin at the expense of aliphatic hydroxyl groups (66). Thus, this work can be used to purify pulp fibres to remove lignin and valorize lignin into platform chemicals.

5.17 TOPOCHEMICALLY CONTROLLED DEPOLYMERIZATION OF CELLULOSE

We have been discussing depolymerization of cellulose and lignin by various chemical and enzymatic treatments. But, can we have control over the depolymerization of cellulose? The answer is, yes, it is possible. The physical and chemical properties of cellulose are largely dependent on the degree of polymerization (DP_v). Hence it is important to have control over depolymerization when we deconstruct cellulose. There are many solvent systems from toxic to eco-friendly nature, available to dissolve cellulose. However, these solvent systems need energy intensive pretreatments like refining and bleaching. Trygg and co-workers found a way to minimize the energy cost by using ethanol-hydrochloric acid (EtOH-HCl) mixture on pulp fibres at 65°C. This caused disruption of remnant primary walls around fibre network (67). Nevertheless aqueous mineral acid conditions are quite reactive; hence controlling the cellulose depolymerization and producing disintegrated cellulose with desired molecular weight are challenging. To solve this, Trygg et al. came up with new formulation in which ethanol (0–96%) was added to mineral acid solution (pK_a 10–4.7). This gave them control over depolymerization and helped them to maintain DP_v to a specific desired value. Though the acid was hydrolyzing cellulose causing decrease in viscosity-averaged degree of polymerization (DP_v) and relative cellulose content, the addition of ethanol preserved the cellulose content from loss. The EtOH-HCl mixture in water yielded DP_v of 291 until ethanol concentration increased to 36% (Figure 5.9). Further increase in [EtOH] made DP_v value to fall drastically in two steps: (i) 36% to 76% and (ii) 76% to 100%. They found that there may be a competing ethanolysis reaction to hydrolysis based on the slope of the curve. Depending on the strength of the mineral acid used, the DP_v was changing from low to high in direct proportion. Strong acids caused disruption in cell walls and reduced the DP_v by 75–80% (68). This method is proven to be energy efficient and a way to produce sugars like glucose, xylose, and mannose from cellulose.

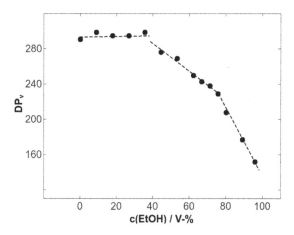

FIGURE 5.9 DP$_v$ profile against ethanol concentrations on pulp fibres. *Reproduced with permission from Romanian Academy Publishing House, the owner of the publishing rights (Ref. 68: Trygg, J.; Trivedi, P.; Fardim. P. Controlled depolymerisation of cellulose to a given degree of polymerisation, Cellulose Chem. Technol. 2016, 50 (5-6), 557-567)*

5.18 CELLULOSE DISSOLUTION AND REGENERATION AS CELLULOSE BEADS

Lignin removed cellulose obtained from Kraft pulping methods can be dissolved in eco-friendly water-based solvent system like NaOH-Water-Urea and further modified chemically and depolymerized by enzymes. But there are other interests also spanning around homogenous cellulose solutions. These solutions can be treated by chemicals once again to regenerate the dissolved cellulose in solid form with new physical and chemical properties. This sold form could be particles, fibre strands, films, or even beads. Cellulose beads are in spherical form with diameter ranging from micro-to millimetre. When the pore size and volume are adjusted in those beads, their internal and external surface will vary. Beads are usually extruded from cellulose solutions in a controlled manner to maintain the geometries of beads uniform. Spinning disk atomization, simple dropping, or dispersion are few methods to produce beads. There are flow techniques and milling as well to produce cellulose beads from solutions and larger sold particles of cellulose. These cellulose beads can be oxidized to introduce new functional groups or subjected to the adsorption of new active groups on their surface to suit our research interests, for example, chromatographic separation of biomolecules (69), drug delivery (70), water and radioactive waste treatment (71, 72), purification and immobilization of bio fluids, enzymes, filtration of heavy metals (73–77).

5.19 ANIONIC CELLULOSE BEADS FOR DRUG ENCAPSULATION AND RELEASE

Trygg et al. reported a method to produce cellulose beads with anionic groups. Their work started with dissolving cellulose in an eco-friendly water-based solvent system (5% cellulose solution in 7 % NaOH-12 % urea-H$_2$O solvent) without

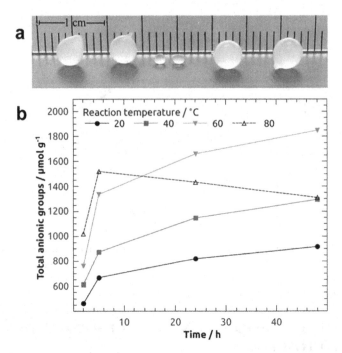

FIGURE 5.10   (a). (Left) Cellulose beads prepared from 5% cellulose-7% NaOH-12% urea-water solution and coagulated in nitric acid and (right) oxidised with Anelli's reaction at 60 ° C for 48 h. (Middle) Beads after drying at room temperature (b) Total anionic groups in CBs oxidised for 2–48 h at 20–80 °C. *Reprinted with permission from Springer Nature (Ref. 81: Trygg, J.; Yildir, E.; Kolakovic, R.; Sandler, N.; Fardim, P. Anionic cellulose beads for drug encapsulation and release. Cellulose 2014, 21(3), 1945–1955)*

producing any toxic waste (78–80). Then this solution was added in an anti-solvent (2 M HNO_3) drop-wise and extruded as neutral cellulose beads (CB) (Figure 5.10a) at room temperature (81). Further AGs were formed on neutral cellulose beads (CB) by modified Anelli's reaction at 20–80°C for 2–48 h. In a typical reaction, the beads were oxidized using TEMPO (2,2,6,6-Tetramethylpiperidin 1-oxyl)/$NaClO_2$/ NaClO (0.1/10/1) with 1.2 moles of $NaClO_2$ for each anhydroglucose (AG) unit. This method helped them to achieve maximum of ~1,850 µmol g^{-1} anhydroglucose units on CBs (degree of substitution: 0.31) (Figure 5.10b). The reaction temperature and time decides the amount of anionic groups formed on the oxidized cellulose beads (OCB).

The oxidation brought new properties like swelling, water retention, and altered pore size in CBs. Solute exclusion technique was used to measure pore size distributions in the beads using dextran as probe (81, 82). The total porosity was found to be 95%. With the oxidation, the mesopores (39–139 Å) increased whereas the macropores (≥560 Å) decreased. Further Swelling of OCBs was measured by soaking the OCBs in sodium phosphate buffered water solution at pH 7.4 and calculating water retention values (WRV) (Table 5.2).

TABLE 5.2
Swelling Ratios of Minimum Diameters Compared to Never-Dried Beads and Water Retention Values (Wrv) of Dried Beads Oxidized at 20, 40, 60°C

	Swelling Ratio %	WRV
CB	48	0.22
OCB20	73	2.50
OCB40	84	4.01
OCB60	88	5.10

Source: Reprinted with permission from Springer Nature (Ref. 81: Trygg, J.; Yildir, E.; Kolakovic, R.; Sandler, N.; Fardim, P. Anionic cellulose beads for drug encapsulation and release. *Cellulose* 2014, 21(3), 1945–1955)

WRV was calculated by the formula,

$$WRV = \left(m_1 / m_2 - 1 \right) \tag{5.1}$$

m_1 was the mass of wet beads after separation by centrifugation and m_2 was the dry mass before swelling. The OCBs showed an increase of 40% in swelling and 500% in water retention value.

Trygg et al. wanted to use these beads in drug delivery applications as the swelling and porous nature of the OCBs are the vital factors for drug encapsulation and delivery. Anionic cellulose beads (OCBs or denoted as A-CB) were studied for ranitidine drug encapsulation and delivery. Ranitidine is an antacid used to cure stomach illness.

According to their method, anionic cellulose beads (A-CB) were soaked in ranitidine HCl solution (20 mg ml^{-1}) at the concentration of 2 beads mL^{-1} solution. Then the beads were filtered and dried for 2 days. These dried beads were used for the controlled release of ranitidine. The released drug was characterized by UV–Vis spectroscopy as ranitidine exhibits absorption peak at 288 nm. The in vitro drug release profile showed that the A-CBs released twice the amount of ranitidine than neutral beads (CB) even though the amounts of accessible water inside the pores same for both the beads (81). Thus, this work promises the application of charged cellulose beads in drug delivery. These anionic beads can also be employed in ion exchange chromatography to purify water, adsorption, and catalysis.

5.20 LIGNIN-CELLULOSE BIO-COMPOSITE BEADS

Plant polymers are potential renewable substrates for the next generation chemical and materials production. Gabov and co-workers have made fascinating lignin-cellulose composite beads from dissolving grade pulp and hydrotropic lignin obtained from birch wood. They used eco-friendly 7% NaOH-12% urea solvent system for mixing cellulose and lignin in different percentage combinations. These solutions were later regenerated in an anti-solvent and extruded into beads. The dimension of the beads was controlled by keeping drop height 5 cm to obtain 1–2 cm diameter beads. With this method one can prepare beads containing lignin from 0 to 40%. Drying methods have great influence on morphology, surface area, and porosity of these beads. The never-dried beads exhibited high porosity and spherical shape with uniform distribution of lignin (Figure 5.11A). The intermolecular hydrogen bonding formed between lignin and cellulose during the composite formation has huge impact on the characteristics of the beads. This can be identified by the – OH stretching band shift to higher wave numbers in IR region. So, where can we use these beads? Gabov et al. tried to employ this material in a completely new application, i.e. antimicrobial activity. They used these beads to inhibit the growth of microorganisms like bacteria (83, 84). They picked gram-positive *Staphylococcus aureus* bacteria and gram-negative *Escherichia coli* and (85). Their finding suggested that lignin free cellulose beads and pure lignin beads of 40% hydrotropic

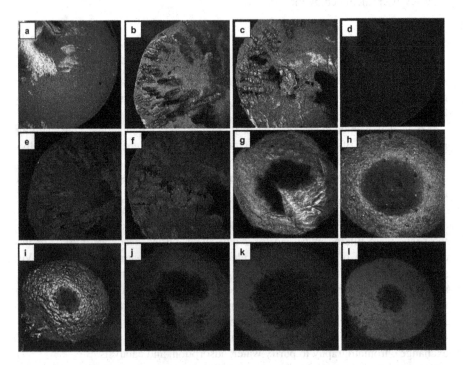

FIGURE 5.11A   Distribution of lignin in cellulose (C)-lignin (L) beads examined by confocal fluorescence microscopy. Cross section of the beads in the reflection channel: (**a**) 90C10L, (**b**) 75C25L, (**c**) 60C40L; surface of the beads in the reflection channel: (**g**) 90C10L, (**h**) 75C25L, (**i**) 60C40L. The cross sections in the fluorescence channel: (**d**) 90C10L, (**e**) 75C25L, (**f**) 60C40L; surface in the fluorescence channel: (**j**) 90C10L, (**k**) 75C25L, (**l**) 60C40L. The colors of the images are artificial. The parameters for all the fluorescence images were adjusted in a same way during processing. 90C10L denotes 90% cellulose, 10% lignin. *Reprinted with permission from Springer Nature (Ref. 85: Gabov, K.; Oja, T.; Deguchi, T.; Fallarero, A.; & Fardim, P. Preparation, characterization and antimicrobial application of hybrid cellulose-lignin beads. Cellulose, 2016, 24(2), 641–658)*

lignin have no antibacterial activity; hence the growth of *E.coli* and *S. aureus* were unabated. On the other hand, lignin-cellulose composite beads showed resistance to the growth of *S. aureus* at the concentration of 40 % lignin. But, still, the composite beads failed to inhibit the growth of *E.coli* (Figure 5.11B). This work paves a way for producing antimicrobial biocomposite material which can be used in the field of biotechnology and medicine.

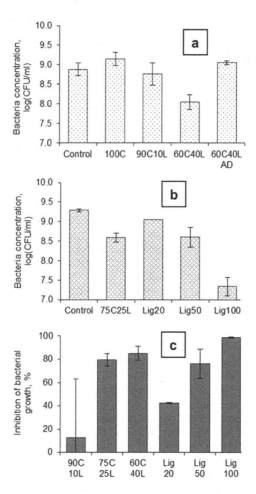

FIGURE 5.11B Concentration of *S. aureus* (a, b) in the broth with different beads and hydrotropic lignin after incubation for 24 h at 37 C; and the inhibition of *S. aureus* growth calculated based on the bacterial concentrations (c, only positive values are presented). a, b runs were performed on two different days. Load of the beads and the dosages of lignin were the same as in the tests with *E. coli*. Initial concentrations of bacteria were 6.38 and 6.26 log(CFU/mL) for a, b runs, respectively. *Reprinted with permission from Springer Nature (Ref. 85: Gabov, K.; Oja, T.; Deguchi, T.; Fallarero, A.; & Fardim, P. Preparation, characterization and antimicrobial application of hybrid cellulose-lignin beads. Cellulose, 2016, 24(2), 641–658)*

5.21 CONCLUSION

Topochemical engineering on bio-based materials is a way forward for sustainability in chemicals and materials production. The advancement of chemistry and characterization techniques is rapidly expanding; hence exploring bioresources as against traditional fossil resources, for technological application is the need of the hour. As we have seen in this chapter, woody biomass in the form of pulp fibres is potential feedstock for the future. The know-how techniques developed in the field of cellulose and lignin chemistry are motivating researchers to engineer those materials at nanostructural and molecular level (86). As we observed from the many reported works, the –OH, -COOH, -O-, RCOO- groups in the biomass are the sites of opportunity for designing and developing new products. When we rightly utilize the interactions like electrostatic, hydrophilic, and hydrophobic forces between host and guest molecules, many new composites can be assembled. These materials could invite commercial-scale production for sustainable smart biomaterials, polymers, paper, renewable fuels, sensors, biomarkers, chromatography, catalysis, etc. Thus, topochemical engineering of bio-based materials possess great scope in circular bio-economy.

REFERENCES

1. Fardim, P.; Holmbom, B.; Ivaska, A.; Karhu, J. Critical comparison and validation of methods for determination of anionic groups in pulp fibres. *Nordic Pulp Paper Res. J.* 2002, 17 (3), 346–351.
2. De Nooy, A.E.J.; Besemer, A. C.; Bekkum, H. van. On the use of stable organic nitroxyl radicals for the oxidation of primary and secondary alcohols. *Synthesis* 1996, 10, 1153–1176.
3. Saito, T.; Isogai, A. TEMPO-mediated oxidation of native cellulose. The effect of oxidation conditions on chemical and crystal structures of the water-insoluble fractions. *Biomacromolecules* 2004, 5 (5), 1983–1989.
4. Kohlshutter, V.; Haenni, P. For the knowledge of graphitic carbon and graphitic acid. *Z. Anorg. Allg. Chem.* 1919, 105 (1), 121–144.
5. Ramamurthy, V.; Venkatesan, K. Photochemical reactions of organic crystals. *Chem. Rev.* 1987, 87(2), 433–481.
6. Langmuir, I. The constitution and fundamental properties of solids and liquids. Part I. Solids. *J. Amer. Chem. Soc.* 1916, 38 (11), 2221–2295.
7. Boldyrev, V.V. Topochemistry and topochemical reactions. *React. Solid.* 1990, 8 (3), 231–246
8. Dence, C.W.; Chemistry of mechanical pulp bleaching. In: Dence CW, Reeve DW (eds) *Pulp bleaching: principles and practice.* TAPPI Press, Atlanta, 1994, 161–181.
9. Pan, G.X.; Spencer, L.; Leary, G.J. A comparative study on reactions of hydrogen peroxide and peracetic acid with lignin chromophores. Part 1. The reaction of coniferaldehyde model compounds. *Holzforschung* 2000, **54(2)**, 144–152.
10. Plesnicar, B. Oxidation with peroxy acids and other peroxides. In: Trahanovsky WS (ed) *Oxidation in organic chemistry. Part C.* Academic Press, New York, 1978, 211.
11. Anderson, J.R.; Amini, B. Hydrogen peroxide bleaching. In: Dence CW, Reeve DW (eds) *Pulp bleaching: principles and practice.* TAPPI Press, Atlanta, 1994, 411–442.
12. James, A.P.; Mackirdy, I. S. The chemistry of peroxygen bleaching. *Chem. Ind (London)* 1990, 20, 641–645.

13. Coucharriere, C.; Mortha, G.; Lachenal, D.; Briois, S.; Larnicol, P. Rationalization of the use of TAED during activated peroxide bleaching. Part II: Bleaching optimization. *J. Pulp. Pap. Sci.* 2004, 30(2), 35–41.

14. Zeinaly, F.; Shakhes, J.; Zeinali, N. Multi stage peroxide and activated peroxide bleaching of kenaf bast pulp. *Carbohydr. Polym.* 2013, 92(2), 976–981.

15. Turner, N.A.; Mathews, A.J. Enhanced delignification and bleaching using TAED activated peroxide. *TAPPI pulping conference, Montre´al, Que´bec, Canada*, 1998, Proceedings book *3*, 1269–1276.

16. Zhao, Q.; Sun, D.; Wang, Z.; Pu, J.; Jin, X.; Xing, M. Effects of different activation processes on H_2O_2/TAED bleaching of *Populus nigra* chemi-thermomechanical pulp. *Bioresources* 2012, 7(4), 4889–4901.

17. Iamazaki, E.T.; Orblin, E.; Fardim, P. Topochemical activation of pulp fibres, *Cellulose* 2013, 20, 2615–2624.

18. Heinze, T.; Koschella, A.; Brackhagen, M.; Engelhardt, J.; Nachtkamp, K. Studies on non-natural deoxyammonium cellulose. *Macromol. Symp.* 2006, 244(1), 74–82.

19. Elegir, G.; Kindl, A.; Sadocco, P.; Orlandi, M. Development of antimicrobial cellulose packaging through laccase-mediated grafting of phenolic compounds. *Enzyme Microb. Technol.* 2008, 43, 84–92.

20. Kim, S.Y.; Zille, A.; Murkovic, M.; Güebitz, G.; Cavaco-Paulo, A. Enzymatic polymerization on the surface of functionalized cellulose fibers. *Enzyme Microb. Technol.* 2007, 40, 1782–1787.

21. Li, X.; Tabil, L. G.; Panigrahi, S. Chemical treatments of natural fiber for use in natural fiber-reinforced composites: A review. *J. Polym. Environ.* 2007, 15(1), 25–33.

22. Persson, P. Strategies for cellulose fiber modification (PhD thesis). Stockholm, Sweden: KTH, STFI-Packforsk AB (April 2004).

23. Qiu, X.; Hu, S. "Smart" materials based on cellulose: A review of the preparations, properties, and applications. *Materials* 2013, 6(3), 738–781.

24. Vega, B.; Wondraczek, H.; Bretschneider, L.; Näreoja, T.; Fardim, P.; Heinze, T. Preparation of reactive fibre interfaces using multifunctional cellulose derivatives. *Carbohydr. Polym.* 2015, 132, 261–273.

25. Yllner, S.; Enstrom, B. Studies of the adsorption of xylan on cellulose fibres during the sulphate cook, Part 1. *Svensk Papperstidning* 1956, 59, 229–232.

26. Yllner, S.; Enstrom, B. Studies of the adsorption of xylan on cellulose fibres during the sulphate cook, Part 2. *Svensk Papperstidning* 1957, 60, 549–554.

27. Linder, A.; Bergman, R.; Bodin, A.; Gatenholm, P. Mechanism of assembly of xylan onto cellulose surfaces. *Langmuir* 2003, 19, 5072–5077.

28. Esker, A.; Becker, U.; Jamin, S.; Beppu, S.; Renneckar, S.; Glasser, W. Self-assembly behavior of some co- and heteropolysaccharides related to hemicelluloses. In P. Gatenholm, & M. Tenkanen (Eds.), *Hemicelluloses: Science and technology* Washington, DC: American Chemical Society 2002, 198–219

29. Vega, B.; Petzold-Welcke, K.; Fardim, P.; Heinze, T. Studies on the fibre surfaces modified with xylan polyelectrolytes, *Carbohydr. Polym.* 2012, 89, 768– 776.

30. Köhnke, T. Adsorption of xylans on cellulosic fibres. Influence of xylan composition on adsorption characteristics and Kraft pulp properties. Ph.D. Thesis, Chalmers University of Technology, 2010.

31. Larsson, P.; Puttaswamaiah, S.; Ly, C.; Vanerek, A.; Hall, C.; Drolet, F. Filtration, adsorption and immunodetection of virus using polyelectrolyte multilayer-modified paper. *Colloids Surf. B* 2013, 101, 205–209.

32. Ankerfors, C.; Lingström, R.; Wågberg, L. A comparison of polyelectrolytes complexes and multilayers: their adsorption behavior and use for enhancing tensile strength of paper. *Nord. Pulp Pap. Res. J.* 2009, 24 (1), 77–86.

33. Gernandt, R.; Wågberg, L.; Gärdlund, L.; Dautzenberg, H. Polyelectrolyte complexes for surface modification of wood fibres. I. Preparation and characterisation of complexes for dry and wet strength improvement of paper. *Colloids Surf. A* 2003, 213, 15–25.

34. Kikuchi, Y.; Noda, A. Polyelectrolyte complexes of heparin with chitosan. *J. Appl. Polym. Sci.* 1976, 20, 2561–2563.

35. Fukuda, H. Polyelectrolyte complexes of chitosan with sodium carboxymethylcellulose. *Bull. Chem. Soc. Jpn.* 1980, 52, 837–840.

36. Hara, M.; Nakajima, A. Formation of polyelectrolyte complex of heparin with aminoacetalized poly(vinyl alcohol). *Polym. J.* 1978, 10 (1), 37–44.

37. Dumitriu, S.; Chornet, E. Inclusion and release of proteins from polysaccharide-based polyion complexes. *Adv. Drug Delivery Rev.* 1998, 31, 223–246.

38. Vega, B.; Wondraczek, H.; Zarth, C.S.P.; Heikkila, E.; Fardim, P.; Heinze, T. Charge-directed fiber surface modification by molecular assemblies of functional polysaccharides. *Langmuir* 2013, 29 (44), 13388–13395.

39. Wondraczek, H.; Pfeifer, A.; Heinze, T. Water soluble photoactive cellulose derivatives: synthesis and characterization of mixed 2-[(4-methyl-2-oxo-2H-chromen-7-yl)oxy]acetic acid–(3-carboxypropyl)trimethylammonium chloride esters of cellulose. *Cellulose* 2012, 19, 1327–1335.

40. Grigoray, O.; Wondraczek, H.; Daus. S.; Kühnöl, K.; Latifi, S.K.; Saketi, P.; Fardim, P.; Kallio, P.; Heinz, T. Photocontrol of mechanical properties of pulp fibers and fiber-to-fiber bonds via self- assembled polysaccharide derivatives. *Macromol. Mater. Eng.* 2014, 300(3), 277–282.

41. Wagberg, L. Polyelectrolyte adsorption onto cellulose fibres – a review. *Nord. Pulp. Pap. Res. J.* 2000, 15 (5), 586–597.

42. Grigoray, O.; Wondraczek, H.; Pfeifer, A.; Fardim, P.; Heinze, T. Fluorescent multifunctional polysaccharides for sustainable supramolecular functionalization of fibers in water, *ACS Sustainable Chem. Eng.* 2017, 5 (2), 1794–1803.

43. Kurrle, F. L.; Parks, C. J. Process of manufacturing authenticatable paper products. U.S. Patent 6,054,021, 2000.

44. Foster, J. J.; Mulcaahy, L. T. Process for making and detecting anti-counterfeit paper. U.S. Patent 6,045,656, 2000.

45. Dou, Y.; Xu, S.; Liu, X.; Han, J.; Yan, H.; Wei, M.; Evans, D. G.; Duan, X. Transparent, flexible films based on layered double hydroxide/cellulose acetate with excellent oxygen barrier property, *Adv. Func. Mater.* 2014, 24, 514–521.

46. Liji Sobhana, S. S.; Mehedi, R.; Mika, M.; Petriina, P.; Mika, L.; Marinela, M. D.; Garcia, Y.; Fardim, P. Heteronuclear nanoparticles supported hydrotalcites containing Ni(II) and Fe(III) stable photocatalysts for Orange II degradation, *Appl. Clay Sci.* 2016, 132–133, 641–649.

47. Liji Sobhana, S.S.; Mohamed, S.; Prevot, V.; Fardim, P. Layered double hydroxides decorated with Au-Pd nanoparticles to photodegradate Orange II from water, *Appl. Clay Sci.* 2016, 134, 120–127.

48. Wang, Q.; O'Hare, D. Recent advances in the synthesis and application of layered double hydroxide (LDH) nanosheets, *Chem. Rev.* 2012, 112(7), 4124–4155.

49. Van Hartman, S.; Heikkila, E.; Lange, C.; Fardim, P. Potential applications of hybrid layered double hydroxide particles in pulp and paper production, *Bioresources* 2014, 9(2), 2274–2288.

50. Yong, C.J.; Bharat B. Mechanically durable carbon nanotube composite hierarchical structures with superhydrophobicity, self-cleaning, and low-drag. *ACS Nano* 2009, 3 (12), 4155–4163.

51. Liji Sobhana, S. S.; Zhang, Xue; Kesavan, L.; Liias, P.; Fardim, P. Layered double hydroxide interfaced stearic acid – Cellulose fibres: A new class of super-hydrophobic hybrid materials. *Colloids Surf A Physicochem. Eng. Asp.* 2017, 522, 416–424.

52. Shu, W.; Zhaojun, T.; Zengfu, J.; Zelong, Y.; Lijuan, W. Preparation and characterization of hydrophobic cotton fibre for water/oil separation by electroless plating combined with chemical corrosion, *Int. J. Environ. Res. Publ. Health.* 2015, 2(10), 144–150.

53. Lange, C.; Lundin, T.; Fardim, P. Hydrophobisation of mechanical pulp fibres with sodium dodecyl sulphate functionalised layered double hydroxide particles. *Holzforschung* 2011, 66, 433–441.

54. Lange, C.; Touaiti, F.; Fardim, P. Hybrid clay functionalized biofibres for composite applications. *Comp. Part B Eng.* 2013, 47, 260–266.

55. Lange, C.; Lastusaari, M.; Reza, M.; Latifi, S.; Kallio, P.; Fardim, P. In situ hybridization of pulp fibers using mg-al layered double hydroxides. *Fibers* 2015, 3, 103–133.

56. Geraud, E.; Prevot, V.; Leroux, F. Synthesis and characterization of macroporous MgAl LDH using polystyrene spheres as template. *J. Phys. Chem. Solids* 2006, 67(5–6), 903–908

57. Mann, S.; Archibaid, D.D.; Didymus, J.M.; Douglas, T.; Heywood, B.R.; Meldrum, F.C.; Reeves, N.J. Crystallization at inorganic-organic interfaces: Biominerals and biomimetic synthesis. *Science* 1993, 261 (5126), 1286–1292.

58. Alan P. I. Thickening and gelling agents for food, 1999, 188.

59. Liu, S.; Tao, D. Cellulose scaffold: A green template for the controlling synthesis of magnetic inorganic nanoparticles. *J. Powder Technol.* 2012, 217, 502–509.

60. Liu, Y.; Goebl, J.; Yin, Y. Templated synthesis of nanostructured materials. *Chem. Soc. Rev.* 2013, 42(7), 2610–2653.

61. Liji Sobhana, S. S.; Bogati, D. R.; Reza, M.; Gustafsson, J.; Fardim, P. Cellulose biotemplates for layered double hydroxides networks, *Micropor. Mesopor. Mat,* 2016, 225, 66–73.

62. Blanch, H.W.; Wilke, C.R. Sugars and chemicals from cellulose. *Rev. Chem. Eng.* 1982, 1, 71–119.

63. Mooney, C.A.; Mansfield, S.D.; Touhy, M.G.; Saddler, J.N. The effect of initial pore volume and lignin content on the enzymatic hydrolysis of softwoods. *Bioresour. Technol.* 1998, 64, 113–119.

64. Rahikainen, J.; Mikander, S.; Marjamaa, K.; Tamminen, T.; Lappas, A.; Viikari, L.; Kruus, K. Inhibition of enzymatic hydrolysis by residual lignins from softwood – study of enzyme binding and inactivation on lignin-rich surface. *Biotechnol. Bioeng.* 2011, 108, 2823–2834.

65. Mou, H.; Elina, O.; Kruus, K.; Fardim, P. Topochemical pretreatment of wood biomass to enhance enzymatic hydrolysis of polysaccharides to sugars. *Bioresour. Technol.* 2013, 142, 540–545.

66. Gabov, K.; Gosselink, R.J.A.; Smeds, A. I.; Fardim, P. Characterization of lignin extracted from birch wood by a modified hydrotropic process, *J. Agric. Food Chem.* 2014, 62 (44), 10759–10767.

67. Trygg, J.; Fardim, P. Enhancement of cellulose dissolution in water-based solvent via ethanol–hydrochloric acid pretreatment, *Cellulose* 2011, 18, 987–994.

68. Trygg, J.; Trivedi, P.; Fardim, P. Controlled depolymerisation of cellulose to a given degree of polymerisation. *Cellulose Chem. Technol.* 2016, 50 (5–6), 557–567.

69. Guile, G. R.; Wong, S.Y.C.; Dwek, R.A. Analytical and preparative separation of anionic oligosaccharides by weak anion-exchange high-performance liquid chromatography on an inert polymer column. *Anal. Biochem.* 1994, 222, 231–235.

70. Bilandi, A.; Mishra, A. K. Ion exchange resins: an approach towards taste making of bitter drugs and sustained release formulations with their patents. *Int. Res. J. Pharm.* 2013, 4, 65–74.

71. Sheldon, A.; Newman, S. Technical reports. *Soc. Sci. Med.* 1967, 1, 441–444.

72. Agency IAE, Vienna (1967). The plutonium-oxygen and uranium-plutonium-oxygen systems: a thermochemical assessment. Technical Reports Series No. 79. Report of a panel on thermodynamics of plutonium oxides held in Vienna, 24–28 Oct 1966.

73. Ettanauer, M.; Loth, F.; Thümmler, K.; Fischer, S.; Weber, V.; Falkenhagen, D. Characterization and functionalization of cellulose microbeads for extraporeal blood purification. *Cellulose* 2011, 18, 1257–1263.

74. Guo, X.; Du, Y.; Chen, F.; Park, H.-S.; Xie, Y. Mechanism of removal of arsenic by bead cellulose loaded with iron oxyhydroxide (β-FeOOH). *J. Colloid Interface Sci.* 2007, 314, 427–433.

75. Štamberg, J.; Peška, J.; Dautzenberg, H.; Phillip B.; Gribnau, T.C.J.; Visser, J.; Nivard R.J.F. (Eds.), *Affinity Chromatography and Related Techniques*, Elsevier Science Publishing Co, Amsterdam 1982, 131–141.

76. Weber, V.; Linsberger, I.; Ettenauer, M.; Loth, F.; Höyhtyä, M.; Falkenhagen, D. Development of specific adsorbents for human tumor necrosis factor-alpha: influence of antibody immobilization on performance and biocompatibility. *Biomacromolecules* 2005, 6, 1864–1870.

77. Zhou, D.; Zhang, L.; Guo, S. Mechanisms of lead biosorption on cellulose/chitin beads *Water Res.* 2005, 39, 3755–3762.

78. Isogai, A.; Atalla, R. Dissolution of cellulose in aqueous NaOH solutions. *Cellulose* 1998, 5, 309–319.

79. Liu, W.; Budtova, T.; Navard, P. Influence of ZnO on the properties of dilute and semi-dilute cellulose-NaOH-water solutions. *Cellulose* 2011, 18, 911–920.

80. Qi, H.; Chang, C.; Zhang, L. Effects of temperature and molecular weight on dissolution of cellulose in NaOH/urea aqueous solution. *Cellulose* 2008, 15, 779–787.

81. Trygg, J.; Yildir, E.; Kolakovic, R.; Sandler, N.; Fardim, P. Anionic cellulose beads for drug encapsulation and release. *Cellulose* 2014, 21(3), 1945–1955.

82. Stone, J.; Scallan, A. A structural model for the cell wall of water-swollen wood pulp fibres based on their accessibility to macromolecules. *Cell. Chem. Technol.* 1968, 2, 343–358.

83. Nada, A.M.A.; El-Diwany, A.I.; Elshafei, A.M. Infrared and antimicrobial studies on different lignins. *Acta. Biotechnol.* 1989, 9, 295–298.

84. Nelson, J.L.; Alexander, J.W.; Gianotti, L.; Chalk, C.L.; Pyles, T. Influence of dietary fiber on microbial growth in vitro and bacterial translocation after burn injury in mice. *Nutrition* 1994, 10, 32–36.

85. Gabov, K.; Oja, T.; Deguchi, T.; Fallarero, A.; & Fardim, P. Preparation, characterization and antimicrobial application of hybrid cellulose-lignin beads. *Cellulose*, 2016, 24(2), 641–658.

86. Liji Sobhana, S. S.; Kesavan, L.; Fardim, P. Topochemical engineering of cellulose-based functional materials, *Langmuir*, 2018, 34 (34), 9857–9878

6 Sustainable Hydrogels from Renewable Resources

Diana Ciolacu

CONTENTS

6.1 INTRODUCTION

In recent years, many industries across various manufacturing sectors have acknowledged the need to adapt to environment-friendly manufacturing technologies and to explore innovative materials, which will be able to meet the growing needs of the population, as well as the environmental concerns of end users (Pandey et al. 2015).

The increasing awareness of the world related to the limited nature of fossil carbon resources required a reassessment of raw materials and thus, the green and renewable resources become a promising alternative (Menon and Rao 2012; Chandel et al. 2018). In addition, the anticipated increase in demand for renewable resources from biomass led to the development of cost-efficient syntheses and production of value-added materials, a crucial requirement for the continued growth of the industry (Ten and Vermerris 2013).

Lignocellulosic biomass, a renewable and abundant feedstock, is available worldwide and holds enormous potential for vast number of applications for human sustainability, such as chemicals, fuels, biofuels, biomolecules, and biomaterials (Isikgor and Becer 2015; Kumar and Sharma 2017).

Lignocellulosic biomass has the crucial advantages over other biomass supplies due to the fact that it can be produced quickly at low cost and represents the non-edible portions of the plants, and therefore it does not interfere with food supplies (Cherubini 2010).

The crucial challenge in converting lignocellulosic biomass is to produce value-added bio-based products at high selectivity and yields at economical cost. The following are main sources of lignocellulosic bio-based products (Ten and Vermerris 2013):

- *The plant cell wall* – from which it can be isolated the natural polymers as cellulose, hemicelluloses, and lignin, that further can be converted to valuable products;
- *Fermentable sugars generated in a biorefinery* – the hexose and pentose sugars obtained through the deconstruction of the cell wall polymers, during the production of second-generation cellulosic biofuels, are converted to fuels such as ethanol or butanol. Furthermore, the use of different microbial strains enables the production of chemical feedstocks that can be used for the production of bio-based products, such as poly(lactic acid) (PLA) (Nampoothiri et al. 2010; Inkinen et al. 2011) and poly(hydroxyalkanoates) (PHA) (Bayer et al. 2017).
- *The waste stream of the biorefinery* – the waste stream which includes a majority of lignin, the resistant part to deconstruction of cellulose, monomeric sugars that can't be converted microbially, products formed during processing from the monomeric sugars, for example, furfural, 5-hydroxymethylfurfural (HMF), and various extractives.
- *The waste stream of the pulp and paper mill* – it consists mainly of lignin and extractives, which can be used for further production of biopolymers and composites.

Renewable resources have gained much attention due to the great importance they have in sustainable development and environmental protection (Mondal and Haque 2019). The direct use of renewable resources or their regenerated forms can develop new generations of engineered biomaterials with designed properties and reduced cost; these are easily producible and are used for a broad array of applications.

Hydrogels made up from renewable resources have attracted attention due to their special structures and fascinating properties, taking into account the biocompatibility, biodegradability, their outstanding mechanical properties and structure similar to extracellular matrix (ECM), and also the possible ability to promote cells growth and to control the release of drugs or bioactive molecules. These properties open up the possibility of their uses in vast areas of applications, such as hygiene (disposable diapers and feminine care products), agriculture (water retention and pesticide delivery), biomedical materials (drug carriers, wound dressings, and tissue engineering scaffolds), pollutant adsorbents (heavy metal ions, dyes, and pesticides), biosensors, etc. (Shen et al. 2016).

This chapter outlines the new developments related to the plant fibre–based hydrogels (flax, bagasse, agave, jute, sugarcane, etc.) and natural polymer–based hydrogels (cellulose, lignin and hemicelluloses) with focus on the preparation methods, physicochemical properties, as well as their various applications.

6.2 POLYMERS FROM RENEWABLE RESOURCES

Polymers obtained from renewable resources are generally known as bio-based polymers and can be obtained from different sources, such as (Thakur and Thakur 2014):

- polymers extracted from biomass: polysaccharides, proteins, and lipids;
- polymers synthesized from bio-derived monomers: polylactides and other polyesters;
- polymers produced directly by the natural or genetically modified organism: PHA and bacterial cellulose.

The most abundant natural polymer from all presented above are polysaccharides, which form the majority of renewable biomass (Eichhorn et al. 2010; Rani et al. 2012; Thakur 2013; Thakur and Singha 2013). These are biorenewable resources, which exhibit biodegradability, biocompatibility, and antibacterial activity (George et al. 2001; Bogoeva-Gaceva et al. 2007; Eichhorn et al. 2010; Rani et al. 2012; Rani et al. 2013). Among these, natural fibres, a natural worldwide-available resource, represent fibrous polymeric materials with numerous advantages; for example, they are environment-friendly/eco-friendly, abundant, easily available, renewable, sustainable, economical, biodegradable, lightweight, non-toxic, and easily processible (Patel and Parsania 2018).

Natural fibres can be classified in function of their origin, such as plant fibres, animal fibres, and mineral fibres (Figure 6.1) (Thakur and Thakur 2014; Pandey et al. 2015; Owonubi et al. 2019). *Plant fibres* can be classified into wood fibres (softwood, hardwood, and recycled wood fibres) and non-wood natural fibres, subdivided into five basic types: bast, leaf, seed, straw, and grasses. These are mainly composed of cellulose, hemicelluloses, lignin, and pectin with a small quantity of extractives. *Animal fibres* are the second most widely used natural fibres after plant fibres and generally consist of proteins, such as wool fibres, silk, hair, and feathers (Puttegowda et al. 2018). Some of the most important *mineral fibres* may be classified into the following: (i) asbestos, which is the only naturally occurring mineral fibre-like serpentine, amphiboles, and anthophyllite (Sapuan et al. 2017) and (ii) basalt fibres, which have superior corrosion resistance and similar mechanical properties to glass fibres (Rousakis 2017).

Lignocellulosic fibres (wood and non-wood natural fibres) may be derived from a wide range of plants and are considered an excellent feedstock for the production of various materials. Even if wood (such as pine, poplar, beech, willow, etc.) has a great potential as biomass feedstock, used as fuel from early times, necessary to produce heat and electricity, it is also used for the production of wood materials and pulp and paper commodities (Sannigrahi et al. 2010; Gonçalves et al. 2018; Zoghlami and Paës 2019). In addition, wood is the most important industrial source of cellulose, together with lignin and hemicelluloses, small amounts of extractives, and inorganic salts. Based on their anatomical features, wood species can be distinguished as hardwoods, with a more complex and heterogeneous structure, and softwoods (Wiedenhoeft and Miller 2005; Nechyporchuk et al. 2016).

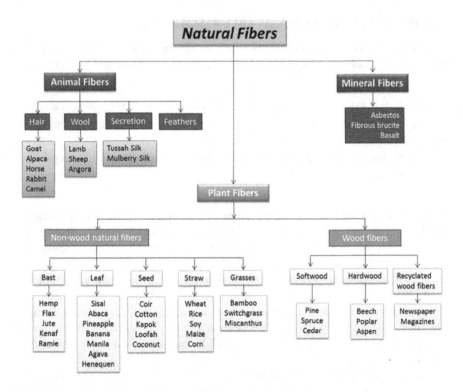

FIGURE 6.1 Classification of natural fibres.

Important amounts of lignocellulose fibres are produced annually in agricultural sector, where the non-wood fibres, such as maize, wheat straw, rice, and sugarcane, are responsible for generating the majority of lignocellulosic materials (Saini et al. 2015; Yuan et al. 2018). Due to their higher lignocellulose productivity and better processability, perennial grass species, such as switchgrass, bamboo, *Erianthus*, etc., have attracted considerable attention as potent lignocellulosic materials (Tye et al. 2016; Bhatia et al. 2017; Umezawa 2018). The grass fibres can be more efficient than wood fibres, for instance sugarcane, corn, and sorghum, which can be used for food, cattle feed, and/or raw materials for fermentation, and at the same time their residual straw, the main source of lignocellulose, can be used for fuels, fibres, materials, bio-chemicals, lignin-derived biofuels, etc. Moreover, sugarcane-based sugar plants use baggage residues as a fuel that can easily cover all the energy requirements of the factory (Bezerra and Ragauskas 2016; Takeda et al. 2019). The other plants, such as corn stover, bagasse, etc., constitute only a minor proportion of the total agriculture waste production in the world.

Due to their strength, flexibility, and easy processability, the natural fibres such as hemp, jute, sisal, banana, coir, and kenaf are extensively used in the production of the lightweight composites (Thakur et al. 2014; Girijappa et al. 2019).

The dimensions of the natural fibres depend mainly on their location within the plant, as for example the fibres from fruits and seeds are few centimetres long, whereas fibres from stems and leaves are much longer (Smole et al. 2013). Some structural characteristics, properties, and applications of different non-wood natural fibres are presented in Table 6.1.

Lignocelluloses are complex assembly mainly composed of carbohydrate polymers (cellulose and hemicelluloses), and non-carbohydrate polymer (lignin), as well

TABLE 6.1
Structural Compositions, Properties, and Applications of Natural Fibres

Type of Fibre	Characteristics	Properties	Applications	References
Cotton *Gossypium*	*Staple* length: 10–80 mm *Elementary fibres* length: 25–60 mm diameter: 12–45 µm	• Hydrophilic (fibres swell considerably in water • Dry tenacity (25–40 cN/tex) • Elongation at break (5–10%) • Low microbial resistance • Doesn't cause skin irritation or other allergies	• Textile (poplins, voiles, clothing, boos, and shoes, carpets, curtains, etc.) • Yarn and cordage • Furniture industry • Automotive (tire cords) • Tarpaulins and marquees	Zou et al. (2011); Al-Oqla et al. (2015); Gupta and Srivastava (2016); Balaji and Senthil Vadivu (2017); Colomban and Jauzein (2018); Elmogahzy and Farag (2018)
Kapok *Ceiba pentandra*	*Plant* length: 90 cm–1 m diameter: 2–3 mm *Elementary fibres* length: 10–35 mm diameter: 20–43 µm	• Light-weight • Hydrophobic • Thermal and acoustic insulating properties • Tensile strength • (0.84 cN/dtex) • Breaking elongation (1.2%)	• Technical textiles • Yachts and boats furnishing • Insulating materials in refrigeration systems • Acoustic insulation • Industrial wastewaters filtration • Reinforcement in composites	Mwaikambo et al. (2000); Prachayawarakorn et al. (2013); Dong et al. (2015); Arumugam (2014); Zheng et al. (2015)
Hemp *Cannabis sativa*	*Plant* length: 1.2–4.5 m diameter: 2 cm *Elementary fibres* length: 2–90 mm	• Excellent moisture resistance • Rot very slowly in water • High tenacity (53–62 cN/tex)	• Technical textiles • Building and isolation materials • Composite materials • Garden mulch • Animal beddings	Kozlowski et al. (2006); Bhoopathi et al. (2014); Panaitescu et al. (2016); Väisänen et al. (2018); Sepe et al. (2018); Réquilé et al. (2018)

(Continued)

TABLE 6.1 (Continued)
Structural Compositions, Properties, and Applications of Natural Fibres

Type of Fibre	Characteristics	Properties	Applications	References
Flax *Linum usitatissimum*	*Plant* length: 90 cm – 1 m diameter: 2–3 mm *Elementary fibres* length: 6–65 mm diameter: 12–20 μm	• Strong strength (55 cN/tex) • Low elongation at break (1.8%) • Moisture regain (12%)	• Automotive or boat components • Twine and rope • Textile (damasks, lace, and sheeting) • Technical textile (fishing nets) • Dyes • Papers • Cosmetics (hair gels and soap)	Sen and Reddy (2011); Zhu et al. (2013); De Prez et al. (2018); Yu et al. (2019); Bourmaud et al. (2019)
Ramie *Boehmeria nivea*	*Plant* length: 1–2 m *Elementary fibres* length: 120–150 mm diameter: 40–60 μm	• Excellent mechanical properties (45–88 cN/tex) • Elongation at break (3–7%)	• Textile (sweaters) • Technical textile (fishing nets, gasmantle) • Marine packings • Papers • Automotive • Furniture • Construction	Cengiz and Babalik (2009); Sen and Reddy (2011); Mather and Wardman (2011); Angelini and Tavarini (2013); Debeli et al. (2018); Rehman et al. (2019)
Jute *Corchorus capsularis*	*Plant* length: 15–20 cm *Elementary fibres* length: 0.5–6 mm diameter: 26–30 μm	• Moderate strength (30–45 cN/tex) • Low extension at break (1–2%)	• Building materials • Furniture industry • Geotextile • Packing materials • Chipboards	Biswas et al. (2015); Das (2017); Gunti et al. (2018); Shahinur and Hasan (2019); Wang et al. (2019)
Bagasse *Saccharum officinarum*	*Plant* length: 6 cm diameter: 3.8 cm *Elementary fibres* length: 0.8–2.8 mm diameter: 10–34 μm	• Good sorption properties (9.21%) • Compressive strength (23–30%) • Modulus of elasticity (50%)	• Energy production – ceramic materials • Pulp, paper, and board • Human and animal food	Guimarães et al. (2009); Xu et al. (2019); Teixeira et al. (2015); Tahir et al. (2016); Samadi et al. (2016)
Wheat straw *Triticum aestivum*	*Plant* length: 6 cm diameter: 8–34 cm *Elementary fibres* length: 0.4–3.2 mm diameter: 10–34 μm	• Tensile strength (58–146 MPa) • Elastic modulus (2.8–7.9 MPa)	• Animal bedding • Animal feed • Construction materials • Lubricants, adhesives • Biochemicals	Hornsby et al. (1997); Guo et al. (2018); Canizares et al. (2020)

as of small amounts of other materials, such as pectin, proteins, or ash, which are present in various percent, depending on the origin source, and are unevenly distributed within the cell walls (Table 6.2) (Saini et al. 2015; Toushik et al. 2017; Tayyab et al. 2018).

Generally, the plant cell wall consists of middle lamella, the primary cell wall, and the secondary cell wall, which is further divided based on their structure and composition into three separate parts: inner layer (S3), middle layer (S2), and outer layer

TABLE 6.2
Chemical Composition of Various Lignocellulosic Fibres

Lignocellulosic Fibres	Cellulose (%)	Hemicelluloses (%)	Lignin (%)	Ash (%)	Pectin (%)
Hardwood					
Beech	45	33	20	0.2	–
Poplar	49	24	20	1	–
Aspen	52.7	21.7	19.5	0.3	–
Cherry wood	46	29	18	0.5	–
Willow	41.7	16.7	29.3	2.5	–
Softwood					
Pine	46.9	20.3	27.3	0.3	–
Pinus armandii Franch	48.4	17.8	24.1	0.2	–
Spruce	43	29.4	27.6	0.6	–
Japanese cedar	38.6	23.1	33.8	0.3	–
Fir	45	22	30	0.5	–
Agricultural by-products					
Wheat straw	37–41	27–32	13–15	11–14	–
Rice straw	32	24	18	14–20	–
Bagasse	32–48	19–24	23–32	1.5–5	–
Barley straw	31–45	27–38	14–19	2–7	–
Cornstalk	39–47	26–31	3–5	12–16	–
Corn stover	38–40	28	7–21	3.6–7	–
Bamboo	39.8	19.49	20.81	1.21	–
Sorghum straw	32	24	13	12	–
Sugarcane bagasse	20–42	19–25	42–48	–	–
Grasses	25–40	25–50	10–30	–	–
Cotton	80–85	5–20	–	–	–
Algae (green)	20–40	20–50	–	–	–
Fruits					
Apple pomace	40–43.6	19–24.4	15–20.4	–	9–11.7
Orange pulp	25.32	5.35	2.2–3.0	–	15.7–16.3
Pine apple shell	40.55–41.57	28.69–29	10–10.39	–	2.49–2.7
Grape pomace	6–17.75	18.0–31	59–64	–	0.25–4
Cherry pomace	16–20	10.7	69.4	–	1.51
Vegetables					
Olive	31.9–36.4	21.9–26.8	26.0–39.2	–	16.2–17.44
Carrot pomace	51.6	12.3	32.2	–	3.88
Tomato pomace	19.02	12.0	36.0	–	7.55
Potato pulp	17.0–21.7	14.0	2.6	–	2.2
Sugar beet pulp	22–30	24–32	3–4	–	24–32

Note: Data from Saini et al. 2015; Toushik et al. 2017; Tayyab et al. 2018.

(S1). The middle lamella contains the highest concentration of lignin, while the main components of the primary cell wall are cellulose, hemicelluloses, and pectin. The secondary wall has more celluloses than the primary wall, but lacks pectin; it is harder and less extended. This is particularly important for the mechanical reinforcement and water transportation. Cell walls mainly consist of cellulose, hemicelluloses, and lignin in a 4:3:3 ratio. This ratio differs from sources such as hardwood, softwood, and herbs (Chen, 2014).

The highest percent of lignin is contained by softwood fibres, while the fibres from hardwood, agriculture by-products, and fruits and vegetables contain cellulose in the highest percentages.

Within the wood secondary cell wall, the S1 layer has the greatest concentration of lignin in the secondary cell wall, while the S2 layer contains the greatest amount of cellulose and hemicelluloses, and the S3 layer forms the boundary of the lamella and has the least amount of lignin (Daniel 2009; Mathews et al. 2015).

The cell wall of vascular plants contains relatively more cellulose and hemicelluloses than pectin, whereas the cell wall of fruits and vegetables contains more pectin than cellulose and hemicelluloses (Chylińska et al. 2014; Toushik et al. 2017).

The strength to the cell wall of fruits and vegetables are given by cellulose and hemicelluloses, which are covalently cross-linked with lignin, and thus, the removal of lignin can accelerate the hydrolysis rates of cellulose and hemicelluloses (Yuan et al. 2013). These are processed via hydrolysis by different lignocellulolytic enzymes, such as bacteria, archaea, fungi, protists, plants, and invertebrate animals (insects and crustaceans), which can potentially transform lignocellulose into glucose and other soluble sugars (Nagar et al. 2012; Sweeney and Xu 2012; Szymanska-Chargot and Zdunek 2013; Iqbal et al. 2013). These lignocellulolytic enzymes are used in various food industries, such as fruit and vegetable juice producing, vegetable oil processing, winemaking, brewing, and baking industries (Toushik et al. 2017).

The constituent of plant fibres, such as wood fibres and non-wood natural fibres, are cellulose, lignin, hemicelluloses, which are shown schematically in Figure 6.2.

Cellulose is a polydisperse linear homopolymer, consisting of D-glucopyranose units that are linked by β-1,4-glycosidic bonds, which are joined by single oxygen atoms (acetal linkages) between the C-1 of one pyranose ring and the C-4 of the next pyranose ring, as anhydroglucose units (AGU) (Ciolacu and Popa 2010). Each of the AGU units presents one primary hydroxyl group (OH) at C6 and two secondary hydroxyl groups at C2, C3, which are capable to be involved in intra- and inter-molecular hydrogen bonds (Figure 6.2.1).

The intra-molecular hydrogen bonds are responsible for the shift and rigid nature of the cellulose molecules and are established between (i) the OH group from C3 of one AGU unit and the pyranose ring oxygen (O5') of an adjacent unit, and (ii) between OH group from C2 of one AGU and the OH group from C6 of adjacent AGU, and are presented on the both side of the chain. The inter-molecular hydrogen bonds in cellulose are responsible for the sheet-like nature of native cellulose and are established between (i) C6 and the C3 hydroxyl groups of an adjacent 020 plane cellulose molecules, and (ii) C6 and the glycosidic oxygen (O4') of a second neighbouring 020 plane (Ciolacu 2018). The presence of the strong bonds between chains determines the formation of a microcrystalline structure with very high packing regions. Cellulose molecules are arranged in elementary fibrils, which are strings of

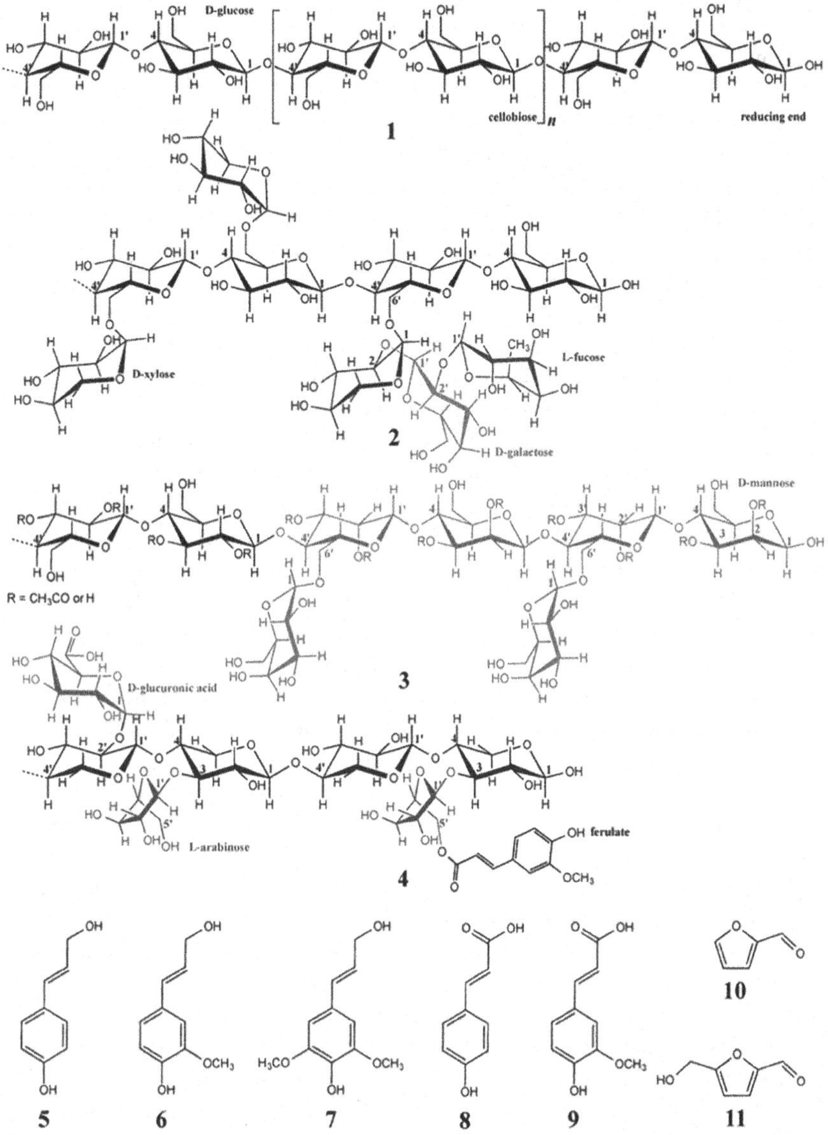

FIGURE 6.2 Structures of the cell wall polymers: cellulose (1), (fucogalacto)xyloglucan (2), galactoglucomannan (3), glucuronoarabinoxylan (4). Sugar residues are colour coded. Also shown are the phenolic compounds *p*-coumaryl alcohol (5), coniferyl alcohol (6), sinapyl alcohol (7), *p*-coumaric acid (8), and ferulic acid (9) and the breakdown products furfural (10) and hydroxymethylfurfural (11) derived from pentose and hexose sugars, respectively. (Reprinted with permission from Ten and Vermerris 2013)

elementary crystallites, and are associated in a more or less random fashion into aggregations. Microfibrils are further associated with each other, forming larger fibrillar (macrofibrils) structures with different diameters. Although the fibrillar structure model is accepted for native cellulose and man-made fibres, there are differences in the structural arrangement between different types of fibres (Krässig 1992; Smole et al. 2013).

The arrangement of the cellulose molecules with respect to each other and to the fibre axis strongly influenced the physical properties of cellulose, as well as their chemical behaviour and reactivity. It is known that the cellulosic materials consist of crystalline and amorphous domains, in varying proportions, depending on their source and history. Most of the reactants penetrate only the amorphous regions, and it is only in these regions with a low level of order and on the surface of the crystallites that the reactions can take place, leaving the intracrystalline regions unaffected (Ciolacu et al. 2011). Interactions between solid cellulosic materials with water, enzymes, or other reactive or adsorptive substances occur first in the noncrystalline domains and/or on the surface of cellulose crystallites (Ciolacu et al. 2012b; Ciolacu et al. 2014). Thus, the secondary and tertiary structures of the noncrystalline domains in cellulose, their properties, and their distribution states should be significant for understanding the behaviour of cellulosic materials under various conditions (Ciolacu et al. 2011).

Hemicelluloses are derived from polysaccharides of plants and include the basic chain containing residues of D-xylose, D-mannose, D-glucose, or D-galactose and other glycosyls as branched chains linked to this basic chain. The category of glycosyl in hemicelluloses varies, such as pyran type, furan type, α-glycoside bond-linked type, β-glycoside bond-linked type, L-configuration type, D-configuration type, etc., and the ways of linkage between glycosyls are various, such as 1–2, 1–3, 1–4, and 1–6 links. Generally, the hemicelluloses are the glucans in the matrix of the cell, and the main components are xylan, xyloglucan, glucomannan, mannan, galactomannan, callose, etc. (Chen 2014). The content and structure of hemicelluloses are different from plant to plant, and the chemical structure of this refers to the composition of the main chain and branched chains of glucose. Thus, the predominant hemicelluloses from (i) monocot angiosperms (grasses) contains mixed-linkage β-glucans, which form as a result of alternating β-1,3 and β-1,4 linkages in the glucan backbone; (ii) dicot angiosperms and non-commelinoid monocots are composed of xyloglucan, which contains a linear backbone of β-1,4-linked D-glucopyranose residues, with substitutions of α-D-xylose on the O-6 position of the glucose (Figure 6.2.2); (iii) softwoods are galactoglucomannans (O-acetyl-galactoglucomannans) (Figure 6.2.3); (iv) the hardwoods contain glucuronoxylans (O-acetyl-4-O-methyl-D-glucuronoxylan) which may contain small amounts of glucomannans (Ten and Vermerris 2013).

Hemicelluloses are imbedded in the plant cell walls to form a complex network of bonds that provide structural strength by linking cellulose fibres into microfibrils and cross-linking with lignin (Isikgor and Becer 2015; Ciolacu 2018).

Lignin is the second most abundant natural polymer after cellulose, a highly branched polymer which contains a diversity of functional groups (aliphatic and phenolic hydroxyls, carboxylic, carbonyl, and methoxyl groups) that vary in function of the morphological location of the lignin and species. This is a complex compound of nonlinearly and randomly linked phenylpropane units (*p*-coumaryl alcohol, coniferyl alcohol, and sinapyl alcohol), which can be divided into the following: syringyl lignin polymerized by syringyl propane, guaiacyl lignin polymerized by guaiacyl propane, and hydroxyphenyl lignin polymerized by hydroxyphenyl propane.

Variation in lignin composition is a function of plant species, age, and tissue type. Monocotyledon (grasses and bulbos plants) contains guaiacyl-syringyl-hydroxyphenyl (GSH) lignin, in proportion of *p*-hydroxyphenyl (H) residues, formed from p-coumaryl alcohol (Figure 6.2.5), guaiacyl (G) residues derived from coniferyl alcohol (Figure 6.2.6), and syringyl (S) residues derived from sinapyl alcohol (Figure 6.2.7). Grasses also contain substantial amounts of cell wall-bound *p*-coumarate (Figure 6.2.8) and ferulate (Figure 6.2.9) (Ralph et al. 2004; Vanholme et al. 2010; Ten and Vermerris 2013). Dicotyledon (hardwood species and most of the herbaceous plants) mainly contains guaiacyl-syringyl (GS) lignin, whereas gymnosperm (softwood species) contains mainly guaiacyl (G) lignin (Lv et al. 2010).

Lignin plays a major role in acting to harden the wall, is responsible for structural rigidity, is a physical barrier that blocks the access to fungi and insects, and facilitates the transport of water through the vascular tissue due to its hydrophobic nature (Ten and Vermerris 2013).

6.3 HYDROGELS BASED ON PLANT FIBRES

During the last decade, hydrogels are considered as a key approach to be used as a soil conditioner in agricultural sector. Hydrogels can retain water, (i) acting as carriers of nutrients in the soil; (ii) improving fertilizers efficiency through reducing the nutrient losses by leaching; (iii) increase the plants life under drought conditions by enhancing the seeds germination, root development, and plant growth; and (iv) decrease the pollution of the environment (Abobatta 2018).

Generally, the hydrogels for agricultural applications are based on synthetic polymers, such as acrylic acid, methacrylic acid, poly(ethylene glycol), vinyl acetate, poly(vinyl alcohol), and various acrylates, due to their better mechanical strength and shelf life or durability (Behera and Mahanwar 2020). The mechanical properties of natural fibres have been found to be inferior to their synthetic counterparts; nevertheless due to the properties such as lower densities, good specific modulus values, better cost per weight, and cost per unit length than synthetic fibres, natural fibres became highly preferable to synthetic fibres, in applications where stiffness and weight are primary concerns (Thakur and Thakur 2014).

In the particular case of the hydrogels with agricultural applications, the problem of mechanical strength is even more pronounced, due to the pressure produced by plants and soil on them, causing the loss of swelling capacity, elasticity, and rigidity (Feng et al. 2010; Guilherme et al. 2015).

In this regard, several approaches were proposed to improve the hydrogel strength, such as (i) increasing the cross-linking density, although this can lead to the gel brittleness and absorbency loss, (ii) increasing the surface cross-linking by using a second cross-linker for a post-treatment on surface, (iii) incorporation of polymers with long macromolecular chains, or (iv) incorporation of inorganic and/or suitable polymers into the hydrogel to prepare nanocomposite structures (Kabiri et al. 2011). One alternative to increase the mechanical properties is the use of natural fibres with excellent characteristics, such as abundance, biodegradability, flexibility, strength, and easy processing (Sannino et al. 2009; Thakur and Thakur 2014; Serna-Cock and Guancha-Chalapud 2017). Thus, the natural fibres, like flax fibres, bagasse, agave, jute, sugarcane, etc. became an optimal alternative to improve the characteristics of a large part of commercial hydrogels, the non-biodegradable synthetic polymer–based hydrogels.

The hydrogels synthesis is carried out by the following methods (Moreau et al. 2016; Che et al. 2016; He et al. 2017; Zain et al. 2018; Behera and Mahanwar 2020):

(i)   Bulk polymerization;
(ii)   Solution polymerization – the solution copolymerization or free radical initiated polymerization of acrylic acid and its salts with acrylamide and a cross-linking agent (N,N'-methylenebisacrylamide, MBA) is the mostly used; the polymerization is initiated by ammonium persulfate (APS) or potassium persulfate (KPS); the advantage of this method over bulk polymerization is the presence of the solvent which helps in controlling viscosity, as well as promotes proper heat transfer;
(iii)   Inverse suspension polymerization;
(iv)   Radiation-induced polymerization.

Generally, the hydrogels synthesis based on natural fibres uses the method of polymerization in solution and several preparation methods of hydrogels for agriculture applications are presented in Table 6.3 (Serna-Cock and Guancha-Chalapud 2017).

Generally, for most of the natural fibre–based hydrogels an increase of the swelling ability and the elastic modulus with the increase of the amount of fibre from hydrogels was recorded.

Thus, Wu et al. synthesized a novel cellulose-poly(acrylic acid-co-acrylamide) superabsorbent composite based on flax yarn waste, with excellent water absorbency and retention capability, with a maximum water absorbency of 875 g g^{-1} distilled water, 490 g g^{-1} natural rainwater, and 90 g g^{-1} 0.9 wt% aqueous NaCl solution (Wu et al. 2012). Feng et al., obtained a superabsorbent hydrogel based on flax shive, with a maximum swelling achieved at a pH of 7.3, the water absorbance in 0.9 % NaCl, KCl, FeCl$_3$ solutions, and urine being 56.47 g g^{-1}, 54.71 g g^{-1}, 9.89 g g^{-1} and 797/21 g g^{-1}, respectively (Feng et al. 2010).

The influence of the swelling degree by the change of pH, the presence of salts (NaCl and CaCl$_2$), and temperature was studied for the hydrogels obtained from sugar cane bagasse. At a pH between 2 and 5, the hydronium ions interact with the hydroxyl groups of cellulose chain, generating a greater presence of hydrogen linking forces, increasing the chain cross-links, and consequently, decreasing the absorption capacity. At a pH of 10–12, the cations interact with the carboxyl group of the polyacrylate and polyacrylamide chain, neutralizing the electrostatic attraction active sites, which causes a decreasing in the swelling ability, at higher saline concentration (Liang et al. 2013). In the pH range of 5–10, the swelling ability of the hydrogels was increased at temperatures between 0 and 50°C (Liang et al. 2013).

The utilization of the kapok fibres for the fabrication of a pH-sensitive superabsorbent hydrogel reveals the greater influence of these fibres on water absorption, gel content, gel strength, and time-dependent swelling behaviours (Shi et al. 2014). The incorporation of 10 wt.% of PKF improved the gel strength to an optimum and the swelling rate constant was enhanced 2.63 folds.

TABLE 6.3

Synthesis of Plant Fibre–based Hydrogels

Type of Natural Fibres	Synthetic Matrices			Method of Polymerization	References
	Monomer/ Polymer	Cross-linker	Initiator		
Natural Fibres					
Flax	• Acrylic acid • Acrylamide	MBA	• Ammonium persulfate	Solution polymerization	Wu et al. (2012)
Flax	• Acrylic acid	MBA	• Potassium persulfate	Microwave-assisted polymerization	Feng et al. (2010)
Flax	• Acrylic acid • Acrylamide	MBA	• Ammonium persulfate	Solution polymerization	Zhang et al. (2013)
Sugar cane bagasse	• Acrylic acid	MBA	• Ammonium persulfate • Sodium sulfite	Solution polymerization	Liang et al. (2013); Huang et al. (2009)
Sugar cane bagasse	• Phosphoric rock • Acrylic acid	MBA	• Potassium persulfate	Solution polymerization	Zhong et al. (2012)
Wheat straw	• Acrylic acid	MBA	• Ammonium persulfate	Solution polymerization	Liang et al. (2009)
Wheat straw	• Acrylic acid • Acrylamide	MBA	• Potassium persulfate • Ceric ammonium nitrate	Solution polymerization	Li et al. (2012)
Kapok	• Acrylic acid	MBA	• Ammonium persulfate	Solution polymerization	Shi et al. (2014)
Cellulose Nanofibres					
Cotton cellulose nanofibres Chitosan	• Acrylic acid	MBA	• Potassium persulfate	Solution polymerization	Spagnol et al. (2012a)
Cotton cellulose nanofibres Cassava starch	• Sodium acrylate	MBA	• Potassium persulfate	Free radical polymerization	Spagnol et al. (2012b)
Cotton cellulose nanofibres	• Acrylic acid • Acrylamide	MBA	• Potassium persulfate	Solution polymerization	Haque and Mondal (2016)
Cellulose nanofibres from kenaf bast	• Acrylic acid	MBA	• Ammonium persulfate	Free radical polymerization	Lim et al. (2015)
Cellulose nanofibres from kenaf bast	• Acrylic acid	MBA	• Ammonium persulfate	Free radical polymerization	Lim et al. (2017)
Cellulose nanofibres from bleached Eucalyptus pulp	• Acrylamide • Acrylic acid	MBA	• Potassium persulfate	Solution polymerization	Mahfoudhi and Bouf (2016)
Cellulose nanofibres from pulp	• Acrylamide	MBS	• Potassium persulfate	Free radical polymerization	Yang et al. (2013)
Cellulose nanofibres from cellulose powder (CF11)	• Acrylamide	MBA	• Sodium persulfate	Solution polymerization	Aouada et al. (2011)
Cellulose nanofibres from microcrystalline cellulose	• Acrylamide	MBA	• Potassium persulfate • Sodium bisulfite	Free radical polymerization	Zhou et al. (2011)

Cellulose nanofibres (CNFs) have achieved a great research attention due to their special properties and have been shown to provide significant reinforcement potential in many polymer matrices.

These may be classified in three main subcategories, as nanofibrillated cellulose (NFC), cellulose nanocrystals (CNC), and bacterial nanocellulose (BNC) and can be extracted from a wide variety of vegetal resources, such as (i) wood fibre (bleached Kraft pulp, bleached sulfite pulp), (ii) plant fibre (cotton, flax, ramie, wheat straw, hemp, etc.), (iii) from marine animals such as tunicate, (iv) from various species of green algae (*Valonia ventricosa, Micrasterias denticulate*) and it can also be synthesized from (v) some bacterial species (*Acetobacter xylinum, Gluconoacetobacter xylinius*) (Ciolacu and Darie 2016). Generally, the main extraction processes in the preparation of cellulose nanoparticles are mechanical treatment and acid hydrolysis. The variation of the lignocellulosic source and the influence of the type and severity of the extraction process bring to CNs differences in the particle size and shape, crystal structure, morphology, crystallinity, and properties. These found diverse applications, including high performance composite materials, packaging, electronics, agriculture, and biomedical applications (Klemm et al. 2011; Nechyporchuk et al. 2016).

As a consequence of the unique properties of the cellulose nanofibres, such as high specific surface areas, high modulus and strength, light-weight, strength, biodegradability, sustainability, the nanofibre-based hydrogels' preparation demonstrated the obtaining of three-dimensional (3D) networks with improved swelling and mechanical properties and elasticity (De France et al. 2017).

The presence of OH groups in the nanofibres facilitates a more precise liquid's diffusion into the hydrogel matrix, more efficient and faster, leading to an increase in the swelling rate (Serna-Cock and Guancha-Chalapud 2017).

Spanol et al. found that the swelling ability of the nanofibre-based hydrogels decreases for concentrations greater than 10 wt% nanofibres (Spagnol et al. 2012a; Spagnol et al. 2012b). These results can be explained by the fact that an excessive increase in the content of CNFs causes their agglomeration in polymer networks and allows an increase in cross-linking, resulting in a decrease in swelling ability (Spagnol et al., 2012b).

Similar results were also obtained by Mahfoudhi and Bouf, who prepared hydrogel nanocomposites based on cellulose nanofibrils (CNF) and poly(acrylic acid-co-acrylamide) cross-linked with bis-acrylamide by radical polymerization in the presence of NFCs. It was demonstrated that the swelling capacity of the hydrogel decreases by 30%, for 10% CNFs loading hydrogel. A huge enhancement in the modulus and compression strength was observed by the addition of NFC with improvement in the strain at break. This strong reinforcing effect was explained by the good dispersion of NFCs within the hydrogel and the grafting of polymer chains on the NFC surface (Mahfoudhi and Bouf 2016). The utilization of an increased content of cellulose nanofibres in hydrogels increases the density of cross-linking points, which improves the resistance to fracture, flexural strength, and compression (Serna-Cock and Guancha-Chalapud 2017). Also, they promote the formation of a porous morphology (Wen et al. 2015), which improves biodegradability, thermal stability, and water retention capacity due to temperature changes in the medium.

Zhou et al. have prepared partially hydrolysed poly(acrylamide)/CNC nanocomposite hydrogels capable of absorbing about 600 times their mass in water and adsorbing more than 90% methylene blue, which can also be used to treat effluents containing toxic metals and other agricultural wastes (Zhou et al. 2014a).

Irrigation and the application of herbicides, insecticides, fungicides, fertilizers, pheromones, and growth regulators only in affected areas of the crop are of great importance for precision agriculture. Moreover, the loss of agrochemicals or other active ingredients by leaching affects their availability for plants, resulting in the need of applying new dosages (Nascimento et al. 2018).

The hydrogels based on natural fibres or cellulose nanofibres are excellent candidates for agriculture because they reduce water consumption in irrigation, irrigation cost, plants and crops death, stabilize fertilizers in the soil, prevent leaching out of active ingredients to the groundwater, and enhance plant growth (Behera and Mahanwar 2020).

6.4 HYDROGELS BASED ON NATURAL POLYMERS

Lignocellulosic biomass, the most common renewable resource, contains as main polymers cellulose, lignin, and hemicelluloses, all of which have great potential of application. Although there has been noticeable progress in cellulose applicability, in case of lignin and hemicelluloses additional investigations are needed to unlock all applicative potential and to increase attractiveness for industry. The widespread use of lignin and hemicelluloses is still limited due to the existence of a underdeveloped marketplace, besides technical and economic aspects related to product development, as well as the existence of competition with petroleum products. In order to achieve high economic efficiency and to minimize market risks, it is necessary to transform them into high-quality products with an increased economic value.

Significant investigations have been focused on the utilization of the polymers from lignocellulosic fibres (cellulose, hemicelluloses, and lignin) for the synthesis of hydrogels with various biomedical applications, such as wound dressing, drug delivery system, and scaffolds.

6.4.1 HYDROGELS BASED ON CELLULOSE

Cellulose-based hydrogels can be obtained *via* either physical or chemical stabilization of aqueous solutions of cellulose products. Additional natural and/or synthetic polymers might be combined with cellulose to obtain composite hydrogels with specific properties. Generally, the preparation of cellulose-based hydrogels consists of the following steps (Ciolacu and Suflet 2018): (i) dissolution of cellulose fibres or powder and (ii) chemical and/or physical cross-linking, in order to obtain a 3D network of hydrophilic polymer chains, which is able to absorb and retain a significant amount of water.

Cellulose is very difficult to be dissolved in common solvents due to its highly extended hydrogen bonded structure, so the major problem for preparing cellulose hydrogel is a lack of appropriate solvents. However, there are a few solvent systems which can dissolve native celluloses, like (i) lithium chloride/dimethylacetamide (LiCl/DMAc), (ii) N-methylmorpholine-N-oxide (NMMO), (iii) some ionic liquids

(ILs), (iv) alkali aqueous system, and (v) alkali/urea (or thiourea) aqueous systems (Chang and Zhang 2011; Shen et al. 2016; Ciolacu and Suflet 2018). Cellulose in solution behaves as random coils, semi-flexible (or semi-rigid) chains, or entangled chains, and the degree of entanglement depends on the polymer.

Physical cross-linking permits the achievement of reversible hydrogels, networks which are held together by physical entanglement of the polymer chains, or other non-covalent inter-actions (hydrogen bonds, van der Waals interactions, and hydrophobic interactions, ionic forces, etc.). This can be realized by two processes (Shen et al. 2016):

- Curing – It consists in the maintaining of the solution in a range of temperatures between 5 and 60°C for a specific period of time; at low temperatures, polymer chains in solution are hydrated and simply entangled with one another, while as temperature increases, macromolecules gradually lose their water , until polymer–polymer hydrophobic associations take place, thus forming the hydrogel network,
- Coagulation – It is achieved in the presence of different anti-solvents such as water, ethanol, methanol, or sulfuric acid/sodium sulfate.

Chemical cross-linking leads to stable and stiff networks with irreversible cross-linking bonds between the cellulosic chains. This can be obtained by using different chemical cross-linkers, such as epichlorohydrin (EPC), citric acid (CA), 1,2,3,4-butanetetracarboxylic dianhydride (BTCA), ethylene glycol diglycidyl ether (EGDE), succinic anhydride (SA), and divinyl sulfone (DVS) (Reddy et al. 2015).

A key step in the formation of any hydrogel is 'gelation', which can be interpreted as the inter-connecting of the macromolecular chains in some manner, such that essentially the whole structure becomes linked together (Xia et al. 2015).

Physical gelation is accompanied by a micro-phase separation and occurs due to cellulose chains self-association because of the preferential cellulose–cellulose and not cellulose–solvent interactions, while chemical gelation (i) perturbs cellulose chains self-association and packing which leads to the decrease of crystallinity, (ii) leads to a more homogeneous morphology which results in transparent swollen coagulated cellulose hydrogels, and (iii) increases swelling in water and adsorption of water vapours due to a more porous structure (Figure 6.3) (Ciolacu et al. 2016).

As a consequence, the difference in the structure and swelling of physical *vs.* chemically cross-linked cellulose influences their release properties. Thus, in chemically cross-linked cellulose, larger amount of drug can be loaded and the release kinetics is faster compared to physically cross-linked matrix. By varying cellulose concentration and the amount of cross-linker it is possible to prepare versatile cellulose hydrogels and dry porous networks with controlled morphology and porosity (Ciolacu et al. 2016).

The utilization of cellulose-based hydrogels as drug delivery systems have been extensively investigated (Shen et al. 2016; De France et al. 2017; Mohammadinejad et al. 2019).

The release of procaine was studied in order to test the physically and chemically cross-linked cellulose as matrices for controlled release applications. It was shown

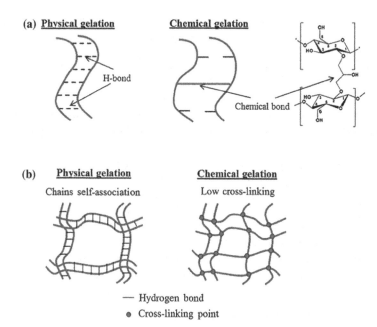

FIGURE 6.3 (a) A sketch of network formation in cellulose solutions: physical gelation via self-association of chains and chemical cross-linking; (b) a schematic presentation of the structures of physical and chemical cellulose gels (Reprinted with permission from Ciolacu et al. 2016).

that the kinetics and the amount of drug released can be tuned by cellulose concentration, type of gelation, and epichlorohydrin (ECH) concentration. It was also shown that in order to understand the kinetics of procaine release, the swelling of cellulose-based hydrogel in water must be taken into account as far as both processes, release and swelling, occur simultaneously (Ciolacu et al. 2016).

The preparation of composite hydrogels is an extremely attractive and advantageous method in order to obtain new and improved polymeric networks (Faroongsarng and Sukonrat 2008; Chang et al. 2008, 2009; Ciolacu and Cazacu 2011; Păduraru et al. 2012; De France et al. 2017; Mohammadinejad et al. 2019).

The introduction of carboxymethyl cellulose (CMC) in cellulose hydrogels increases the pore size, the swelling rate, and consequently the drug release rate significantly (Chang et al. 2010).

The influence of CNC on the properties of cellulose physical gels was evaluated (Wang and Chen 2011). The results revealed that CNC could act as the 'bridge' to facilitate the cross-linking of cellulose chains during gel formation process. Moreover, they can significantly improve the dimensional stability and mechanical strength, and they demonstrate capacity to steadily release bovine serum albumin (BSA) in the simulated body fluid.

Poly(N-isopropylacrylamide) (NIPAAm) – CNC hydrogels were studied and the release of dimethyl methylene blue demonstrated that these systems are promising smart materials for drug delivery (Wang et al. 2013). In addition, these hydrogels showed pH-responsiveness, having a higher water uptake capacity at alkaline pHs than at acidic pHs (Zhou et al. 2014b).

The use of cellulose-based hydrogels for tissue engineering has been intensively studied due to the excellent bio- and cyto-compatibility of these materials.

Biocompatible and non-toxic microspheres were developed by using CNF in the presence of Kymene, as cross-linking agent, which showed no toxicity at cultured 3T3 NIH cells and facilitated the cell attachment, differentiation, and proliferation (Cai et al. 2014).

Nanofibrillar cellulose (CNF) has been radiolabelled with technetium-99m, allowing for the tracking/monitoring of the localization of CNF hydrogels and evaluation of drug delivery in vivo as a function of time, further showing that CNF hydrogels are promising for a variety of tissue engineering applications (Lauren et al. 2014; De France et al. 2017).

Recently, by using 3D printing technologies, it was developed hydrogels based on gelatin methacrylamide and cellulose nanofibres, without cytotoxicity and high cell viability, which contain living cells, human chondrocytes (Markstedt et al. 2015).

6.4.2 HYDROGELS BASED ON HEMICELLULOSES

Hemicelluloses-based hydrogels exhibit different physical properties because of changed structure, composition, and amounts of the hemicelluloses in different biomass, as a function of the source of biomass (Mohammadinejad et al. 2019). Hemicelluloses from various sources such as aspen wood, birch wood spruce, straw, etc. have been used for the development of hydrogels.

Weak hydrogels with improved shear modulus and swelling ratio were produced from hemicelluloses derived from spruce chips, O-acetyl-galactoglucomannan (AcGGM) (Maleki et al. 2017), while hydrogels produced from acetylated galacto-glucomannan-rich wood hydrolysate (WH) showed a high swelling ratio of up to 270 g g^{-1} (Maleki et al. 2014).

Hydrogels prepared from xylan extracted from birch wood and chitosan have been studied, when an increase of water uptake with chitosan content was observed (Gabrielii and Gatenholm 1998). The swelling behaviour was studied at various pH levels and salt concentrations, and the hydrogels responded in a reversible manner to various stimuli. Moreover, highly swellable hydrogels were prepared by using hemicelluloses obtained from aspen wood (*Populus tremula*) and chitosan (up to 20 wt.%) (Gabrielii et al. 2000). Hemicelluloses were characterized by ^{1}H and ^{13}C-NMR spectroscopy and were found to be composed of a linear $(1\rightarrow4)$-β-linked D-xylose main chain with a 4-O-methyl-α-D-glucuronic acid substituting the 2-position of approximately every eighth xylose unit. The cohesive forces of the hydrogels are suggested to be the result of the crystalline arrangement of the polymers and of electrostatic interactions between acidic groups in the hemicelluloses and amino groups of chitosan.

A pH-sensitive and biodegradable hemicelluloses-based hydrogel was prepared by grafting acrylic acid (AAc) into hemicelluloses (HC) by using different ratios of hemicelluloses to monomer (Sun et al. 2013). The swelling degree of the hydrogels increased from 33 to 79%, when the AAc:HC ratio increased from 6:1 to 8:1. The release dynamics of the acetylsalicylic acid-loaded hydrogels closed to zero-order drug release kinetics for 6 h, and the cumulative release rate of 85% was achieved.

A stimuli-responsive hemicelluloses-based hydrogels were developed consisting of the insitu formation of magnetic iron oxide (Fe_3O_4) nanoparticles during the covalent cross-linking of O-acetyl-galactoglucomannan (AcGGM) (Zhao et al. 2015). In addition, the magnetic field-responsive hemicelluloses hydrogels exhibited excellent adsorption and controlled release profiles with BSA as the model drug.

A very recent CMC and xylan-based homopolymerized as well as copolymerized hydrogels were synthesized, using an ethylene glycol diglycidyl ether (EGDGE) cross-linker, in alkaline medium (Kundu and Banerjee 2019). In vitro release of vitamin B_{12} (VB12) in artificial gastric fluid (AGF, pH = 1.2), artificial intestinal fluid (AIF, pH = 6.8), and phosphate-buffered saline (PBS, pH = 7.4) was studied. It was observed that the release of vitamin B_{12} in physiological buffers of the gastrointestinal tract was greater than 90%, for 10h.

6.4.3 HYDROGELS BASED ON LIGNIN

Lignin presents interesting properties such as, biodegradability, antioxidant/antimicrobial activity, high availability as a by-product of industrial activities, higher resistance to most of the biological attacks, which make them the ideal candidate to polymer composites and hydrogels with various applications (Mohammadinejad et al. 2019).

Generally, the preparation of lignin-based hydrogels can be done by using several procedures, but the most common method involves the formation of covalently cross-linked hydrogels (Sathawong et al. 2018). For this purpose, lignin is combined either with (i) synthetic polymers such as, acrylic polymers (AA), poly(vinyl alcohol) (PVA), poly(ethylene glycol) (PEG) derivatives, or with (ii) different biopolymers: cellulose, starch, xanthan, etc. (Thakur and Thakur 2015).

The lignin-based hydrogels have been developed for (i) removal of heavy metal and cationic compounds from aqueous solutions, (ii) controlled release of different substances, such as food ingredients, fertilizers, pesticides, etc., or (iii) for drug release, wound healing, and tissue engineering applications (Peñaranda and Sabino 2010; Thakur and Thakur 2015; Yu et al. 2016; Thakur et al. 2017; Domínguez-Robles et al. 2018; Witzler et al. 2018; Zmejkoski et al. 2018; Chen et al. 2019).

Lignin-based hydrogels containing pine kraft and spruce organosolv lignin were prepared by cross-linking with poly (ethylene glycol) diglycidyl ether (PEGDGE) in alkaline media (Passauer 2012). It was demonstrated that the cross-linking reaction mainly occurred by an etherification reaction between PEGDGE and the phenolic OH groups of lignin.

Feng et al. developed a temperature-sensitive lignin hydrogel by using a graft copoly-merization reaction, in the presence of acetic acid lignin (cross-linker), N-isopropyl acrylamide, NIPAAm (thermoresponsive polymer), methylenebisacrylamide, MBA (cross-linker), and H_2O_2 (radical initiator) (Feng et al. 2011). It was reported that the pore size in the hydrogel structure increased with the increase in the acetic acid lignin content and that the material presented a lower critical solution temperature (LCST) at 31°C; thus, the resulting hydrogels were thermo-responsible.

Chemically cross-linked cellulose–lignin hydrogels, in the presence of epichlorohydrin, were obtained with improved swelling capacities, due to the presence of lignin (Ciolacu et al. 2012a; Ciolacu et al. 2013). The success of cross-linking reaction between lignin and cellulose was proved by FTIR spectroscopy, when a new ether absorption band was highlighted. Moreover, a less dense hydrogel structure was shown, due to increased lignin content, allowing a higher polyphenols release rate.

Larraneta et al. developed hydrogels containing lignin and poly(ethylene glycol) (PEG) by using poly(methyl vinyl ether-co-maleic acid) (PMVE-MA) as cross-linker (Larraneta et al. 2018). The lignin-based hydrogels showed sustained curcumin release profiles and, in addition, a reduction in adherence of *Proteus mirabilis* and *Staphylococcus aureus* relative to poly(vinyl chloride) (PVC), which is the most commonly employed medical material; in fact this proved that these hydrogels are ideal candidates to be used as medical materials.

Alginate/lignin hybrid aerogels were developed as scaffolds for tissue engineering (Quraishi et al. 2015). The results showed that the aerogels were non-cytotoxic and presented enhanced cell adhesion, which makes them proper candidates for an extensive range of applications in tissue engineering/regenerative medicine.

6.5 CONCLUSIONS

During the last years, the sustainable development imposed the capitalization of the raw materials from renewable sources and their utilization in the production of environment-friendly high value-added products. Due to their advantages like biodegradability, biocompatibility, low cost, antibacterial activity, antioxidant properties, the materials synthesized from renewable resources become highly desired for various applications.

In this context, the development of hydrogels from renewable resources is therefore of first interest in modern chemistries including a wide range of applications in various areas, from agriculture, biotechnology to biomedicine, tissue engineering, and many others.

In this chapter, an overview was provided on the latest developments on synthesis, characterization, and potential applications of green hydrogels from bio-based polymers (polymers extracted from biomass) and plant fibres (non-wood natural fibres and wood fibres), from basic research to practical applications.

Although a significant number of researches has been achieved related to the biomaterials from renewable resources, there are still many potentials in the development and application of these materials, and therefore, it can be assumed that there is a need for greater research efforts, with considerable promise for the future.

ACKNOWLEDGEMENTS

This work was supported by a grant of the Romanian Ministry of Research and Innovation, CCCDI – UEFISCDI, project number PN-III-P1-1.2-PCCDI-2017-0697/13PCCDI/2018, within PNCDI III.

LIST OF ABBREVIATIONS

AAc	acrylic acid
APS	ammonium persulfate
AGU	anhydroglucose units
BNC	bacterial nanocellulose
BSA	bovine serum albumin
BTCA	1,2,3,4-butanetetracarboxylic dianhydride
CMC	carboxymethyl cellulose
CNC	cellulose nanocrystals
CNFs	cellulose nanofibres
CA	citric acid
DVS	divinylsulfone
ECM	extracellular matrix
EPC	epichlorohydrin
EGDE	ethylene glycol diglycidyl ether
G	guaiacyl residues
HMF	5-hydroxymethylfurfural
LiCl/DMAc	lithium chloride/dimethylacetamide
LCST	lower critical solution temperature
MBA	N,N′-methylenebisacrylamide
NFC	nanofibrillated cellulose
NIPAAm	N-isopropyl acrylamide
NMMO	N-methylmorpholine-N-oxide
AcGGM	O-acetyl-galactoglucomannan
H	p-hydroxyphenyl units
PEGDGE	poly(ethylene glycol) diglycidyl ether
PEG	poly(ethylene glycol)
PMVE-MA	poly(methyl vinyl ether-co-maleic acid)
NIPAAm	poly(N-isopropylacrylamide)
PVC	poly(vinyl chloride)
PHA	polyhydroxyalkanoates
PLA	polylactic acid
PVA	polyvinyl alcohol
KPS	potassium persulfate
SA	succinic anhydride
S	syringyl units

REFERENCES

Abobatta, W. 2018. Impact of hydrogel polymer in agricultural sector. Advances in agriculture and environmental science. *Adv Agr Environ Sci* 1(2):59–64.

Al-Oqla, F. M., Sapuan, S. M., Anwer, T. et al. 2015. Natural fiber reinforced conductive polymer composites as functional materials: A review. *Synth Met* 206:42–54.

Angelini, L. G., and S. Tavarini. 2013. Ramie [*Boehmeria nivea* (L.) Gaud.] as a potential new fibre crop for the Mediterranean region: Growth, crop yield and fibre quality in a long-term field experiment in Central Italy. *Ind Crops Prod* 51:138–144.

Aouada, F. A., de Moura, M. R., Orts, W. J. et al. 2011. Preparation and characterization of novel micro- and nanocomposite hydrogels containing cellulosic fibrils. *J Agric Food Chem* 59:9433–9442.

Arumugam, V. 2014. A preliminary investigation on Kapok/polypropylene nonwoven composites for sound absorption. *Ind J Fibre Text Res* 37:385–388.

Balaji, V., and K. Senthil Vadivu. 2017. Mechanical characterization of coir fiber and cotton fiber reinforced unsaturated polyester composites for packaging applications mechanical characterization of coir fiber and cotton fiber reinforced. *J Appl Packag Res* 9:12–19.

Bayer, J., Granda, L.A., Méndez, J.A. et al. 2017. Cellulose polymer composites (WPC). In *Advanced high strength natural fibre composites in construction*, ed. M. Fan and F. Fu, chapter 5, 115–139. Woodhead Publishing.

Behera, S., and P. A. Mahanwar. 2020. Superabsorbent polymers in agriculture and other applications: A review. *Polym-Plast Technol Mat* 59:341–356.

Bezerra, T. L., and A. J. Ragauskas. 2016. A review of sugarcane bagasse for second-generation bioethanol and biopower production. *Biofuels Bioprod Bioref* 10:634–647.

Bhatia, R., Gallagher, J. A., Gomez, L. D. et al. 2017. Genetic engineering of grass cell wall polysaccharides for biorefining. *Plant Biotechnol J* 15:1071–1092.

Bhoopathi, R., Ramesh, M., and C. Deepa. 2014. Fabrication and property evaluation of banana-hemp-glass fiber reinforced composites. *Proc Eng* 97:2032–2041.

Biswas, S., Shahinur, S., Hasan, M. et al. 2015. Physical, mechanical and thermal properties of jute and bamboo fiber reinforced unidirectional epoxy composites. *Proc Eng* 105:933–939.

Bogoeva-Gaceva, G., Avella, M., Malinconico, M. et al. 2007. Natural fiber eco-composites. *Polym Composite* 28(1):98–107.

Bourmaud, A., Siniscalco, D., Foucat, L. et al. 2019. Evolution of flax cell wall ultrastructure and mechanical properties during the retting step. *Carbohydr Polym* 206:48–56.

Cai, H. L., Sharma, S., Liu, W. Y. et al. 2014. Aerogel microspheres from natural cellulose nanofibrils and their application as cell culture scaffold. *Biomacromolecules* 15:2540–2547.

Canizares, D., Angers, P., and C. Ratti. 2020. Flax and wheat straw waxes: Material characterization, process development, and industrial applications. *Biomass Convers Bior* 10:555–565.

Cengiz, T. G., and F. C. Babalik. 2009. The effects of ramie blended car seat covers on thermal comfort during road trials. *Int J Ind Ergon* 39:287–294.

Chandel, A. K., Garlapati, V. K., Singh, A. K. et al. 2018. The path forward for lignocellulose biorefineries: Bottlenecks, solutions, and perspective on commercialization. *Bioresour Technol* 264:370–381.

Chang, C., Lue, A., and L. Zhang. 2008. Effects of crosslinking methods on structure and properties of cellulose/PVA hydrogels. *Macromol Chem Phys* 209:1266–1273.

Chang, C., and L. Zhang. 2011. Cellulose-based hydrogels: Present status and application prospects. *Carbohydr Polym* 84:40–53.

Chang, C., Duan, B., and L. Zhang. 2009. Fabrication and characterization of novel macroporous cellulose-alginate hydrogels. *Polymer* 50:5467–5473.

Chang, C., Duan, B., Cai, J. et al. 2010. Super absorbent hydrogels based oncellulose for smart swelling and controllable delivery. *Eur Polym J* 46:92–100.

Che, A. N., Jamari, S. S., and Y. W. S. N. Wan. 2016. Effect of cross-linker concentration on the synthesis and swelling behaviour of superabsorbent polymers (SAP) using graft polymerization techniques. *Key Eng Mater* 719:62–66.

Chen, H. 2014. Chemical composition and structure of natural lignocellulose In *Biotechnology of lignocellulose*, 25–71. Springer, Dordrecht.

Chen, Y., Zheng, K., Niu, L. et al. 2019. Highly mechanical properties nanocomposite hydrogels with biorenewable lignin nanoparticles, *Int J Biol Macromol* 128:414–420.

Cherubini, F. 2010. The biorefinery concept: Using biomass instead of oil for producing energy and chemicals. *Energ Convers Manage* 51(7):1412–1421.

Chylińska, M., Szymánska-Chargot, M., and A. Zdunek. 2014. Imaging of polysaccharides in the tomato cell wall with Raman microspectroscopy. *Plant Methods* 10:14.

Ciolacu D., Doroftei, F., Cazacu G. et al. 2013. Morphological and surface aspects of cellulose-lignin hydrogels. *Cell Chem Technol* 47(5–6):377–386.

Ciolacu, D. 2018. Biochemical modification of lignocellulosic biomass. In *Biomass as renewable raw material to obtain bioproducts of high-tech value*, ed. V. I. Popa, I. Volf, chapter 9, 315–350. Elsevier.

Ciolacu, D., Chiriac, A. I., Pastor, F. I. J. et al. 2014. The influence of supramolecular structure of cellulose allomorphs on the interactions with cellulose-binding domain, CBD3b from *Paenibacillus barcinonensis*. *Bioresour Technol* 157:14–21.

Ciolacu, D., Oprea, A. M., Anghel, N. et al. 2012a. New cellulose–lignin hydrogels and their application in controlled release of polyphenols. *Mater Sci Eng C* 32:452–463.

Ciolacu, D., and M. Cazacu. 2011. Synthesis of new hydrogels based on xanthan and cellulose allomorphs. *Cell Chem Technol* 45(3–4):163–169.

Ciolacu, D., and R. N. Darie. 2016. Nanocomposites based on cellulose, hemicelluloses, and lignin. In *Nanomaterials and nanocomposites: Zero- to three-dimensional materials and their composites*, ed. P. M. Visakh and M. J. M. Morlanes, 391–424. Wiley-VCH Verlag GmbH.

Ciolacu, D., and V. I. Popa. 2010. Cellulose allomporphs – overview and perspectives. In *Cellulose: Structure and properties, derivatives and industrial uses*, ed. A. Lejeune and T. Deprez, chapter 1, 1–38. New York: Nova Science Publishers.

Ciolacu, D., Rudaz, C., Vasilescu, M. et al. 2016. Physically and chemically cross-linked cellulose cryogels: Structure, properties and application for controlled release. *Carbohydr Polym* 151:392–400.

Ciolacu, D., and D. Suflet. 2018. Cellulose-based hydrogels for medical/pharmaceutical applications, In *Biomass as renewable raw material to obtain bioproducts of high-tech value*, ed V. I. Popa and I. Volf, chapter 11, 401–440. Elsevier.

Ciolacu, D., Ciolacu, F., and V. I. Popa. 2011. Amorphous cellulose structure and characterization. *Cell Chem Technol* 45(1–2):13–21.

Ciolacu, D., Pitol-Filho, L., and F. Ciolacu. 2012b. Studies concerning the accessibility of different allomorphic forms of cellulose. *Cellulose* 19:55–68.

Colomban, P., and V. Jauzein. 2018. Silk: Fibers, films, and composites-types, processing, structure, and mechanics. In *Handbook of properties of textile and technical fibres*, ed. A. R. Bunsell, 137–183. Woodhead Publishing.

Daniel, G. 2009. Wood fibre morphology. In: *Wood chemistry and wood biotechnology*. ed. M. Ek. D. Gellerstedt, and G. Henrikson, chapter 3, 45–70, Berlin: De Gruyter and Co.

Das, S. 2017. Mechanical properties of waste paper/jute fabric reinforced polyester resin matrix hybrid composites. *Carbohydr Polym* 172:60–67.

De France, K. J., Hoare, T., and E. D. Cranston. 2017. Review of hydrogels and aerogels containing nanocellulose. *Chem Mater* 29:4609–4631.

De Prez, J., Van Vuure, A. W., Ivens, J. et al. 2018. Enzymatic treatment of flax for use in composites. *Biotechnol Rep* 20:e00294.

Debeli, D. K., Qin, Z., and J. Guo. 2018. Study on the pre-treatment, physical and chemical properties of ramie fibers reinforced poly (lactic acid) (PLA) biocomposite. *J Nat Fibers* 15:596–610.

Domínguez-Robles, J., Peresin, M.S., Tamminen, T. et al. 2018. Lignin-based hydrogels with 'super-swelling' capacities for dye removal. *Int J Biol Macromol* 115:1249–1259.

Dong, T., Xu, G., and F. Wang. 2015. Adsorption and adhesiveness of kapok fiber to different oils. *J Hazard Mater* 296:101–111.

Eichhorn, S. J., Dufresne, A., Aranguren, M. et al. 2010. Review: Current international research into cellulose nanofibers and nanocomposites. *J Mater Sci* 45(1):1–33.

Elmogahzy, Y., and R. Farag. 2018. Tensile properties of cotton fibers: Importance, research, and limitations. In *Handbook of properties of textile and technical fibres*, ed. A. R. Bunsell, chapter 7, 223–273. Woodhead Publishing.

Faroongsarng, D., and P. Sukonrat. 2008. Thermal behavior of water in the selected starch-and cellulose-based polymeric hydrogels. *Int J Pharmaceut* 352:152–158.

Feng, H., Li, J., and L. Wang. 2010. Preparation of biodegradable flax shive cellulose-based superabsorbent polymer under microwave irradiation. *BioResources* 5(3):1484–1495.

Feng, Q., Chen, F., and H. Wu. 2011. Preparation and characterization of atemperature-sensitive lignin-based hydrogel. *BioResources* 6:4942–4952.

Gabrielii, I., and P. Gatenholm. 1998. Preparation and properties of hydrogels based on hemi-cellulose. *J Appl Polym Sci* 69(8):1661–1667.

Gabrielii, I., Gatenholm, P., Glasser, W. G. et al. 2000. Separation, characterization and hydrogel-formation of hemicellulose from aspen wood. *Carbohydr Polym* 43:367–374.

George, J., Sreekala, M. S., and S. Thomas. 2001. A review on interface modification and characterization of natural fiber reinforced plastic composites. *Polym Eng Sci* 41(9):1471–1485.

Girijappa, Y. G. T., Rangappa, S. M., Parameswaranpillai, J. et al. 2019. Natural fibers as sustainable and renewable resource for development of eco-friendly composites: A comprehensive review. *Front Mater* 6: 226.

Gonçalves, A. C., Malico, I., and A. M. O. Sousa. 2018. Solid biomass from forest trees to energy: A review. In *Renewable resources and biorefineries*, ed. E. Jacob-Lopes and L. Q. Zepka, chapter 3, 23–46. IntechOpen.

Guilherme, M. R., Aouada, F. A., Fajardo, A. R. et al. 2015. Superabsorbent hydrogels based on polysaccharides for application in agriculture as soil conditioner and nutrient carrier: A review. *Eur Polym J* 72:365–385.

Guimarães, J. L., Frollini, E., da Silva C. G. et al. 2009. Characterization of banana, sugarcane bagasse and sponge gourd fibers of Brazil. *Ind Crop Prod* 30:407–415.

Gunti, R., Prasad, A. V. R., and A. Gupta. 2018. Mechanical and degradation properties of natural fiber-reinforced PLA composites: Jute, sisal, and elephant grass. *Polym Compos* 39:1125–1136.

Guo, Y., Zhu, W., Gao R. et al. 2018. Research on wheat straw application in the preparation of superplasticizer. *Adv Mater Sci Eng* 2018:Article ID 2092383.

Gupta, M. K., and R. K. Srivastava. 2016. Mechanical properties of hybrid fibers-reinforced polymer composite: A review. *Polym Plast Technol Eng* 55:626–642.

Haque, M. O., and M. I. H. Mondal. 2016. Synthesis and characterization of cellulose-based eco-friendly hydrogels. *Rajshahi Uni J Sci Eng* 44: 45–53.

He, G., Ke, W., Chen X. et al. 2017. Preparation and properties of quaternary ammonium chitosan-g-poly(acrylic acid-co-acrylamide) superabsorbent hydrogels. *React Funct Polym* 111:14–21.

Hornsby, P.R., Hinrichsen, E., and K. Tarverdi. 1997. Preparation and properties of polypropylene composites reinforced with wheat and flax straw fibres. 1. Fibre characterization. *J Mater Sci* 32(2):443–449.

Huang, Z., Liang, X., Hu, H. et al. 2009. Influence of mechanical activation on the graft copolymerization of sugarcane bagasse and acrylic acid. *Polym Degrad Stabil* 94(10):1737–1745.

Inkinen, S., Hakkarainen, M., Albertsson, A. et al. 2011. From lactic acid to poly(lactic acid) (PLA): Characterization and analysis of PLA and its precursors. *Biomacromolecules* 12:523–532.

Iqbal, H. M. N., Kyazze, G., and T. Keshavarz. 2013. Advances in the valorization of lignocel-lulosic materials by biotechnology: An overview. *BioResources* 8:3157–3176.

Isikgor, F. H., and C. R. Becer. 2015. Lignocellulosic biomass: A sustainable platform for the production of bio-based chemicals and polymers. *Polym Chem* 6:4497–4559.

Kabiri, K., Omidian, H., Zohuriaan-Mehr, M. J. et al. 2011. Superabsorbent hydrogel composites and nanocomposites: A review, *Polym Composite* 32(2):277–289.

Klemm, D., Kramer, F., Moritz, S. et al. 2011. Nanocelluloses: A new family of nature-based materials. *Angew Chem Int Ed.* 50:5438–5466.

Kozlowski, R., Baraniecki P., and J. Barriga-Bedoya. 2006. Bast fibres (flax, hemp, jute, ramie, kenaf, abaca). In *Biodegradable and sustainable fibres*, ed. R. S. Blackburn, chapter 2, 36–88, Cambridge: Woodhead Publishing.

Krässig, H. H. 1992. *Cellulose, structure, accessibility and reactivity*. Philadelphia, USA: Gordon and Breach Publishers.

Kumar, A.K., and S. Sharma. 2017. Recent updates on different methods of pretreatment of lignocellulosic feedstocks: A review. *Bioresour Bioprocess* 4:7:1–19.

Kundu, D., and T. Banerjee. 2019. Carboxymethyl cellulose–xylan hydrogel: Synthesis, characterization, and in vitro release of vitamin B12, *ACS Omega* 4:4793–4803.

Larraneta, E., Imizcoz, M., Toh, J. X. et al. 2018. Synthesis and characterization of lignin hydrogels for potential applications as drug eluting antimicrobial coatings for medical materials. *ACS Sustain Chem Eng* 6(7):9037–9046.

Lauren, P., Lou, Y. R., Raki, M. et al. 2014. Technetium-99m-Labeled Nanofibrillar Cellulose Hydrogel for in Vivo Drug Release. *Eur J Pharm Sci* 65:79–88.

Li, Q., Ma, Z., Yue, Q. et al. 2012. Synthesis, characterization and swelling behavior of superabsorbent wheat straw graft copolymers. *Bioresource Technol* 118:204–209.

Liang, X., Huang, Z., Zhang, Y. et al. 2013. Synthesis and properties of novel superabsorbent hydrogels with mechanically activated sugarcane bagasse and acrylic acid. *Polym Bull* 70(6):1781–1794.

Liang, R., Yuan, H., Xi, G. et al. 2009. Synthesis of wheat straw-g-poly(acrylic acid) superabsorbent composites and release of urea from it, *Carbohydr. Polym.* 77: 181–187.

Lim, L. S., Rosli, N. A., Ahmad, I. et al. 2017. Synthesis and swelling behavior of pH-Sensitive semi-IPN superabsorbent hydrogels based on poly(acrylic acid) reinforced with cellulose nanocrystals, *Nanomaterials-Basel* 7(11):399.

Lim, S. L., Ahmad, I., and A. M. Lazim. 2015. pH Sensitive hydrogel based on poly(acrylic acid) and cellulose nanocrystals, *Sains Malays* 44(6):779–785.

Lv, W. J., Xue, C. Y., Cao, C. Y. et al. 2010. Lignin distribution in wood cell wall and its testing methods. *J Beijing Univ* 32(1):136–141.

Mahfoudhi, N., and S. Bouf. 2016. Poly (acrylic acid-co-acrylamide)/cellulose nanofibrils nanocomposite hydrogels: Effects of CNFs content on the hydrogel properties, *Cellulose* 23:3691–3701.

Maleki, L., Edlund, U., and A. C. Albertsson. 2014. Unrefined wood hydrolysates are viable reactants for the reproducible synthesis of highly swellable hydrogels. *Carbohydr Polym* 108:281–290.

Maleki, L., Edlund, U., and A. C. Albertsson. 2017. Synthesis of full interpenetrating hemicellulose hydrogel networks. *Carbohydr Polym* 170:254–263.

Markstedt, K., Mantas, A., Tournier, I. et al. 2015. 3D bioprinting human chondrocytes with nanocellulose–alginate bioink for cartilage tissue engineering applications. *Biomacromolecules* 16:1489–1496.

Mather R.R., and R.H. Wardman. 2011. *The chemistry of textile fibres*. Cambridge: RSC Publishing.

Mathews, S. L., Pawlak, J., and A. M. Grunden. 2015. Bacterial biodegradation and bioconversion of industrial lignocellulosic streams. *Appl Microbiol Biotechnol* 99(7):2939–2954.

Menon, V., and M. Rao. 2012. Trends in bioconversion of lignocellulose: Biofuels, platform chemicals and biorefinery concept. *Prog Energy Combust Sci* 38:522–550.

Mohammadinejad, R., Larrañeta, H. M. E., Fajardo, A. R. et al. 2019. Status and future scope of plant-based green hydrogels in biomedical engineering. *Appl Mater Today* 16:213–246.

Mondal M. I. H., and M. O. Haque. 2019. Cellulosic hydrogels: A greener solution of sustainability. In: *Cellulose-based superabsorbent hydrogels*. ed. M. Mondal, chapter 1, 1–33. Springer International Publishing.

Moreau, D., Chauvet, C., Etienne, F. et al. 2016. Hydrogel films and coatings by swelling-induced gelation. *PNAS*, 113(47):13295–13300.

Mwaikambo, L. Y., Martuscelli, E., and M. Avella. 2000. Kapok/cotton fabric-polypropylene composites. *Polym Test* 19:905–918.

Nagar, S., Mittal, A., and V. K. Gupta. 2012. Enzymatic clarification of fruit juices (apple, pineapple, and tomato) using purified *Bacillus pumilus* SV-85S xylanase. *Biotechnol Bioproc E* 17:1165–1175.

Nampoothiri, K. N., Nair, N. R., and R. P. John. 2010. An overview of the recent developments in polylactide (PLA) research. *Bioresour Technol* 101:8493–8501.

Nascimento, D. M., Nunes, Y. L., Figueirêdo M. C. B. et al. 2018. Nanocellulose nanocomposite hydrogels: Technological and environmental issues, *Green Chem* 20:2428–2448.

Nechyporchuk, O., Belgacem, M. N., and J. Bras. 2016. Production of cellulose nanofibrils: A review of recent advances. *Ind Crop Prod* 93(25):2–25.

Owonubi, J. S., Stephen, A. C., Chioma, A. G. et al. 2019. Fiber-matrix relationship for composites preparation. In *Renewable and sustainable composites*, ed. A. Pereira and F. Fernandes, chapter 2, 1–30. IntechOpen.

Păduraru, O. M., Ciolacu, D., Darie, R. N. et al. 2012. Synthesis and characterization of polyvinyl alcohol/cellulose cryogels and their testing as carriers for a bioactive component. *Mat Sci Eng C* 32(8):2508–2515.

Panaitescu, D. M., Nicolae, C. A., Vuluga, Z. et al. 2016. Influence of hemp fibers with modified surface on polypropylene composites. *J Ind Eng Chem* 37:137–146.

Pandey, J. K., Nagarajan, V., Mohanty, A. K. et al. 2015. Commercial potential and competitiveness of natural fiber composites. In *Biocomposites: Design and mechanical performance*, ed. M. Misra, J. K. Pandey, and A. K. Mohanty, chapter 1, 1–15. Woodhead Publishing.

Passauer, L. 2012. Highly Swellable lignin hydrogels: Novel materials with interesting properties. *Funct Mater Renew Sources* 1107:211–228.

Patel, J. P., and P. H. Parsania. 2018. Characterization, testing, and reinforcing materials of biodegradable composites. In *Biodegradable and biocompatible polymer composites*, ed. N. G. Shimpi, chapter 3, 55–79, Woodhead Publishing.

Peňaranda, A. J. and M. Sabino. 2010. Effect of the presence of lignin or peat in IPN hydrogels on the sorption of heavy metals. *Polym Bull* 65:495–508.

Prachayawarakorn, J., Chaiwatyothin, S., Mueangta, S. et al. 2013. Effect of jute and kapok fibers on properties of thermoplastic cassava starch composites. *Mater Des* 47:309–315.

Puttegowda, M., Rangappa, S. M., Jawaid, M. et al. 2018. Potential of natural/synthetic hybrid composites for aerospace applications. *In Sustainable composites for aerospace applications*, ed. M. Jawaid and M. Thariq, chapter 15, 315–351. Woodhead Publishing

Quraishi, S., Martins, M., Barros, A. A. et al. 2015. Novel non-cytotoxic alginate–lignin hybrid aerogels as scaffolds for tissue engineering. *J Supercrit Fluids* 105:1–8.

Ralph, J., Lundquist, K., Brunow, G. et al. 2004. Lignins: Natural polymers from oxidative coupling of 4-hydroxyphenylpropanoids. *Phytochem Rev* 3:29–60.

Rani, G. U., Mishra, S., Sen, G. et al. 2012. Polyacrylamide grafted Agar: Synthesisand applications of conventional and microwave assisted technique. *Carbohydr Polym* 90(2):784–791.

Rani, P., Mishra, S., and G. Sen. 2013. Microwave based synthesis of polymethylmethacrylate grafted sodium alginate: Its application as flocculant. *Carbohydr Polym* 91(2):686–692.

Reddy, N., Reddy, R., and Q. Jiang. 2015. Crosslinking biopolymers for biomedical applications. *Trends Biotechnol* 33:362–369.

Rehman, M., Gang, D., Liu, Q. et al. 2019. Ramie, a multipurpose crop: Potential applications, constraints and improvement strategies. *Ind Crops Prod* 137:300–307.

Réquilé, S., Le Duigou, A., Bourmaud, A. et al. 2018. Peeling experiments for hemp retting characterization targeting biocomposites. *Ind Crops Prod* 123:573–580.

Rousakis, T. 2017. Natural fibre rebar cementitious composites. In *Advanced high strength natural fibre composites in construction*, ed. M. Fan and F. Fu, chapter 9, 215–234. Woodhead Publishing.

Saini J. K., Saini, R., and L. Tewari. 2015. Lignocellulosic agriculture wastes as biomass feedstocks for second-generation bioethanol production: Concepts and recent developments. *3 Biotech* 5(4):337–353.

Samadi, S., Wajizah, S., Usman, Y. et al. 2016. Improving sugarcane bagasse as animal feed by ammoniation and followed by fermentation with *Trichoderma harzianum* (in vitro study), *Anim Prod* 18(1):14–21.

Sannigrahi, P., Ragauskas, A. J., and G. A. Tuskan. 2010. Poplar as a feedstock for biofuels: A review of compositional characteristics. *Biofuels Bioprod Bioref* 4:209–226.

Sannino, A., Demitri, C., and M. Madaghiele. 2009. Biodegradable cellulose-based hydrogels: Design and applications. *Materials* 2(2):353–373.

Sapuan, S. M., Tamrin, K. F., Nukman, Y. et al. 2017. Natural fiber-reinforced composites: Types, development, manufacturing process, and measurement, In *Comprehensive materials finishing*, ed. M. S. J. Hashmi, chapter 1.8, 203–230. Elsevier.

Sathawong, S., Sridach, W., and K. A. Techato. 2018. Lignin: Isolation and preparing the lignin based hydrogel. *J Environ Chem Eng* 6:5879–5888.

Sen, T., and H. N. J. Reddy. 2011. Various industrial applications of hemp, kinaf, flax and ramie natural fibres. *Int J Innov Manag Technol.* 2(3):192–198.

Sepe, R., Bollino, F., Boccarusso, L. et al. 2018. Influence of chemical treatments on mechanical properties of hemp fiber reinforced composites. *Compos Part B Eng* 133:210–217.

Serna-Cock, L., and M. A. Guancha-Chalapud. 2017. Natural fibers for hydrogels production and their applications in agriculture. *Acta Agron* 66 (4

Shahinur, S., and M. Hasan. 2019. Jute/coir/banana fiber reinforced biocomposites: Critical review of design, fabrication, properties and applications, In *Reference Module in Materials Science and Materials Engineering*, ed. S. Hashmi and I. A. Choudhury, vol. 2, 751–756. Elsevier.

Shen, X., Shamshina, J. L., Berton, P. et al. 2016. Hydrogels based on cellulose and chitin: Fabrication, properties, and applications. *Green Chem* 18:53–75.

Shi, X., Wang, W., Zheng, Y. et al. 2014. Utilization of hollow kapok fiber for the fabrication of a pH-sensitive superabsorbent composite with improved gel strength and swelling properties. *RSC Adv* 4(92):50478–50485.

Smole, M. S., Hribernik, S., Stana Kleinschek, K. et al. 2013. Plant fibres for textile and technical applications, In *Advances in agrophysical research*, ed. S. Grundas, A. Stępniewski, chapter 15, 369–397. IntechOpen.

Spagnol, C., Rodrigues, F. H. A., Neto, A. G. V. C. et al. 2012a. Nanocomposites based on poly(acrylamide-co-acrylate) and cellulose nanowhiskers. *Eur Polym J* 48(3):454–463.

Spagnol, C., Rodrigues, F., Pereira, A. et al. 2012b. Superabsorbent hydrogel nanocomposites based on starch-g-poly (sodium acrylate) matrix filled with cellulose nanowhiskers. *Cellulose* 19:1225–1237.

Sun, X. F., Wang, H. H., Jing, Z. X. et al. 2013. Hemicellulose-based pH-sensitive and biodegradable hydrogel for controlled drug delivery. *Carbohydr Polym* 92:1357–1366.

Sweeney, M. D., and F. Xu. 2012. Biomass converting enzymes as industrial biocatalysts for fuels and chemicals: Recent developments. *Catalysts* 2:244–263.

Szymanska-Chargot, M., and A. Zdunek. 2013. Use of FT-IR spectra and PCA to the bulk characterization of cell wall residues of fruits and vegetables along a fraction process. *Food Biophys* 8:29–42.

Tahir, H., Sultan, M., Akhtar, N. et al. 2016. Application of natural and modified sugar cane bagasse for the removal of dye from aqueous solution. *J Saudi Chem Soc* 20:S115-S121.

Takeda, Y., Tobimatsu, Y., Yamamura, M. et al. 2019. Comparative evaluations of lignocellulose reactivity and usability in transgenic rice plants with altered lignin composition. *J Wood Sci* 65:6.

Tayyab, M., Noman, A., Islam, W. et al. 2018. Bioethanol production from lignocellulosic biomass by environment-friendly pretreatment methods: A review. *Appl Ecol Env Res* 16(1):225–249.

Teixeira, S. R., Arenales, A., de Souza, A. E. et al. 2015. Sugarcane bagasse: Applications for energy production and ceramic materials, *J Solid Waste Technol Manag* 41(3):229–238.

Ten, E., and W. Vermerris. 2013. Functionalized polymers from lignocellulosic biomass: State of the art. *Polymers-Basel* 5:600–642.

Thakur, S., Govender, P. P., Mamo, M. A. et al. 2017. Progress in lignin hydrogels and nanocomposites for water purification: Future perspectives. *Vacuum* 146:342–355.

Thakur, V. K. 2013. *Green composites from natural resources*. CRC Press/Taylor & Francis.

Thakur, V. K., and M. K. Thakur. 2015. Recent advances in green hydrogels from lignin: Are view. *Int J Biol Macromol* 72:834–847.

Thakur, V. K., and A. S. Singha. 2013. *Biomass-based biocomposites*. Smithers Rapra Technology.

Thakur, V. K., and M. K. Thakur. 2014. Processing and characterization of natural cellulose fibers/thermoset polymer composites. *Carbohyd Polym* 109:102–117.

Thakur, V. K., Thakur, M. K., and R. K. Gupta. 2014. Review: Raw natural fiber-based polymer composites. *Int J Polym Anal Charact* 19:256–271.

Toushik, S. H., Lee, K. T., Lee, J. S. et al. 2017. Functional applications of lignocellulolytic enzymes in the fruit and vegetable processing industries. *J FOOD SCI* 82(3):585–593.

Tye, Y. Y., Lee, K. T., Abdullah, W. N. W. et al. 2016. The world availability of non-wood lignocellulosic biomass for the production of cellulosic ethanol and potential pretreatments for the enhancement of enzymatic saccharification. *Renew Sustain Ener Rev* 60:155–172.

Umezawa, T. 2018. Lignin modification in planta for valorization. *Phytochem Rev* 17:1305–1327.

Väisänen, T., Batello, P., Lappalainen, R. et al. 2018. Modification of hemp fibers (Cannabis sativa L.) for composite applications. *Ind Crops Prod* 111:422–429.

Vanholme, R., Demedts, B., Morreel, K. et al. 2010. Lignin biosynthesis and structure. *Plant Physiol* 153:895–905.

Wang, H., Memon, H., Hassan, E. A. M. et al. 2019. Effect of jute fiber modification on mechanical properties of jute fiber composite. *Materials* 12:E1226.

Wang, J., Zhou, X., and H. Xiao. 2013. Structure and properties of cellulose/poly(N-isopropylacrylamide) hydrogels prepared by SIPN strategy. *Carbohydr Polym* 94:749–754.

Wang, Y., and L. Chen. 2011. Impacts of nanowhisker on formation kinetics andproperties of all-cellulose composite gels. *Carbohydr Polym* 83:1937–1946.

Wen, Y., Zhu, X., Gauthier, D. E. and An, X. 2015. Development of poly (acrylic acid)/nanofibrillated cellulose superabsorbent composites by ultraviolet light induced polymerization. *Cellulose* 22(4):2499–2506.

Wiedenhoeft, A. C., and R. B. Miller. 2005. Structure and function of wood. In: *Handbook of wood chemistry and wood composites*, ed. R. M. Rowell, 9–33. CRC Press.

Witzler, M., Alzagameem, A., Bergs, M. et al. 2018. Lignin-derived biomaterials for drug release and tissue engineering. *Molecules* 23:1885.

Wu, F., Zhang, Y., Liu, L. et al. 2012. Synthesis and characterization of a novel cellulose-g-poly(acrylic acid-co-acrylamide) superabsorbent composite based on flax yarn waste. *Carbohyd Polym* 87(4):2519–2525.

Xia, Z., Patchan, M., Maranchi, J. et al. 2015. Structure and relaxation in cellulose hydrogels. *J Appl Polym Sci* 132(24):42071.

Xu, Q., Ji, T., Gao, S. J. et al. 2019. Characteristics and applications of sugar cane bagasse ash waste in cementitious materials. *Materials* 12:39.

Yang, J., Han, C. R., Duan, J. F. et al. 2013. Synthesis and characterization of mechanically flexible and tough cellulose nanocrystals–polyacrylamide nanocomposite hydrogels, *Cellulose* 20:227–237.

Yu, C., Wang, F., Zha, C. et al. 2016. The synthesis and absorption dynamics of a lignin-based hydrogel for remediation of cationic dye-contaminated effluent, *React Funct Polym* 106:137–142.

Yu, H., Wang, X., and M. Petru. 2019. The effect of surface treatment on the creep behavior of flax fiber reinforced composites under hygrothermal aging conditions. *Constr Build Mater* 208:220–227.

Yuan, T. Q., You, T. T., Wang, W. et al. 2013. Synergistic benefits of ionic liquid and alkaline pretreatments of poplar wood. Part 2: Characterization of lignin and hemicelluloses. *Bioresource Technol* 136:345–350.

Yuan, Z., Wen, Y., and G. Li. 2018. Production of bioethanol and value added compounds from wheat straw through combined alkaline/alkaline-peroxide pretreatment. *Bioresour Technol* 259:228–236.

Zain, G., Nada, A. A., El-Sheikh, M. A. et al. 2018. Superabsorbent hydrogel based on sulfonated-starch for improving water and saline absorbency. *Int J Biol Macromol* 115:61–68.

Zhang, Y., Wu, F., Liu, L. et al. 2013. Synthesis and urea sustained-release behavior of an eco-friendly superabsorbent based on flax yarn wastes. *Carbohyd Polym* 91(1): 277–283.

Zhao, W., Odelius, K., Edlund, U., Zhao, C. and Albertsson, A.C. 2015. In situ synthesis of magnetic field-responsive hemicellulose hydrogels for drug delivery, *Biomacromolecules* 16:2522–2528.

Zheng, Y., Wang, J., Zhu, Y. et al. 2015. Research and application of kapok fiber as an absorbing material: A mini review. *J Environ Sci* 27:21–32.

Zhong, K., Zheng, X. L., Mao, X. Y. et al. 2012. Sugarcane bagasse derivative based superabsorbent containing phosphate rock with water-fertilizer integration. *Carbohyd Polym* 90(2):820–826.

Zhou, C., Wu, Q., Yue, Y. et al. 2011. Application of rod-shaped cellulose nanocrystals in polyacrylamide hydrogels. *J Colloid Interf Sci* 353(1):116–123.

Zhou, C., Wu, Q., Lei, T. et al. 2014a. Adsorption kinetic and equilibrium studies for methylene blue dye by partially hydrolyzed polyacrylamide/cellulose nanocrystal nanocomposite hydrogels, *Chem Eng J* 251:17–24.

Zhou, H., Zhu, H., Yang, X. et al. 2014b. Temperature/pH sensitive cellulose-based hydrogel: Synthesis, characterization, loading, and release of model drugs for potential oral drug delivery. *Bioresources* 10:760–771.

Zhu, J., Zhu, H., Njuguna, J. et al. 2013. Recent development of flax fibres and their reinforced composites based on different polymeric matrices. *Materials* 6:5171–5198.

Zmejkoski, D., Spasojević, D., Orlovska, I. et al. 2018. Bacterial cellulose-lignin composite hydrogel as a promising agent in chronic wound healing. *Int J Biol Macromol* 118:494–503.

Zoghlami, A., and G. Paës. 2019. Lignocellulosic biomass: Understanding recalcitrance and predicting hydrolysis. *Front Chem* 7:874.

Zou, Y., Reddy, N., and Y. Yang. 2011. Reusing polyester/cotton blend fabrics for composites. *Compos Part B Eng* 42:763–770.

7 Production of Cellulosic Membranes from Rice Husks for Reverse Osmosis Applications

Ana Carolina de Oliveira, Franciélle Girardi-Alves, and Luizildo Pitol-Filho

CONTENTS

7.1 INTRODUCTION

Economically, rice is viewed as the most important product in many developing countries. During its processing, a significant amount of husks is produced, equivalent to approximately 22 wt.% of the grain. The accumulation of rice husks may generate an environmental problem owing to its low rate of biodegradation (Netto, 2006; Conab, 2020). The main organic components are cellulose, hemicelluloses, and lignin. By applying a chemical treatment, it is possible to obtain cellulose acetate that is used in the production of dense membranes. (Nitzke and Biedrzycki, 2012). In order to improve the properties of membranes, such as selectivity, density, and mechanical resistance, cross-linking agents may be incorporated (Costa Jr and Mansur, 2008; Farrugia, 2013). Cross-linking agents are low weight molecules that present at least two reactive carbonyl groups, as in aldehydes or carboxylic acids that are capable to bond polymeric chains. This phenomenon is known as polymer cross-linking. By using cross-linking agents, it is possible therefore to process agricultural waste into value added products (Costa Jr and Mansur, 2008).

In this chapter, membranes were prepared by acetylation of rice husks, and permeation experiments were performed in order to estimate the salt retention and recommend this material for reverse osmosis applications. Results compared the effect of cross-linking agents on permeation experiments.

7.2 THEORETICAL BACKGROUND

Waste management is considered a worldwide challenge, because of the high costs, variety of treatment, and choice of an appropriate place for disposal. Therefore, projects focusing on the reduction of waste production, recycling strategies, and reuse of products have gained a lot of attention globally in the last decades, because of their impact on the reduction of costs and their environmental improvement (Lorenzett et al., 2012).

The solid waste generated in the rice production is approximately equal to 20 wt.% of the grain. Rice husks do not decompose easily and occupy a significant volume in landfill. To reduce this environmental issue, some rice companies use husks as a complementary energy source, owing to their high calorific content, around 16,720 kJ kg^{-1} (Della et al., 2001; Conab, 2020; Nunes et al., 2017). The volume of ashes is significantly lower than the volume of the raw husks. Rice husk ashes are widely used to stabilize soils and landfills in the production of glasses and ceramics, and also may enter the composition of thermal insulators, bricks, and refractant materials (Bezerra et al., 2011).

However, when unappropriately disposed, rice husks may cause serious environmental problems, since its average absorption takes around five years, and also may be easily spread by the wind, owing to their low specific weight (Lorenzett et al., 2012).

Rice husks are composed of 32 wt.% cellulose, 28 wt.% lignin, 20 wt.% hemicelluloses, and 20 wt.% inorganic matter, approximately. About the inorganic matter, 96 wt.% is silica, and oxides of potassium, magnesium, sodium, and calcium. All the rest is iron, manganese, and aluminium (Johar et al., 2012). Therefore, rice husks also may be used as a substrate in agriculture, being widely available all around the globe (Walter and Rossato, 2010).

On the other hand, chemical industries could process rice husks, providing a noble destination to such a rich product. For example, they could be used to produce dense membranes composed of cellulose acetate (Della et al., 2001). Separation processes based on synthetic membranes (microfiltration/MF, ultrafiltration/UF, nanofiltration/NF, and reverse osmosis/RO) have a high selectivity and demand a reduced amount of energy, if compared to standard distillation or evaporation processes (Habert et al., 2006; Vigneswaran et al., 2012). For example, cellulose acetate membranes were first used in the USA for water desalination in a reverse osmosis unit (Habert et al., 2006). Those membranes became a viable alternative not only on laboratory scale but also for industries. The separation of components crossing a membrane may take place by a difference of a variety of properties, such as pressure, temperature, chemical potential, or even electrical potential (Ribeiro and Síntese, 2013).

7.3 MATERIALS AND METHODS

To verify if the rice husks had the potential to become a raw material for desalination processes, two sets of membranes were produced: one with commercial cellulose acetate (CCA; Sigma–Aldrich, 100 wt.%), and the other through acetylation of rice husks. Dichloromethane (Reatec) was used to dissolve CCA. To synthesize cellulose acetate from the rice husks, sodium hydroxide (Sigma–Aldrich), acetic acid (Alphatec, 99.7 wt.%), sulfuric acid (Cinética, 98 wt.%), and acetic anhydride (Dinamica, 97 wt.%) were used. Rice husks were kindly supplied by 'Tio Urbano' Rice Company, located in Jaraguá do Sul (Santa Catarina State). Cross-linking was performed by using oxalic acid (Aldrich) and an aqueous solution of glutaraldehyde (Exodo Cientifica, 50 wt.%). Purification and acetylation of the rice husks are described below.

7.3.1 RICE HUSKS PURIFICATION

Rice husks were purified as described by Laureano (2014), being weighed and rinsed with distilled water during one hour. After that, the husks were filtered and dried at 45°C during 24h, then submersed in a 1.25 M NaOH solution and then autoclaved during one hour at 120°C and 1.25 atm. Husks were then extensively rinsed with distilled water, to get a neutral pH, and then were dried again during 24h at 45°C.

7.3.2 SYNTHESIS OF CELLULOSE ACETATE

Once the husks were purified, cellulose acetate was synthesized by following the methodology described by Meireles (2007) and Laureano (2014). In one Erlenmeyer, 15ml of acetic acid were added for each 1g of purified rice husks, and the mixture was vigorously mixed at 37.5 rpm during 30 minutes. Then, a solution composed by 0.1ml sulfuric acid and 6.6 ml acetic acid was added. Agitation continued during 15 minutes. The supernatant liquid was removed, and more 15ml acetic anhydride were added, followed by more 30 minutes of mixing. The solution was stored without agitation during 24h, being rinsed with distilled water, to precipitate the cellulose acetate that was filtered and dried at 45°C for 24h. The cellulose acetate was stored in polyethylene recipients.

7.3.3 MEMBRANE PREPARATION

The membranes of CCA were prepared by dissolving the polymer in dichloromethane in a ratio 1:50 (m/v) at vigorous mixing during 30 minutes in sealed Erlenmeyers to avoid the evaporation of the solvent. The solution was then distributed in Petri plates and evaporation took place at room temperature to form the membranes. To obtain the membranes from the rice husks (ARH), the same procedure was followed, by using the polymer obtained through acetylation, as described in the previous topic. In Figure 7.1, images of the membranes of CCA (right, transparent) and of ARH (left, brownish) are shown. Probably the visual differences occurred since the membranes of rice husks had impurities that were not removed by the purification after acetylation.

7.3.4 Infrared Spectroscopy

Attenuated total reflection (ATR), used in conjunction with infrared spectroscopy (FTIR) allowed to analyse the solid materials without further preparation. The principle of ATR lies on the specific infrared absorbance of each functional group in an organic molecule (Souza, 2009). As a reference, the FTIR of cellulose acetate published by Santos et al. (2020), shown in Figure 7.2, was used. Dashed lines represent cellulose samples, and solid lines represent cellulose acetate samples. Infrared spectroscopy was also used to assess if cross-linking effectively occurred, when ARH samples were treated with cross-linking agents.

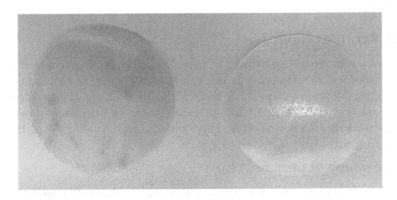

FIGURE 7.1   Samples of ARH membrane (left) and CAC membrane (right).

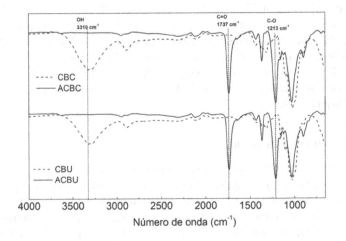

FIGURE 7.2   Reference FTIR spectra of cellulose and cellulose acetate.

7.3.5 HETEROGENEOUS CROSS-LINKING

Heterogeneous cross-linking was performed by using two different cross-linking agents: oxalic acid and glutaraldehyde. Stock solutions of 8.8 wt.% oxalic acid and 50 v/v% glutaraldehyde were prepared and subsequently diluted in the following proportions: 4/5, 3/5, 2/5, 1/5. CCA and ARH membranes were stored in each solution during 30 minutes and then extensively rinsed with distilled water to remove the excess of cross-linking agent, and then were dried at approximately 25 °C.

7.3.6 PERMEATION EXPERIMENTS

To assess the salt retention, permeation experiments were performed, as described by Bueno (2013) and Guollo (2017). A membrane was placed in a syringe filled with a concentrated saline solution (5M) and the syringe was partially immersed on a recipient with distilled water. The electric conductivity of the recipient was monitored each 5 minutes during one hour to determine if salt was permeating the membrane. As an effect of comparison, the conductivities of both distilled water and saline solution were determined before each experiment. All the membranes produced in this work were used to study the salt permeation for each cross-linking agent.

Results were analysed in terms of a normalized conductivity θ, as expressed by Eq. (7.1):

$$\theta = \frac{C(t) - C_{ref}}{C_s - C_{ref}} \tag{7.1}$$

where $C(t)$ is the conductivity measured at each time (μS cm^{-1});C_{ref} is the conductivity of the distilled water (μS cm^{-1});C_s is the conductivity of the saline solution (μS cm^{-1}).

7.4 RESULTS

7.4.1 FTIR RESULTS

Figure 7.3 shows the FTIR spectra for the ARH membranes, indicating the occurrence of the acetylation reaction. The high intensity band in 1747 cm^{-1} indicated C=O stretching present in the cellulose acetate and the band in 1228 cm^{-1} indicated the C–O stretching of the acetyl group. Also, the spectra show no evidence of the cellulosic hydroxylic groups that should appear at 3310 cm^{-1}, suggesting that the hydroxyl groups were substituted by acetyl groups. However, in all the membranes some impurities appear, from 2800 cm^{-1} onwards. Besides, there is evidence of reactions between cellulose acetate and dichloromethane, as bands between 700 and 600 cm^{-1} appear, showing a region of chlorides, as suggested by Silverstein et al. (2010). As reported by Pinto et al. (2013), in the best scenario, the conversion of cellulose into cellulose acetate may reach up to 75%, because factors such as high temperature, high pressure, and alkaline medium could disrupt some cellulosic fibres and limit the acetylation reaction.

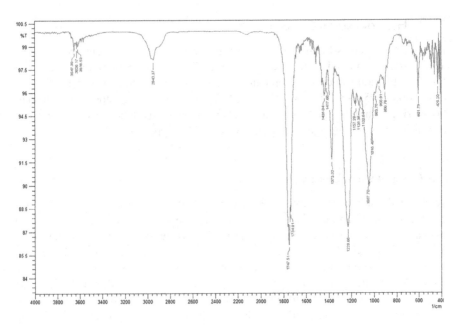

FIGURE 7.3 FTIR spectra of ARH membranes.

7.4.2 PERMEATION EXPERIMENTS

In the permeation experiments, the conductivity was measured each 5 minutes for each sample, treated with different cross-linking agent solutions. Figure 7.4 shows the variation of the normalized conductivity on time for the samples treated with oxalic acid with different concentrations and also for the sample that was not cross-linked (S/E). As it is evident by the graph, the permeation experiments where cross-linked membranes were used showed a reduced electrical conductivity of the distilled water recipient, meaning that the cross-linking reaction effectively improved the salt retention by the membranes. The best results were obtained when membranes were treated with 8.8 wt.% oxalyic acid.

Figure 7.5 shows the analogue experiment, where membranes treated with glutaraldehyde as cross-linking agent were tested, compared to the blank membrane (S/E) that did not undergo cross-linking. However, in this case, glutaraldehyde concentrations lower than 40% (v/v) show a reduction of the salt permeation, when compared to the untreated membrane. Therefore, a minimal concentration of 40% (v/v) is recommended if glutaraldehyde is the chosen cross-linking agent, as also observed by Bolto et al. (2009), who report an optimal cross-linking degree for each agent. Chen et al. (2004) confirmed an improvement of the membrane properties by using cross-linking agents. According to the authors, the size of pores reduces significantly and also interconnectivity is enhanced. However, there is a risk of a phenomenon known as over cross-linking that may lead to a drastic reduction in performance since the pressure drop through the membrane is magnified.

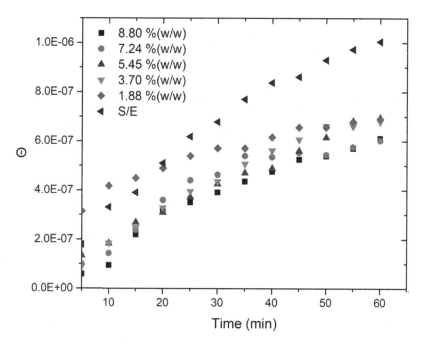

FIGURE 7.4   Permeation experiments of membranes cross-linked with oxalic acid.

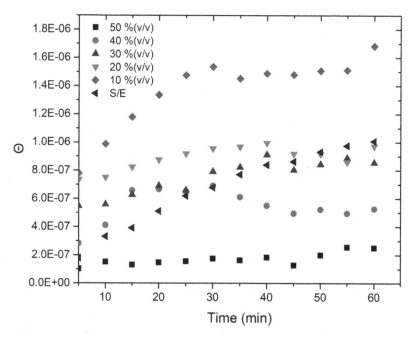

FIGURE 7.5   Permeation experiments of membranes cross-linked with glutaraldehyde.

7.5 CONCLUSIONS

As confirmed by FTIR, rice husks were effectively converted into cellulose acetate, which was precipitated into membranes that could be used in desalination processes. To improve the salt retention, cross-linking agents (oxalic acid and glutaraldehyde) were used to cross-link the membranes. The permeation experiments indicate oxalic acid as the best cross-linking agent studied, with increasing salt retention as the concentration of the oxalic acid increased. Therefore, cross-linked membranes made from rice husks are recommended as promising materials for desalination membrane processes, such as reverse osmosis.

REFERENCES

Bezerra, I. M. T.; Souza, J.; Carvalho, J. P. Q.; Neves, G. Aplicação da cinza da casca do arroz em argamassas de assentamento. *Revista Brasileira de Engenharia Agrícola e Ambiental* v.15, n.6, p.639–645, 2011 Campina Grande, PB, UAEA/UFCG.

Bolto, B; Tran, T; Hoang, M; Xie, Z. Crosslinked poly(vinyl alcohol) membranes. *Progress in Polymer Science*, v. 34, p. 969–981, 2009.

Bueno, M. Z. *Nanofiltração e Osmose Inversa aplicadas a Remoção de Agrotóxicos (Carbamatos) em águas de abastecimento: avaliação em escala de bancada.* Pós-Graduação em Engenharia Ambiental, Universidade Federal de Santa Catarina, Florianópolis, 2013.

Chen, Z; Deng, M; Chen, Y; He, G; Wu, M; Wang, J. Preparation and performance of cellulose acetate/polyethyleneimine blend microfiltration membranes and their applications. *Journal of Membrane Science*, v. 235, p. 73–86. 2004.

CONAB *Companhia Nacional de Abastecimento*, 2020. Available in: www.conab.gov.br

Costa Jr, E. de S; Mansur, H. S. Preparação e caracterização de blendas de quitosana/poli(álcool vinílico) reticuladas quimicamente com glutaraldeído para aplicação em engenharia de tecido. *Química Nova*, v. 31, n. 6, 2008.

Della, V.P.; Kühn, I.; Hotza, D. Characterization of rice husk ash for use as raw material in the manufacture of silica refractory. *Química Nova*, v. 24, n. 6, 2001.

Farrugia, B. Membranas de filtração, tecnologia eficiente em aplicações diversas. *Revista Meio Filtrante*, v. 12, n. 64, Setembro – Outubro, 2013.

Guollo, L. C. *Produção de membranas de acetato de celulose a partir da casca de arroz.* Graduação em Engenharia Química, Universidade do Vale do Itajaí, Itajaí, 2017.

Habert, A. C.; Borges, C. P.; Nobrega, R. *Processos de separação com membranas.* E-papers, Rio de Janeiro, 2006.

Johar, N.; Ahmad, I.; Dufresne, A. Extraction, preparation and characterization of cellulose fibres and nanocrystals from rice husk. *Industrial Crops and Products*,v. 37, p. 93–99, 2012.

Laureano, P. *Obtenção do acetato de celulose a partir da casca de arroz.* 2014. 32p. Trabalho de Conclusão de Curso em Engenharia de Produção – Centro Universitário Católica de Santa Catarina em Jaraguá do Sul.

Lorenzett. D. B.; Neuhaus, M.; Schwab, N. T. S. *Gestão de Resíduos e a Indústria de Beneficiamento de Arroz. Revista Gestão Industrial.* Universidade Tecnológica Federal do Paraná – UTFPR, Campus Ponta Grossa, Paraná, v. 08, n. 01: p. 219–232, 2012.

Meireles, C.S. *Síntese e caracterização de membranas de acetato de celulose, obtido do bagaço da cana –de –açúcar e blendas de acetato de celulose com poliestireno de copos plásticos descartáveis.* 2007. 80 f. Dissertação (Mestrado em Química) – Curso de Química, Instituto de Química, Universidade Federal de Uberlândia, Uberlândia, 2007.

Netto, R. M. *Materiais pozolânicos*. 2006. 148 f. Dissertação (Especialização em Construção Civil) - Universidade Federal de Minas Gerais, Belo Horizonte, 2006.

Nitzke, J. A.; Biedrzycki, A. *Terra de arroz*. 2012. Available in: http://www.ufrgs.br/alimentus1/terradearroz/.

Pinto, B.; Calloni, G.; Silva, S. A. Obtenção de acetato de celulose a partir da casca de arroz (Oryza sativa). *Liberato*, Novo Hamburgo, v. 14, n. 21, p.01–112, jan/jun 2013.

Ribeiro, E. A. M. *Síntese, caracterização e aplicação de membranas de acetato de celulose a partir da reciclagem de palha de milho em processo de ultrafiltração*. Pós Graduação em Química, Universidade Federal de Uberlândia, Minas Gerais – MG, 2013.

Santos, M. S. Amaral, H. R. Cipriano, D. F. Ferreti, J. V. T. Meireles, C. S. Freitas, J. C. C. Santos. Aproveitamento de precursores lignocelulósicos para produção de acetato de celulose. *Química Nova*, v. 15, n. 00, p. 1–7, 2020.

Silverstein, R. M.; Webster, F. X.; Kiemle, D. J. *Identificação Espectrométrica de Compostos Orgânicos*. 7ed. Rio de Janeiro: Ltc, 2010. 490 p.

Souza, R. M. ATR: avanço da espectrofotometria de infravermelho na análise de materiais plásticos. *Instituto de Tecnologia de Alimentos*, v. 21, n. 3, p. 1–3, 2009.

Vigneswaran, S.; Sathananthan, S.; Shon, H. K.; Kandasamy, J.; Visvanathan, C. Delineation of membrane process. p 45–58. In: *Membrane technology and environmental applications*. American Society of Civil Engineers, 2012.

Walter, J. P.; Rossato, M. V. Destino do resíduo casca de arroz na microrregião de Restinga Seca – RS: um enfoque à responsabilidade sócio ambiental. In: *VI CONGRESSO NACIONAL DE EXCELÊNCIA EM GESTÃO, 05 a 07 de Ago. 2010, Niterói, RJ. Proceedings of the conference Niterói, RJ*, 2010.

8 Morphological Aspects of Sustainable Hydrogels

Daniela Rusu, Diana Ciolacu, and Roxana Vlase

CONTENTS

8.1 INTRODUCTION

Three-dimensional (3D) cross-linked networks are capable to absorb a large volume of water and have tailorable swelling–deswelling cycles appropriate to many nature-mimicking purposes, especially the ones related to the medical, pharmacological, and agricultural fields. Hydrogels, the materials which include a huge diversity of the porous, soft 3D systems, can perform a lot of the particular requirements of diverse applications. These are found in a broad range of physical forms, from micro- and nanoparticles to films, sponges, slabs, and beads.

The most intense researches of hydrogels are related to biomedical applications, such as drug and gene delivery, tissue engineering, cell cultures, regenerative medicine, biosensors, and others (Abdelaal and Darwish 2011). This tremendous interest comes from a unique combination of benefits like biocompatibility, controllable functionality and biodegradability, tunable mechanical features, and tissue-like flexibility. Many of them impart 'smart' features based on the sensitivity towards various stimuli. Most of these attractive characteristics are directly or indirectly dependent on the porous architecture of hydrogels, starting with the water sorption mechanism and swelling kinetics and continuing with other above-mentioned features. At the same time, porosity-specific characteristics like stability, homogeneity, pore arrangement, dimensions, interconnectivity, and overall surface area are directly responsible for the final use and applicative potential of the hydrogel. They dictate the loading and release capacity in drug delivery, cell attachment, and tissue formation in tissue engineering, or cell filtration and nutrients and oxygen permeability in wound healing (Ahuja and Pathak 2009).

The porous nature of hydrogels is regarded by many researchers as a crucial feature of these soft materials, starting from design and structure up to the desired performance and final usage. As a consequence, the morphological assessment of hydrogels represents one of the most important steps in their characterization. Visual particulars and overall information obtained at the nano- and micro-scale are paramount in the comprehension of morphological details and their connection to the structure, behaviour, and applicative promises of any material, and hydrogels, in particular. Moreover, morphological analysis is very useful in elucidating the results acquired by other investigative methods, from swelling analysis and mechanical properties up to biological tests and incorporation of active principles.

Scanning electron microscopy (SEM) is the most used method in the morphological appraisal of hydrogels, based on its distinct ability to swiftly deliver reliable, detailed information regarding morphology, porous topology, homogeneity, size, shape, etc. (Breton 1999; Stokes 2003). Its usefulness is incomparable especially when it comes to the porous-directed appearance, integrity, organization, quality, and uniformity of such materials. The obtained data are multiple and allow the construction on several hierarchical levels of a true image of the porous assembly, in a wide dimensional range.

While much of the research on hydrogels is centred on identification of advanced pathways to create novel materials of superior quality, there will always be a primary need to obtain detailed and complex information about the porous architecture, supramolecular organization, and morphology of these materials.

This chapter highlights the important role of scanning electron microscopy in the study of hydrogels morphology, with a focus on biopolymer-based hydrogels (cellulose, hemicellulose and lignin), designed and employed in biomedical applications.

8.2 HYDROGELS

Hydrogels are three-dimensional polymer networks established either by chemical (permanent) or physical (reversible) means, capable of absorbing and releasing large amounts of water or biological fluids (at least 20% of the total weight). They can also be considered colloidal materials (gels) in which the dispersion medium is water or another biological fluid (Buchholz, and Graham 1998; Hoarea and Kohane 2008; Ahmed et al. 2013). The water retention capacity of hydrogels derives from the very large number of hydrophilic functional groups attached to the polymeric skeleton, while the resistance to dissolution is determined by the presence in the network structure of numerous cross-linking points between polymer chains (Singh et al. 2010; Laftaha et al. 2011). The ability to retain water imparts flexibility, durability, and (selective) permeability towards water, metabolites and other small molecules, which unlocks a plethora of practical opportunities. The physically and chemically driven operating principles of hydrogels are depicted in Figure 8.1.

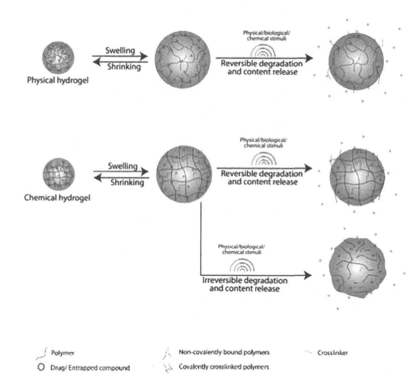

FIGURE 8.1 Schematic diagram representing the operating basics of physical- and chemical-derived hydrogels. (Reproduced from Onaciu et al. 2019).

A very wide range of materials, of both natural and synthetic nature, fall into the above definition. Since the seminal work of Wichterle and Lim in 1960 on hydrogels based on 2-hydroxyethyl methacrylate (Wichterle and Lim 1960), academic and industrial research in this field has known an exponential development due to the innumerable possibilities in terms of design, structure, features, and implementation options (Hoffman 2002; Langer and Peppas 2003; Kashyap et al. 2005). Nevertheless, these possibilities are strictly dependent on structuring methods (Maolin et al. 2000; Yang et al. 2002; Gil and Hudson 2004; Dadsetan et al. 2007; Lee and Tsao 2007; Takashi et al. 2007). While the available literature provides several collections of minimal requirements for the ideal hydrogel for a specific application, most empirical approaches rely on the optimization of production strategies to achieve an appropriate balance between the properties of the final material.

In general, hydrogels provide an unsurpassed set of benefits: (i) accessible preparation, (ii) easily tailorable properties, (iii) similarity with natural tissues (in terms of biological interactions at the molecular level and flexibility), (iv) access to either inert or functional surfaces, (v) high biocompatibility or biodegradability, (vi) tunable and acceptable transport properties (Ulijin et al. 2007).

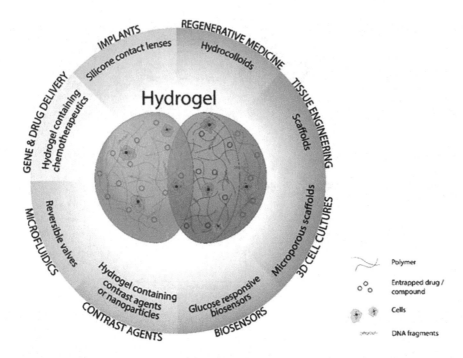

FIGURE 8.2　Major biomedical application areas of hydrogels. (Reproduced from Onaciu et al. 2019).

These set the basis for a high practical relevance in a very broad range of applicative fields, the medical realm being, by far, the one that enjoys the most attention. The literature abounds in success stories related to hydrogels used as (bio)materials in specific biomedical sectors. The most important biomedical application areas of hydrogels are represented in Figure 8.2.

A major part of this success is directly or indirectly connected to the porous architecture of the material; several studies prove that the porous morphology and topology are key elements in the final usage and biomedical potential of the hydrogel-based material (De et al. 2002; Pal et al. 2009; Das et al. 2012; Nokhodchi et al. 2012; Simões et al. 2012; Lee et al. 2013; Sharpe et al. 2014; Caló and Khutoryanskiy 2015; Buwalda et al. 2017).

8.2.1 Hydrogels Based on Sustainable Polymers (Cellulose, Hemicellulose, and Lignin)

Research in the broad area of hydrogels (especially those with biomedical applications) has managed in recent decades to set the course for a new goal: the use of materials from renewable, sustainable raw materials. This research direction is enabled by multifaceted considerations. On one side, there is a set of features of high interest for the research: controlled biodegradability, extraordinary biocompatibility, biomimicry, and overall non-toxicity of bio-based polymers. On the other, there are some traits highly appealing to the profit-driven industry: cost efficiency, large bio-availability, high stability, eco-friendliness (Lin and Metters 2006; Grassi et al. 2009; Miller 2011).

In this respect, numerous investigations have been focused on structure–properties studies of natural polymers, especially plant-based ones (Yang et al. 2015; Ajdary et al. 2018; Nadimi et al. 2018). In biomedical research, the challenge is to discover the perfect plant-based hydrogel that can imitate the physicochemical properties of human tissues in terms of structure, function, and performance.

Cellulose is the most abundant natural macromolecule (the major constituent of plant fibres) and therefore an important performer in the naturally derived hydrogels area. From the chemical point of view, it is a homopolymer with a linear backbone based on β-D-glucopyranose units (up to 1500 per macromolecule) connected through β-1,4-glycosidic bonds (Ciolacu 2018). One key chemical feature is the abundance in hydroxyl functionalities since each β-glucose unit has three hydroxyls in the 2, 3, and 6 positions. These hydroxyl moieties impart multiple functions within the supramolecular structure, like chemically driven modifications or inter- and intra-molecular interactions (via hydrogen bonds) with other cellulose chains and with any absorbed water. They also provide tight molecular packing which results in a partially crystalline structure, and solubility or processability issues.

This carbohydrate is the first choice in biomedical applications due to its biocompatibility, biodegradability, hydrophilicity, availability, and complex physico-mechanical characteristics (Lee et al. 2014). Cellulose-based hydrogels combine the incontestable benefits of cellulose within a tailorable porous structure and thus represent a very actual research theme in the broad domain of material science (Ciolacu et al. 2012; Fu et al. 2019). Moreover, they assure structural integrity to tissue assemble and manage the formation and growth of desired biological support and the diffusion of molecules through the porous architecture (Babensee et al. 1998; Drury and Mooney 2003). As a consequence, the 3D soft networks derived from cellulose are widely used in different configurations in the pharmaceutical industry, offering the possibility to optimize the viscosity of gel formulations, a decisive parameter when it comes to material erosion and bioactive species' diffusion. They are important players in biosensors (Adhikari and Majumdar 2004; Pourjavadi et al. 2007), drug delivery (Zhang et al. 2002; Rodríguez et al. 2003), tissue engineering (Kim et al. 2007), and wound dressing (Mogoşanu and Grumezescu 2014).

Hemicellulose is the second fundamental component of plant cell walls and is a heteropolymer with a mixed chemical blueprint, composed of five- and six-membered saccharide units. They aggregate into a random, amorphous structure of lower mechanical strength, water and enzymatic stability, and a commercial value as a bulk product (Al-Rudainy et al. 2019). The singular form usually refers to the homopolymer mainly containing xylose-derived sugars, while the plural form describes a class of homologues polysaccharides based on the above blueprint. The variation in structure and overall composition is a function of the biomass source.

Hemicellulose-based hydrogels not only impart the high biocompatibility and non-toxicity of cellulose but also display peculiar features (different biodegradability, optical features, and stimuli-responsiveness are the most important), depending on the hemicellulose amount used in formulations (Hu et al. 2018; Liu et al. 2020). They are mostly studied for drug delivery applications due to the possibility to control the swelling behaviour when different functional groups are introduced on to the hemicellulose structure (Gabrielii and Gatenholm 1998; Gabrielii et al. 2000; Sun et al. 2013; Zhao et al. 2014). Tissue engineering, wound healing, and cell encapsulation are other topics of interest for porous networks developed from hemicellulose (Ferreira et al. 2009; Venugopal et al. 2013; Liu et al. 2017b).

Lignin is the second most abundant biopolymer on earth after cellulose, with a key function as structural materials that provide mechanical support and high resistance to microbial attacks (Doherty et al. 2011; Thakur et al. 2013; Oyarce et al. 2019; Vanholme et al. 2019). The chemical structure of this amorphous biopolymer is highly complex and heterogeneous, strongly depending on biosource and isolation procedure (since it is also the main by-product of cellulose and hemicellulose conversions and separations). It lacks a well-defined primary structure and is widely viewed as a long-chain, hyperbranched, phenolic polymer largely composed of phenylpropane-derived moieties mostly connected through ether bridges.

Due to its particular (relatively) hydrophobic nature and richness in aromatic subunits, lignin represents a vital renewable resource with various chemically driven modification opportunities. (Laurichesse and Avérous 2014; Naseem et al. 2016; Tarrés et al. 2017). As a biopolymer, lignin is associated with several advantages like bioavailability, biocompatibility, distinctive mechanical features, and antifungal, antibacterial, and antioxidant activity (Sánchez et al. 2016). There are several studies dealing with lignin-based hydrogels as smart materials for drug delivery (hydrophobic drugs in particular) and tissue engineering (high cytocompatibility and enhanced cell adhesion) applications (Wang et al. 2016; Bian et al. 2018; Sun et al. 2018; Witzler et al. 2018; Zmejkoski et al. 2018; Chen et al. 2019).

8.3 MORPHOLOGICAL ASSESSMENT OF HYDROGELS

While most of the research in the field of hydrogels is focused on the development of novel 3D polymeric networks or advanced methods for generating new porous materials with superior or specific properties, there will be always a primary need to get detailed and complex information related to the characteristics of these 3D porous assemblies, such as their supramolecular organization and morphology.

The design and formulation of hydrogels to attain the envisaged porous architecture, specific features, and overall performance come with a versatile assortment of characterization tools. A typical investigation protocol involves several analytical techniques that follow one of two categories: functional and structural (Onaciu et al. 2019).

The functional route is fundamental in grasping the relationship between the network's structure and any external factors (solute, stimuli) and usually involves studies regarding swelling/deswelling, absorption and diffusion behaviour (rate, in bulk or under load), pH- and thermal-responsiveness, stability (degradability), and other specific techniques required by a given application.

Structural assessment is inevitably based on a collection of microscopy-based tools, with SEM being the golden standard in the field. Other useful alternatives or complementary methods are laser scanning confocal microscopy, scanning probe microscopy, and atomic force microscopy, and they are used in parallel or correlatively.

8.3.1 Scanning Electron Microscopy

SEM is an impressively powerful tool that delivered a major contribution in materials science, biology, and medicine by giving researchers the possibility to observe a specimen at the nanometre scale and to build complex micrographs of a material's surface in a 'pixel by pixel' manner (Nico et al. 2007). Although the data generated by this technique are limited to a certain area, the information obtained is varied and complex and construct a detailed picture of the fine morphology generated by the supramolecular structure.

SEM-based assessment of a sample represents the point-to-point construction of a two- or three-dimensional micrograph. It starts with the use of an electron source to emit a finely focused electron beam (energy between 0.2 and 40 keV) that is directed to the surface of the investigated material with the help of a complex system of electromagnetic lenses (Joy 1991). The electron beam penetrates the sample to a depth that varies between a few tens of nanometres up to about 5 µm, depending on the electrons' energy at surface contact and the density, shape, and composition of the sample.

The specimen is examined in a raster pattern from one side to the other, and the interactions between the focused beam's electrons and the surface atoms generate various signals that carry information concerning the morphology, topography, and composition. The signals are generated by backscattered electrons, secondary electrons, and X-rays. Each of them is identified, captured, and quantified by a specific detector to obtain a virtual image of the analysed surface. The standard detection method is based on the imaging of secondary electrons that are emitted through an inelastic scattering mechanism close to the surface of the analysed specimen. They provide details regarding several high-impact features of the studied material like size, shape, texture, uniformity, and elemental composition (Hayes and Pease 1968; Sant'Anna et al. 2005; Zhou et al. 2006).

Backscattered electrons detection addresses greater depths of the sample and provides information concerning phases, composition, topography, crystallography, or magnetic field. X-ray detection is used to evaluate the composition and measure the abundance of the elements in the sample's surface, thus enabling a highly accurate, qualitative elemental analysis for elements from 5B to 92U (Pease and Nixon 1965).

The obtained micrographs can be only viewed in black and white (tones of grey) because the electrons detected by this method do not have a colour spectrum similar to the incandescent light in optical microscopes. An interesting possibility to colour SEM-derived images is to generate pseudo colored micrographs through specialized image processing software.

The technology behind SEM is constantly being upgraded, thus maintaining its key role in materials science and assuring an important investigation tool for emerging technologies and applications (Minkoff 1967; Johari and Bhattacharyya 1969; Boyde et al. 1986; Hazelton and Gelderblom 2003; De Jonge et al. 2009; Golding et al. 2016).

8.3.2 SCANNING ELECTRON MICROSCOPY IN HYDROGELS' EVALUATION

The morphological investigation of hydrogels is one interesting and spectacular research direction in materials science. The porosity (the average pore size, the pore size distribution, and the pore interconnections) is one essential feature of these 3D porous materials and empowers a broad range of applications, especially in biotechnology and medicine. Freezing and lyophilization of hydrogels are the most common physical methods in generating the characteristic porous network structure, and the evaluation of their achievement or subsequent tailoring involves the constant use of detailed morphological investigations (Zhang and Cooper 2007). The structuring of the porous architecture is a decisive factor in shaping the final properties of the material, from swelling and mechanical properties to transport kinetics and the release of active principles (Ren et al. 2002; Chiriac et al. 2014).

SEM became the go-to technique in the morphological evaluation of hydrogels based on its distinct ability to swiftly deliver reliable, detailed information regarding morphology, porous topology, cross-linking status, homogeneity, size, shape, and others (Liu et al. 2014). Its usefulness is unparalleled especially when it comes to the porous-directed appearance, integrity, organization, quality, and uniformity of such materials. The obtained data are multiple and allow the construction of a true image of the porous assembly on several hierarchical levels, in a wide dimensional range.

The preparation of hydrogel-based samples prior to SEM investigation is relatively straightforward and standardized for each gel types. The swelled hydrogel is dehydrated and freeze-dried (usually in liquid nitrogen or by common freezing) and the surface of the sample is covered with a thin conductive layer by sputtering Au, Pd or combinations therefrom (Shen et al. 2012; Luo et al. 2016). Some samples do not need the additional conductive layer, while other protocols refrain from freeze-drying since it was shown to determine gel shrinkage in some cases and, therefore, an incorrect evaluation of pore size (McMahon et al. 2010; Aston et al. 2016).

One primary role of the SEM evaluation is to respond to a relatively simple question: 'how big is this?' (the term 'this' having a lot of meanings). The quality of this response has been greatly improved in recent years, especially due to the needs of pharmaceutical and biomedical research that frequently rely on SEM investigations. The ability of hydrogels to be structure-modified allows them to be tailored in different shapes and sizes that fulfil the needs of various drug delivery applications (Narayanaswamy and Torchilin 2019). This leads to a wide collection of hydrogel-derived dosage forms with distinct designs and shapes, as a function of drug administration route (Table 8.1). Certainly, SEM is one of the go-to methods in certifying their overall appearance and specific, in-depth attributes.

8.3.3 MORPHOLOGICAL ASPECTS OF HYDROGELS BASED ON SUSTAINABLE POLYMERS

A first direction in using SEM in the morphological evaluation of hydrogels is the assessment of the formulation technique's success. For example, Chang et al. appealed to distinct post-formulation methods to compare the morphology of

TABLE 8.1
Typical Shapes and Dimensions for Main Types of Hydrogel-Derived Products Administered via Different Routes

Administration Route	Shape	Typical Dimension
Peroral	Spherical beads	1 μm to 1mm
	Discs	Diameter of 0.8 cm and thickness of 1 mm
	Nanoparticles	10–1000 nm
Rectal	Suppositories	Conventional adult suppositories dimensions (length ≈ 32 mm) with a central cavity of 7 mm and wall thickness of 1.5 mm
Vaginal	Vaginal tablets	Height of 2.3 cm, width of 1.3 cm and thickness of 0.9 cm
	Torpedo-shaped pessaries	Length of 30 mm and thickness of 10 mm
Ocular	Contact lenses	Conventional dimensions (typical diameter ≈ 12 mm)
	Drops	Hydrogel particles present in the eye drops must be smaller than 10 μm
	Suspensions	N/A
	Ointments	
Transdermal	Circular inserts	Diameter of 2 mm and total weight of 1 mg
	Dressings	Variable
Implants	Discs	Diameter of 14 mm and thickness of 0.8 mm
	Cylinders	Diameter of 3 mm and length of 3.5 cm

Source: (Data from Caló and Khutoryanskiy 2015)

chemically cross-linked, cellulose-based hydrogels, by comparing the effects of heating and freezing treatment, respectively (Chang et al. 2010b). SEM micrographs showed a macroporous inner structure and a decrease in the pore size with an increase in cellulose concentration in the case of the heating post-treatment. However, cellulose-based hydrogels structured *via* the freezing treatment revealed a completely different morphology, with a fibre-like aspect, as it can be observed in Figure 8.3.

This type of structure is most likely determined by the slow, yet powerful self-association of cellulose chains at lower temperature (Chang et al. 2010b). The different morphological assemblies determined by the two methods result in distinct swelling ratios, reswelling water uptakes, light transmittance, and mechanical strength and pave the way for relatively facile tailoring of the networks' porous architecture.

Ciolacu et al. (Ciolacu et al. 2016) obtained highly porous cellulose-based matrices *via* physical and chemical gelation (gelling agent: epichlorohydrin, ECH). It was observed that the content of cellulose, gelation type, and ECH concentration have an important impact over the swelling capacity, the drug release behaviour and the morphology of the cellulose-based hydrogels. The swelling degree of the studied hydrogels increased with the increase of epichlorohydrin concentration, meanwhile a decrease in the density of gels with the increase of ECH concentration was noticed. These results were confirmed by SEM measurements that followed the influence of cross-linking type on hydrogels morphology and porosity, as detailed in in Figure 8.4.

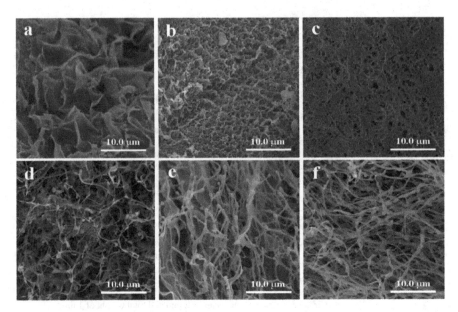

FIGURE 8.3 Morphology of hydrogels prepared by heating (50° C for 20 h; top row) and freezing (-20° C for 20 h; bottom row) methods applied to various cellulose compositions: (a), (d) 2 wt%; (b), (e) 3 wt%; (c), (f) 4 wt. (Reproduced from Chang et al. 2010)

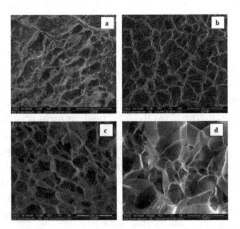

FIGURE 8.4 SEM micrographs of cellulose-based (5% wt), physically (a) and chemically (b-d) cross-linked cryogels by various molar ratios of ECH (b) 1, (c) 2 and (d) 4 (with respect to anhydroglucose units). (Reproduced from Ciolacu et al. 2016)

The surface morphology for physically cross-linked formulations appeared extremely heterogeneous as compared to the chemically derived counterparts. Higher pores' diameter and the disappearance of denser regions were observed with the increase in ECH concentration, together with the formation of a foam-like morphological structure. These structural differences have a direct influence on the release properties of the cellulose-based hydrogels and evidence the possibility to control the morphology of the porous networks to obtain versatile cellulose-based materials (Ciolacu et al. 2016).

The physical cross-linking pathway was also used in the preparation of bio-based matrices with tunable morphology starting from cellulose and various amounts of glycine (Palantöken et al. 2019). Three-dimensional network architectures with evenly distributed pores and higher homogeneity were obtained. This type of material displays a micro-roughness that can prove useful in tissue engineering applications, since cellular adhesion, proliferation, and tissue formation are required for the scaffolds.

Another study employed ECH in the development of new superabsorbent hydrogels based on cellulose and carboxymethyl cellulose (CMC) (Chang et al. 2010a). The cross-section SEM analysis of hydrogels revealed a macroporous structure, as presented in Figure 8.5. Furthermore, it determined an increase in the pore size at higher CMC amounts, resulting in more relaxed architectures, with improved equilibrium swelling ratio and BSA (bovine serum albumin) release behaviour. The micrographs also revealed that cellulose acted as the strong backbone of the hydrogels offering support, while the hydrophilic CMC provided the increased pores and higher swelling capacity.

Niu et al. introduced a new strategy to obtain cellulose-based dual network (DN) hydrogels with improved mechanical strength: cellulose is firstly chemically cross-linked with ECH, and the resulting material is treated with diluted acid to generate porous matrices (Niu et al. 2020). SEM analysis observed that the utilization of diluted acid conducted to a uniform porous structure with increased pore size and distribution also provides enhanced mechanical stability of the 3D networks.

In another study, Kentaro et al. described hydrogels preparation from cellulose nanofibres following an alkaline treatment (Kentaro and Hiroyuki 2012). It was found that morphological differences (specific structures and crystalline formations) in hydrogel formulations strongly depend on the concentration of alkaline solutions (9 wt% NaOH and 15 wt% NaOH) and can be pursued by varying the magnification degree of the investigation technique.

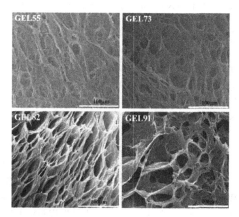

FIGURE 8.5   SEM images (cross-sections) of cellulose-CMC hydrogels (GEL). The numbers represent the CMC:cellulose ratio. (Reproduced from Chang et al. 2010a).

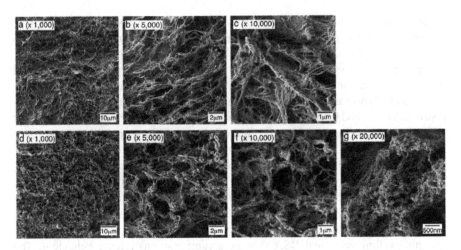

FIGURE 8.6 SEM micrographs of freeze-dried hydrogels prepared by using NaOH solutions of different concentrations: (a–c) 9 wt%, and (d–g) 15 wt%. (Reproduced from Abe and Yano 2011).

As it can be observed in Figure 8.6, at low magnification, both hydrogel formulations revealed a micro- and nano-porous structure (Abe and Yano 2011).

When the morphological analysis was performed at higher magnifications, different morphological features between the two were demonstrated. The hydrogel formed at lower alkaline concentration showed a three-dimensional structure formed by the aggregation of individual cellulose nanofibres (cellulose I form).the formulation obtained at the higher NaOH concentration presented a network with a sponge-like agglomeration (cellulose II structure), due to the longitudinal shrinkage of cellulose nanofibres in response to the increased alkaline environment (Abe and Yano 2011).

Liu et al. investigated smart hydrogels prepared from cellulose acetoacetate and cystamine dihydrochloride (Liu et al. 2017a). The morphological changes of the hydrogels under various physiological conditions (pH = 3.5 and 7.4) and at different times were assessed by SEM and are depicted in Figure 8.7.

SEM showed a pH-dependent, homogeneous, macroporous structure of the cellulose-based hydrogels and proved their ability to facilitate drug loading and release capacity, an important property for several biomedical applications (Liu et al. 2017a).

Hydroxyethyl cellulose (HEC) of various molar masses and acrylic acid (AAc) in different concentrations were used to obtain superabsorbent hydrogel formulations with variable porous architecture, as it can be observed in the micrographs presented in Figure 8.8 (Fekete et al. 2017).

In the case of hydrogels with low AAc concentrations (5% and 10%, respectively), SEM evaluation showed a smooth, porous network with large pores of similar sizes. When the AAc concentration was raised to 20% and 30%, the surface became heterogeneous and of granular structure, the biggest inconsistencies being observed at the highest AAc content (Fekete et al. 2017). The molar mass of HEC also determined some changes in the morphological appearance of the formulations. As expected, an increase in chains length determines a reduction in pore size coming from the higher

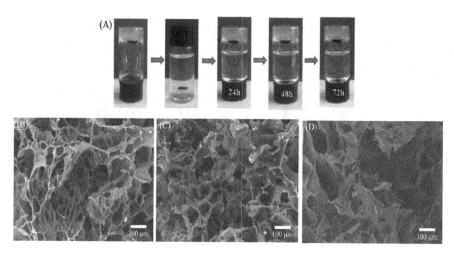

FIGURE 8.7 Stability of cellulose-based multi-responsive hydrogel in PBS (A); SEM images of the hydrogel at: (B) pH 7.4, 0 h; (C) pH 7.4, 72 h; (D) 3.5, 72 h. (Adapted from Liu et al. 2017a).

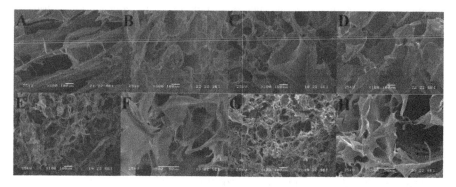

FIGURE 8.8 Morphology of hydrogels based on HEC of different molar mass (A-D: 90 kDa; E-H: 1300 kDa) and various amounts of AAc: (A, G, H) 5%; (B) 10%; (C) 20%; (D) 30%; (E, F) 0%. (Reproduced from Fekete et al. 2017).

cross-linking density that generates a smaller water uptake. The gelation modifications determined by a 5% AAc content leads to an even denser architecture and smaller pores. SEM micrographs connect the high heterogeneity of these formulations (above the one observed at a 30% AAc concentration in the lower molar mass derivative) with the superior viscosity of the higher molar mass cellulose derivative.

Zhao et al. fabricated composite hydrogels with a semi-interpenetrating network (semi-IPN) structure, based on the cross-linking of the aforementioned HEC and a lignosulfonate-graft-poly(acrylic acid) network (Zhao et al. 2017). These formulations presented good mechanical features and a superabsorbent character coming from their highly porous (86% porosity), honeycomb architecture, as evidenced in Figure 8.9.

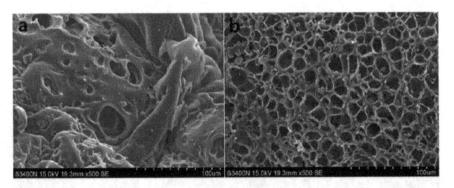

FIGURE 8.9 Morphology of semi-IPNs based on(a): hydroxyethyl cellulose, and (b): hydroxyethyl cellulose-lignosulfonate-*graft*-poly(acrylic acid). (Reproduced from Zhao et al. 2017).

Detailed SEM analysis showed interconnected pores with an average diameter of 11.2 μm that build capillary channels that empower a high accessibility of aqueous solutions towards the amorphous zones of the networks. In the absence of lignosulfonate, a less homogeneous morphology with different non-porous regions is obtained that further impacts the superabsorbent feature (Zhao et al. 2017).

Biocompatible and biodegradable hydrogels with superabsorbent capacity and high equilibrium swelling ratios were successfully obtained by using different mixtures of chitin and CMC (Tang et al. 2014). SEM evaluation confirmed that chitin offered a strong support to the hydrogels pore walls, while CMC contributed to improve the water absorption capacity of hydrogels. The pores' size increased with the CMC amount, turning on to a more open and relaxed structure. The three-dimensional porous arrangement, with well-defined, interconnected micropores is shown in Figure 8.10.

This porous structure facilitates the water permeation into the inner structure of the hydrogels' formulations (Tang et al. 2014).

Nayak and Kundu fabricated porous hydrogel matrices based o CMC and silk sericin (SS) with real potential in tissue engineering applications, especially as wound dressing materials (Nayak and Kundu 2014). As denoted by the SEM images shown in Figure 8.11, the microstructure of the SS-CMC matrices is based on a highly porous surface with interconnected and uniformly distributed pores.

In hydrogels based only on CMC, the mechanical strength and the pore size is observed to be higher than in the SS-CMC analogues that present a decreased pore size and higher pores once the SS ratio is higher in the blend matrices. Regardless of the application, an optimal pore size of the scaffolds is very important for cell attachment, migration, and proliferation. In this case, an 1:1 ratio of the SS:CMC combination determines good performances regarding mechanical properties, swelling capacity, and cell proliferation (Nayak and Kundu 2014).

In another study, CMC-SS hydrogels were obtained by varying CMC's molar mass (Siritientong and Aramwit 2015). The molecular weight influenced the biodegradability and sericin release mechanism from the hydrogels, while the

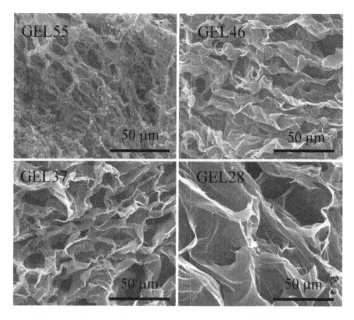

FIGURE 8.10 SEM images (cross-sections) of chitin-carboxymethyl cellulose hydrogels (GEL). The numbers represent the ratio between the two components. (Reproduced from Tang et al. 2014).

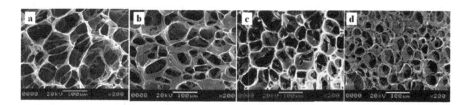

FIGURE 8.11 Morphology of silk sericin-carboxymethyl cellulose blend matrices: (a) pure CMC; (b) 1:2 wt %; (c) 1:1 wt %; (d) 2:1 wt %. (Adapted from Nayak and Kundu 2014).

porosity and the swelling capacity of the CMC matrices were maintained. The micrographs obtained from hydrogels with different CMC chain lengths revealed a porous architecture with the same well-defined, interconnected pores. It was noticed that the CMC's molecular weight did not have an effect over the hydrogels' morphology, the pore diameter varying between 205 and 291 μm (Siritientong and Aramwit 2015).

Zhang et al. successfully combined cellulose and β-cyclodextrin (β-CD) in hydrogels as carriers for the controlled delivery of different bioactive principles (Zhang et al. 2013). The SEM images of the prepared hydrogels displayed in Figure 8.12 showed a homogeneous microporous structure, indicating adequate miscibility between the two components.

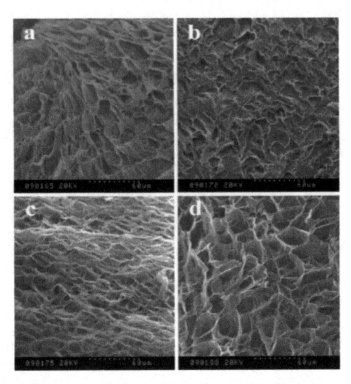

FIGURE 8.12 SEM micrographs of hydrogels prepared from different weight ratios of β-CD:cellulose: (a) pure cellulose; (b) 1:4; (c) 1:2; (d) 1:1. (Reproduced from Zhang, et al. 2013).

Also, a constant average pore diameter of the hydrogels was demonstrated during the increase in β-CD content, while the swelling degree of the hydrogels decreased. The complementary investigation techniques showed that the hydrogels qualify as proper vehicles for the release of hydrophobic drugs (Zhang et al. 2013).

Another study reported interpenetrating polymer network (IPN) hydrogels based on cellulose nanocrystals (CNC) in a sodium alginate-gelatin matrix (Naseri et al. 2016). CNC enhanced the structural integrity and mechanical strength of the hydrogel formulations, as was confirmed by SEM structural morphology studies. The CNC were uniformly distributed within matrices shown in Figure 8.13 and generated a highly porous microstructure based on interconnected and evenly distributed pores with diameters in the range of 16–113 μm.

The rigid hydrogel networks are therefore considered appropriate for cell adhesion and the production of extracellular matrix and can be used in cartilage substitution studies (Naseri et al. 2016).

Ghorbani and Roshangar prepared and investigated a series of hydrogels based on collagen by incorporating different amounts of CNC (Ghorbani and Roshangar 2019). The effects of CNC integration were evidenced from the hydrogels' morphology to swelling ratio and degradation resistance. SEM assessment indicated in all cases porous surfaces, with well-defined and uniform pores, representative micrographs being presented in Figure 8.14.

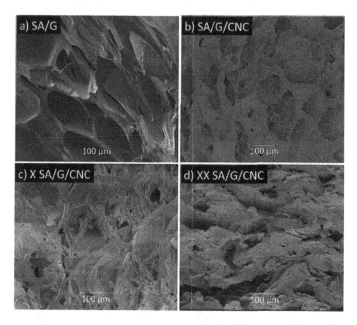

FIGURE 8.13 Morphology of sodium alginate (SA)-gelatin (G) hydrogels incorporating cellulose nanocrystals (CNC) (X: crosslinked with $CaCl_2$; XX: crosslinked with $CaCl_2$ and genipin). (Reproduced from Naseri et al. 2016).

FIGURE 8.14 SEM images of hydrogels prepared from collagen and nanocrystalline cellulose at different ratios (from left to right): 1:0.5; 1:1; 1:2. (Reproduced from Ghorbani and Roshangar 2019).

The formulations based on a 1:0.5 ratio of collagen:CNC proved to be optimal in obtaining highly porous hydrogels with good biodegradability rate and swelling degree (Ghorbani and Roshangar 2019).

The detailed study of the scientific research literature revealed that there are only a few studies dealing with the morphological assessment of hemicellulose-derived hydrogels.

For example, Sun et al. prepared pH-responsive networks by grafting acrylic acid to hemicellulose to obtain new vehicles for controlled drug delivery (Sun et al. 2013).

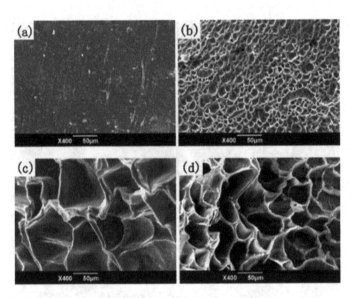

FIGURE 8.15 SEM micrographs of pH-responsive hemicellulose-acrylic acid (1:8) hydrogel after swelling in various buffer solutions: (a) pH 1.5 (vacuum oven drying); (b) pH 1.5; (c) pH 7.4 (d) pH 10.0; (b-d) freeze drying. (Adapted from Sun et al. 2013).

SEM technique was used to evaluate the porous, honeycomb-like structure in which the pore size varied with the pH of the buffer solutions, as it can be observed in Figure 8.15.

The hydrogels swollen at pH 7.4 presented a homogeneous, macroporous structure with higher pore sizes at lower hemicellulose content and cross-linking density, due to the formation of a more relaxed, open architecture. In all cases, SEM images revealed that hemicellulose acted as a rigid backbone of the porous matrix. As in the morphology case, the equilibrium swelling degree and the biodegradability of the networks were considerably influenced by the hemicellulose content and cross-linking density. A decrease in cross-linking led to a smaller number of cross-linking points, thus allowing for a higher amount of water to penetrate the hydrogels (Sun et al. 2013).

A novel hydrogel based on hemicellulose was prepared by using polyethylene glycol (PEG) as a porogen agent for the development of a proper dye adsorbent (Jing et al. 2012). SEM micrographs revealed a highly porous surface morphology with uniformly distributed, interconnected pores. This porosity is a very important characteristic, providing a higher adsorption capacity and a rapid adsorption rate for the matrix. The study confirmed that the methylene blue dye was successfully absorbed by the network.

A synthetic pathway towards hemicellulose-based single-network gels and subsequent IPNs was developed from two different cross-linking mechanisms based on the consecutive synthesis of full IPNs from O-acetyl-galactoglucomannan (AcGGM) and employing a free-radical polymerization or a thiol-ene click reaction (Maleki et al. 2017). A higher swelling degree for the full IPN formulations was confirmed by the SEM micrographs which are shown in Figure 8.16.

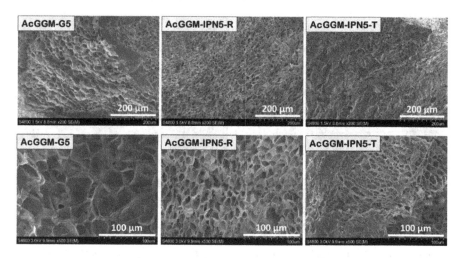

FIGURE 8.16 Morphology of single-network AcGGM hydrogels (G: gel; 5: 50% maleic anhydride (w/w)) and corresponding IPN formulations (R: free-radical cross-linking; T: thiol-ene cross-linking). (Reproduced from Maleki et al. 2017).

The analysis revealed a denser, homogeneous microstructure with uniformly distributed pores that have a significantly smaller diameter as in the case of single-network gel formulation. As a result of the second cross-linking process, the degree of interconnectivity is outstanding, offering the means for proper behaviour as absorbents or membranes (Maleki et al. 2017).

The recent scientific literature is also not so generous in research studies dedicated to lignin as the principal component of hydrogel formulations, and morphology-centred papers are even fewer.

In one such example, Ciolacu et al. prepared cellulose-lignin hydrogels (CL) with superabsorbent characteristics, for the release of polyphenols (Ciolacu et al. 2012). It was found that the lignin content from matrices has a significant influence on the hydrogels' morphology, water retention capacity, and drug release behaviour. SEM micrographs indicated a macroporous structure strongly dependent on the cellulose-lignin ratio, larger pores being obtained while increasing the lignin content in the hydrogel formulations. The large pores induce higher swelling degrees and an increased amount of released polyphenols (Ciolacu et al. 2012).

The same group investigated new hydrogels based on lignin (L) or lignin epoxy-resin (LE) and poly(vinyl alcohol) (PVA), by chemically cross-linking them with epichlorohydrin (Ciolacu and Cazacu 2018).

Interesting morphological aspects of the lignin-based hydrogels were evidenced by SEM, especially for the L-PVA ones. The micrographs revealed a homogeneous porous structure, with well-defined, small-sized pores. By increasing the lignin content within three-dimensional architecture, the formulations become more relaxed, revealing bigger, interconnected pores that are capable to uptake higher amounts of water and determine a superabsorbent character. Based on these morphological differences, L-PVA hydrogels showed higher swelling degrees as compared to the LE-PVA formulations (Ciolacu and Cazacu 2018).

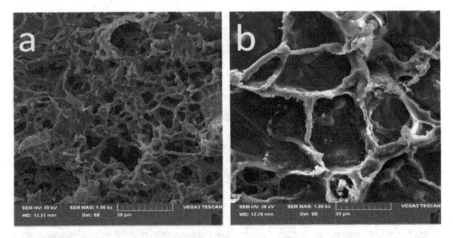

FIGURE 8.17 Morphology of the superabsorbent lignin hydrogel: (a) before Cu (II) adsorption; (b) Cu (II) after adsorption. (Reproduced from Sun et al. 2019).

In another study, lignin-based hydrogels were synthesized and labelled as good candidates for water purification applications (Domínguez-Roblesa et al. 2018). Different types of bio-based lignin were cross-linked with poly (methyl vinyl ether co-maleic acid) (PMVE/MA) *via* a simple, green thermal process. SEM investigations showed a direct connection between the pore size distribution of the highly porous surface and the lignin structure or the base employed during preparation. SEM images confirmed the 'super-swelling' capacity disclosed by swelling capacity measurements: three-dimensional architectures with some quite large pores that are able to incorporate high amount of water. Adsorption and release mechanism of methylene blue confirmed the potential of these eco-friendly hydrogels as organic dyes adsorbents (Domínguez-Roblesa et al. 2018).

The superabsorbent feature of lignin-based hydrogels has also been explored in the removal of Cu (II) ions (Sun et al. 2019). Eco-friendly montmorillonite (MMT) was introduced in lignin-based matrices to improve the adsorption capacity and mechanical properties of resulting hydrogels. SEM was used to observe the composites' morphology before and after the adsorption of Cu (II) ions, two typical images being depicted in Figure 8.17.

It revealed a microporous architecture before adsorption, with interconnected and unevenly distributed pores that completely changed after Cu adsorption: the fine porous structure disappeared and the pore walls became thicker. The superabsorbent feature was confirmed by swelling degree studies, varying pH, contact time and initial ions concentration (Sun et al. 2019).

8.4 CONCLUSIONS

Some of the appealing traits of hydrogels are directly or indirectly connected to their porous architecture, starting with the water sorption mechanism and swelling kinetics and continuing with biocompatibility, biodegradability, and the mechanical features. Porosity-specific characteristics like stability, homogeneity, pore arrangement, dimensions, interconnectivity, and overall surface area are straightly responsible for the final usage and applicative potential of the hydrogel-based material.

Therefore, it is natural that the porous nature of a hydrogel is regarded by many researchers as a crucial feature of these soft materials, starting from design and structure up to the desired performance and final usage. As a consequence, the morphological assessment of hydrogels represents one of the most important steps in their evaluation.

Scanning electron microscopy is the most used method in the morphological appraisal of hydrogels, based on its distinct ability to swiftly deliver reliable, detailed information regarding morphology, porous topology, homogeneity, size, shape, etc. on a multidimensional scale.

Visual particulars and overall information attainable at the nano- and micro-scale are paramount in the comprehension of morphological details and their connection to the structure, behaviour, and applicative promises. Moreover, morphological analysis is very useful in elucidating the results acquired by other investigative methods, from swelling analysis and mechanical properties up to biological tests and incorporation of active principles.

LIST OF ABBREVIATIONS

AAc	acrylic acid
CMC	carboxymethyl cellulose
C	Cellulose
CNC	cellulose nanocrystals
β-CD	β-cyclodextrin
DN	dual network
ECH	Epichlorohydrin
HEC	hydroxyethyl cellulose
IPN	interpenetrating polymer network
L	Lignin
LE	lignin epoxy-resin
MMT	Montmorillonite
AcGGM	O-acetyl-galactoglucomannan
PMVE/MA	poly (methyl vinyl ether co-maleic acid)
PVA	poly(vinyl alcohol)
PEG	polyethylene glycol
SEM	scanning electron microscopy
semi-IPN	semi-interpenetrating network
SS	silk sericin
3D	three-dimensional

REFERENCES

Abdelaal, O. A., and S. M. Darwish. 2011. Fabrication of tissue engineering scaffolds using rapid prototyping techniques. *World Acad Sci Eng Technol Int J Mec Aerosp Ind Mechatron Manuf Eng* 5:2317–2325.

Abe, K., and H. Yano. 2011. Formation of hydrogels from cellulose nanofibers. *Carbohydr Polym* 85:733–737.

Adhikari, B., and S. Majumdar. 2004. Polymers in sensor applications. *Prog Polym Sci* 29:699–766.

Ahmed, E. M., Aggor, F. S., Awad, A. M., and A. T. El-Aref. 2013. An innovative method for preparation of nanometal hydroxide superabsorbent hydrogel. *Carbohydr Polym* 91:693–698.

Ahuja, G., and K. Pathak. 2009. Porous carriers for controlled/modulated drug delivery. *Indian J Pharm Sci* 71:599–607.

Ajdary, M., Moosavi, M., Rahmati, M. et al. 2018. Health concerns of various nanoparticles: A review of their in vitro and in vivo toxicity. *Nanomaterials* 8:634

Al-Rudainy, B., Galbe, M., Arcos Hernandez, M., Jannasch, P., and O. Wallberg. 2019. Impact of lignin content on the properties of hemicellulose hydrogels. *Polymers* 11:35

Aston, R., Sewell, K., Klein, T., Lawrie, G., and L. Grøndahl. 2016. Evaluation of the impact of freezing preparation techniques on the characterisation of alginate hydrogels by cryo-SEM. *Eur Polym J* 82:1–15.

Babensee, J. E., Anderson, J. M., McIntire, L. V., and A. G. Mikos. 1998. Host response to tissue engineered devices. *Adv Drug Delivery Rev* 33:111–139.

Bian, H., Wei, L., Lin, C., Ma, Q., Dai, H., and J. Zhu. 2018. Lignin-containing cellulose nano-fibril-reinforced polyvinyl alcohol hydrogels. *ACS Sustain Chem Eng* 6:4821–4828.

Boyde, A., Maconnachie, E., Reid, S. A., Delling, G., and G. R. Mundy. 1986. Scanning electron microscopy in bone pathology: Review of methods, potential and applications. *Scanning Electron Microsc* 4:1537–1554.

Breton, P. J. 1999. From microns to nanometres: Early landmarks in the science of scanning electron microscope imaging. *Scanning Microsc* 13:1–6.

Buchholz, F. L., and A. T. Graham. 1998. *Modern superabsorbent polymer technology*. New York, Wiley-VCH

Buwalda, S. J., Vermonden, T., and W. E. Hennink. 2017. Hydrogels for therapeutic delivery: Current developments and future directions. *Biomacromolecules* 18:316–330.

Caló, E., and V. V. Khutoryanskiy. 2015. Biomedical applications of hydrogels: A review of patents and commercial products. *Eur Polym J* 65:252–267.

Chang, C., Duan, B., Cai, J., and L. Zhang. 2010a. Superabsorbent hydrogels based on cellulose for smart swelling and controllable delivery. *Eur Polym J* 46:92–100.

Chang, C., Zhang, L., Zhou, J., Zhang, L., and J. F. Kennedy. 2010b. Structure and properties of hydrogels prepared from cellulose in NaOH/urea aqueous solutions. *Carbohydr Polym* 82:122–127.

Chen, Y., Zheng, K., Niu L. et al. 2019. Highly mechanical properties nanocomposite hydrogels with biorenewable lignin nanoparticles. *Int J Biol Macromol* 128:414–420.

Chiriac, A. P., Nistor, M. T., and L. E. Niță. 2014. An investigation on multi-layered hydrogels based on poly(n, n-dimethylacrylamide – co – 3, 9-divinyl-2, 4, 8, 10-tetraoxaspiro (5.5) undecane). *Rev Roum Chim* 59:1059–1068.

Ciolacu, D. and Cazacu, G. 2018. New green hydrogels based on lignin. *J Nanosci Nanotechnol* 18:2811–2822.

Ciolacu, D., Oprea, A. M., Anghel, N., Cazacu, G., and M. Cazacu. 2012. New cellulose–lignin hydrogels and their application in controlled release of polyphenols. *Mater Sci Eng C* 32:452–463.

Ciolacu, D., Rudaz, C., Vasilescu, M., and T. Budtova. 2016. Physically and chemically cross-linked cellulose cryogels: Structure, properties and application for controlled release. *Carbohydr Polym* 151:392–400.

Ciolacu, D. E. 2018. Structure-property relationships in cellulose-based hydrogels. In: *Cellulose-based superabsorbent hydrogels*, ed. Md. I. H. Mondal, chapter 3, 65-95. Springer International Publishing.

Dadsetan, M., Szatkowski, J. P., Yaszemski, M. J., and L. Lu. 2007. Characterization of photo-cross-linked oligo[poly(ethylene glycol) fumarate] hydrogels for cartilage tissue engineering. *Biomacromolecules* 8:1702– 1709.

Das, N., Bera, T., and A. Mukherjee. 2012. Biomaterial hydrogels for different biomedical applications. *Int J Pharm Bio Sci* 3:586–595.

De Jonge, N., Peckys, D. B., Kremers, G. J., and D. W. Piston. 2009. Electron microscopy of whole cells in liquid with nanometer resolution. *Proc Natl Acad Sci USA* 106:2159–2164.

De, S. K., Aluru, N., Johnson, B., Crone, W., Beebe, D. J., and J. Moore. 2002. Equilibrium swelling and kinetics of pH-responsive hydrogels: Models, experiments, and simulations. *J Microelectromech Syst* 11:544–555.

Doherty, W. O. S., Mousavioun, P., and C. M. Fellows. 2011. Value-adding to cellulosic ethanol: Lignin polymers. *Ind Crops Prod* 33:259–276.

Domínguez-Roblesa, J., Peresin, M. S., Tamminen, T., Rodríguez, A., Larrañeta, E., and A. S. Jääskeläinen. 2018. Lignin-based hydrogels with 'super-swelling' capacities for dye removal. *Int J Biol Macromol* 115:1249–1259.

Drury, J. L., and D. J. Mooney. 2003. Hydrogels for tissue engineering: Scaffold design variables and applications. *Biomaterials* 24:4337–4351.

Fekete, T., Borsa, J., Takács, E., and L. Wojnárovits. 2017. Synthesis and characterization of superabsorbent hydrogels based on hydroxyethylcellulose and acrylic acid. *Carbohydr Polym* 166:300–308.

Ferreira, L. M., Blanes, L., Gragnani et al. 2009. Hemicellulose dressing versus rayon dressing in the re-epithelialization of split-thickness skin graft donor sites: A multicenter study. *J Tissue Viability* 18:88–94.

Fu, L. H., Qi, C., Ma, M. G., and P. Wan. 2019. Multifunctional cellulose-based hydrogels for biomedical applications. *J Mater Chem B* 7:1541–1562.

Gabrielii, I., and P. Gatenholm. 1998. Preparation and properties of hydrogels based on hemicellulose. *J Appl Polym Sci* 69:1661–1667.

Gabrielii, I., Gatenholm, P., Glasser, W. G., Jain, R. K., and L. Kenne. 2000. Separation, characterization and hydrogel-formation of hemicellulose from aspenwood. *Carbohydr Polym* 43:367–374.

Ghorbani, M. and L. Roshangar. 2019. Construction of collagen/nanocrystalline cellulose based-hydrogel scaffolds: Synthesis, characterization, and mechanical properties evaluation. *Int J Polym Mater Polym Biomater*

Gil, E. S., Hudson, S. M. 2004. Stimuli-reponsive polymers and their bioconjugates. *Prog Polym Sci* 29:1173–1222.

Golding, C. G., Lamboo, L. L., Beniac, D. R., and T. F. Booth. 2016. The scanning electron microscope in microbiology and diagnosis of infectious disease. *Sci Rep* 6: 26516.

Grassi, M., Sandolo, C., Perin, D., Coviello, T., Lapasin, R., and G. Grassi. 2009. Structural characterization of calcium alginate matrices by means of mechanical and release tests. *Molecules* 14:3003–3017.

Hayes, T. L., and R. F. W. Pease. 1968. The scanning electron microscope: Principles and applications in biology and medicine. *Adv Biol Med Phys* 12:85–137.

Hazelton, P. R., and H. R. Gelderblom. 2003. Electron microscopy for rapid diagnosis of infectious agents in emergent situations. *Emerg Infect Dis* 9:294–303.

Hoarea, T. R., and D. S. Kohane. 2008. Hydrogels in drug delivery: Progress and challenges. *Polymer* 49:1993–2007.

Hoffman, A. S. 2002. Hydrogels for biomedical applications. *Adv Drug Deliv Rev* 54:3–12.

Hu, L., Du, M., and J. Zhang. 2018. Hemicellulose-based hydrogels present status and application prospects: A brief review. *Open J For* 8:15–28.

Jing, Z., Sun, X. F., Ye, Q., and Y. Li. 2012. Hemicellulose-based porous hydrogel for methylene blue adsorption. *Adv Mater Res* 560–561:482–487.

Johari, O., and S. Bhattacharyya. 1969. The application of scanning electron microscopy for the characterization of powders. *Powder Technol* 2: 335–348.

Joy, D. C. 1991. The theory and practice of high-resolution scanning electron microscopy. *Ultramicroscopy* 37:216–233.

Kashyap, N., Kumar, N., and M. N. V. Ravi Kumar. 2005. Hydrogels for pharmaceutical and biomedical applications. *Crit Rev Ther Drug Carrier Syst* 22:107–149.

Kentaro, A., and Y. Hiroyuki. 2012. Cellulose nanofiber-based hydrogels with high mechanical strength. *Cellulose* 19:1907–1912.

Kim, J., Lee, K.W., Hefferan, T. E., Currier, B. L., Yaszemski, M. J., and L. Lu. 2007. Synthesis and evaluation of novel biodegradable hydrogels based on poly(ethylene glycol) and sebacic acid as tissue engineering scaffolds. *Biomacromolecules* 9:149–157.

Laftaha, W. A., Hashima, S., and A. N. Ibrahima. 2011. Polymer hydrogels: A review. *Polym Plast Technol Eng* 50:1475–1486.

Langer, R., and N. A. Peppas. 2003. Advances in biomaterials, drug delivery, and bionanotechnology. *AIChE J* 49:2990–3006.

Laurichesse, S., and L. Avérous. 2014. Chemical modification of lignins: Towards biobased polymers. *Top Issue Biomater* 39:1266–1290.

Lee, H. V., Hamid, S. B. A., and S. K. Zain. 2014. Conversion of lignocellulosic biomass to nanocellulose: Structure and chemical process. *Sci World J* 2014:1–20.

Lee, S. C., Kwon, K., and K. Park. 2013. Hydrogels for delivery of bioactive agents: A historical perspective. *Adv Drug Deliv Rev* 65:17–20.

Lee, W. F., and K. T. Tsao. 2007. Effect of intercalant content of mica on the various properties for the charged nanocomposite poly(N-isopropyl acrylamide) hydrogels. *J Appl Polym Sci* 104:2277–2287.

Lin, C. C., and A. T. Metters. 2006. Hydrogels in controlled release formulations: Network design and mathematical modeling. *Adv Drug Deliv Rev* 58:1379–1408.

Liu, H., Chen, T., Dong, C., and X. Pan. 2020. Biomedical applications of hemicellulose-based hydrogels. *Curr Med Chem* 27(28): 4647–4659.

Liu, H., Rong, L., Wang, B., Xie et al. 2017a. Facile fabrication of redox/pH dual stimuli responsive cellulose hydrogel. *Carbohydr Polym* 176:299–306.

Liu, J., Li, Q., Su, Y., Yue, Q., and B. Gao. 2014. Characterization and swelling–deswelling properties of wheat straw cellulose based semi-IPNs hydrogel. *Carbohydr Polym* 107:232–240.

Liu, X., Lin, Q., Yan, Y., Peng, F., Sun, R., and J. Ren. 2017b. Hemicellulose from plant biomass in medical and pharmaceutical application: A critical review. *Curr Med Chem* 24:1

Luo, X., Zhang, H., Cao, Z., Cai, N., Xue, Y., and F. Yu. 2016. A simple route to develop transparent doxorubicin-loaded nanodiamonds/cellulose nanocomposite membranes as potential wound dressings. *Carbohydr Polym* 143:231–238.

Maleki, L., Edlund, U., and A. C. Albertsson. 2017. Synthesis of full interpenetrating hemicellulose hydrogel networks. *Carbohydr Polym* 170:254-263.

Maolin, Z., Jun, L., Min, Y., and H. Hongfei. 2000. The swelling behavior of radiation prepared semi-interpenetrating polymer networks composed of polyNIPAAm and hydrophilic polymers. *Radiat Phys Chem* 58:397–400.

McMahon, R., Hahn, M., Pendleton, M., and E. Ellis. 2010. A simple preparation method for mesh fibrin hydrogel composites for conventional SEM. *Microsc Microanal* 16:1030–1031.

Miller, J. S. 2011. The discovery of medicines from plants: A current biological perspective. *Econ Bot* 65: 396–407.

Minkoff, I. 1967. Applications of the scanning electron microscope in materials science. *J Mater Sci* 2:388–394.

Mogoşanu, G. D., and M. A. Grumezescu. 2014. Natural and synthetic polymers for wounds and burns dressing. *Int J Pharm* 463:127–136.

Nadimi, A. E., Ebrahimipour, S. Y., Afshar, E. G. et al. 2018. Nano-scale drug delivery systems for antiarrhythmic agents *Eur J Med Chem* 157:1153–1163.

Narayanaswamy, R. and V. P. Torchilin. 2019. Hydrogels and their applications in targeted drug delivery. *Molecules* 24:603.

Naseem, A., Tabasum, S., Zia, K. M., Zuber, M., Ali, M., and A. Noreen. 2016. Lignin-derivatives based polymers, blends and composites: A review. *Int J Biol Macromol* 93: 296–313.

Naseri, N., Bhanumathyamma, D., Mathew, A. P., Oksman, K., and L. Girandon. 2016. Nanocellulose based interpenetrating polymer network (IPN) hydrogels for cartilage applications. *Biomacromolecules* 17:3714–3723.

Nayak, S., and S. C. Kundu. 2014. Sericin–carboxymethyl cellulose porous matrices as cellular wound dressing material. *J Biomed Mater Res Part A* 102:1928–1940.

Nico, B., Crivellato, E., and D. Ribatti. 2007. The importance of electron microscopy in the study of capillary endothelial cells: An historical review. *Endothelium* 14:257–264.

Niu, L., Zhang, D., Liu, Y. et al. 2020. Combination of acid treatment and dual network fabrication to stretchable cellulose based hydrogels with tunable properties. *Int J Biol Macromol* 147:1–9.

Nokhodchi, A, Raja, S, Patel, P, and K. Asare-Addo. 2012. The role of oral controlled release matrix tablets in drug delivery systems. *Bioimpacts* 2:175–187.

Onaciu, A., Munteanu, R. A., Moldovan, A. I., Moldovan, C. S., and I. Berindan-Neagoe. 2019. Hydrogels based drug delivery synthesis, characterization and administration. *Pharmaceutics* 11:432.

Oyarce, P., De Meester, B., Fonseca, F. et al. 2019. Introducing curcumin biosynthesis in Arabidopsis enhances lignocellulosic biomass processing. *Nat Plants* 5:225.

Pal, K., Banthia, A. K., and D. K. Majumdar. 2009. Polymeric hydrogels: Characterization and biomedical applications. *Des Monomers Polym* 12:197–220.

Palantöken, S., Bethke, K., Zivanovic, V., Kalinka, G., Kneipp, J., Rademann, K. 2019. Cellulose hydrogels physically crosslinked by glycine: Synthesis, characterization, thermal and mechanical properties. *J Appl Polym Sci* 137:48380.

Pease, R. E. W., and W. C. Nixon. 1965. High resolution scanning electron microscopy. *J Sci Instr* 42:81.

Pourjavadi, A., Ghasemzadeh, H., and R. Soleyman. 2007. Synthesis, characterization, and swelling behavior of alginate-g-poly(sodium acrylate)/kaolin superabsorbent hydrogel composites. *J Appl Polym Sci* 105:2631–2639.

Ren, L., Tsuru, K., Hayakawa, S., and A. Osaka. 2002. A novel approach to fabricate porous gelatin-siloxane hybrids for bone tissue engineering. *Biomaterials* 23:65–73.

Rodríguez, R. A., Alvarez-Lorenzo, C., and A. Concheiro. 2003. Cationic cellulose hydrogels: Kinetics of the cross-linking process and characterization as pH-/ion-sensitive drug delivery systems. *J Control Release* 86:253–265.

Sánchez, R., Espinosa, E., Domínguez-Robles, J., Loaiza, J. M., and A. Rodríguez. 2016. Isolation and characterization of lignocellulose nanofibers from different wheat straw pulps. *Int J Biol Macromol* 92:1025–1033.

Sant'Anna, C., Campanati, L., Gadelha, C. et al. 2005. Improvement on the visualization of cytoskeletal structures of protozoan parasites using high-resolution field emission scanning electron microscopy (FESEM). *Histochem Cell Biol* 124:87–95.

Sharpe, L. A., Daily, A. M., Horava, S. D., and N. A. Peppas. 2014. Therapeutic applications of hydrogels in oral drug delivery. *Expert Opin Drug Deliv* 11:901–915.

Shen, J., Yan, B., Li, T., Long, Y., Li, N., and M. Ye. 2012. Study on graphene-oxide-based polyacrylamide composite hydrogels. *Compos Part A Appl Sci Manuf* 43:1476–1481.

Simões, S., Figueiras, A., and F. Veiga. 2012. Modular hydrogels for drug delivery. *J Biomater Nanobiotechnol* 3:185–199.

Singh, A., Sharma, P. K., Garg, V. K., and G. Garg. 2010. Hydrogels: A review. *Int J Pharm Sci Rev Res* 4:106–110.

Siritientong, T., and P. Aramwit. 2015. Characteristics of carboxymethyl cellulose/sericin hydrogels and the influence of molecular weight of carboxymethyl cellulose. *Macromol Res* 23: 861–866

Stokes, D. J. 2003. Recent advances in electron imaging, image interpretation and applications: Environmental scanning electron microscopy. *Phil Trans R Soc Lond A* 361:2771–2787.

Sun, X. F., Hao, Y., Cao, Y., and Q. Zeng. 2019. Superadsorbent hydrogel based on lignin and montmorillonite for Cu (II) ions removal from aqueous solution. *Int J Biol Macromol* 127:511–519.

Sun, X. F., Wang, H. H., Jing, Z. X., and R. Mohanathas. 2013. Hemicellulose-based pH-sensitive and biodegradable hydrogel for controlled drug delivery. *Carbohydr Polym* 92:1357–1366.

Sun, Z., Fridrich, B., de Santi, A., Elangovan, S., and K. Barta. 2018. Bright side of lignin depolymerization: Toward new platform chemicals. *Chem Rev* 118:614–678.

Takashi, L., Hatsumi, T., Makoto, M., Takashi, I., Takehiko, G., and J. Shuji. 2007. Synthesis of porous poly(N -isopropylacrylamide) gel beads by sedimentation polymerization and their morphology. *J Appl Polym Sci* 104:842–850.

Tang, H., Chen, H., Duan, B., Lu, A., and L. Zhang. 2014. Swelling behaviors of superabsorbent chitin/carboxymethylcellulose hydrogels. *J Mater Sci* 49:2235–2242.

Tarrés, Q, Espinosa, E., Domínguez-Robles, J., Rodríguez, A., Mutjé, P., and M. Delgado-Aguilar. 2017. The suitability of banana leaf residue as raw material forthe production of high lignin content micro/nano fibers: From residue tovalue-added products. *Ind Crops Prod* 99:27–33.

Thakur, V. K., Thakur, M. K., and R. K. Gupta. 2013. Graft copolymers from cellulose: Synthesis, characterization and evaluation. *Carbohydr Polym* 97:18–25.

Ulijin, R. V., Bibi, N., Jayawarna, V. et al. 2007. Bioresponsive hydrogels. *Mater Today* 10:40–48.

Vanholme, R., De Meester, B., Ralph, J., and W. Boerjan. 2019. Lignin biosynthesis and its integration into metabolism. *Curr Opin Biotechnol* 56:230–239.

Venugopal, J., Rajeswari, R., Shayanti et al. 2013. Xylan polysaccharides fabricated into nano-fibrous substrate for myocardial infarction. *Mater Sci Eng C* 33:1325–1331.

Wang, X. H., Zhou, Z. L., Guo, X. W., He, Q., Hao, C., and C. W. Ge. 2016. Ultrasonic-assisted synthesis of sodium lignosulfonate-grafted poly(acrylic acid-co-poly(vinylpyrrolidone)) hydrogel for drug delivery. *RSC Adv* 6:35550–35558.

Wichterle, O., and D. Lim. 1960. Hydrophilic gels for biological use. *Nature* 185:117–118.

Witzler, M., Alzagameem, A., Bergs et al. 2018. Lignin-derived biomaterials for drug release and tissue engineering. *Molecules* 23:1885.

Yang, L., Chu, J. S., and J. A. Fix. 2002. Colon-specific drug delivery: New approaches and in vitro/in vivo evaluation. *Int J Pharm* 235:1–15.

Yang, M., Zhou, G., Castano-Izquierdo, H., Zhu, Y., and C. Mao. 2015. Biomineralization of natural collagenous nanofibrous membranes and their potential use in bone tissue engineering. *J Biomed Nanotechnol* 11:447–456.

Zhang, H., and A. I. Cooper. 2007. Aligned porous structures by directional freezing. *Adv Mater* 19:1529–1533.

Zhang, L., Zhou, J., and L. Zhang. 2013. Structure and properties of β -cyclodextrin/cellulose hydrogels prepared in NaOH/urea aqueous solution. *Carbohydr Polym* 94:386–393.

Zhang, X. Z., Yang, Y. Y., and T. S. Chung. 2002. The influence of cold treatment on properties of temperature-sensitive poly(n-isopropylacrylamide) hydrogels. *J Colloid Interface Sci* 246:105–111.

Zhao, J., Zheng, K., Nan, J., Tang, C., Chen, Y., and Y. Hu. 2017. Synthesis and characterization of lignosulfonate-graft-poly (acrylic acid)/hydroxyethyl cellulose semi-interpenetrating hydrogels. *React Funct Polym* 115:28–35.

Zhao, W., Glavas, L., Odelius, K., Edlund, U., and A. C. Albertsson. 2014. Facile and green approach towards electrically conductive hemicellulose hydrogels with tunable conductivity and swelling behavior. *Chem Mater* 26:4265–4273.

Zhou, W., Apkarian, R. P., Wang, Z. L., and D. Joy. 2006. Fundamentals of scanning electron microscopy. In *Scanning microscopy for nanotechnology.* 1–40, New York, Springer.

Zmejkoski, D., Spasojevic, D., Orlovska et al. 2018. Bacterial cellulose-lignin composite hydrogel as a promising agent in chronic wound healing. *Int J Biol Macromol* 118:494–503.

9 Bio-based Stimuli-responsive Hydrogels with Biomedical Applications

Raluca Nicu and Diana Ciolacu

CONTENTS

9.1 INTRODUCTION

In recent years, materials from renewable resources have received special attention in a multitude of fields as a way to achieve sustainable development (Alvarez-Lorenzo et al. 2013). The concept 'sustainable development' refers to 'development that meets the needs of the present, without compromising the ability of future generations to meet their own needs' (Emas 2015). Transition to the use of renewable resources (biomass) eliminates the dependence on limited resources and, at the same time, reduces the contribution to the greenhouse effect and allows preserving mineral resources for future generations (Voevodina and Kržan 2013). Biomass feedstock can be converted into raw materials called 'bio-materials' (Nakajima et al. 2017), and this new generation of 'bio-materials' (bio-based, bio-compatible, bio-degradable) offers the possibilities to reduce the dependence on fossil fuels, as well as environmental impacts (Kimura 2009; Babu et al. 2013; Salerno and Pascual 2015; Gicquel 2017). Bio-based materials are obtained through biological, chemical, and physical methods from renewable biomass including crops, trees, or other plants. They have special characteristics such as environmental-friendly, renewable, biodegradable, and other properties that traditional polymers do not have (Gao et al. 2019).

Recently, much research has been focused on obtaining hydrogels from biopolymers due to their unique properties, such as biocompatibility and biodegradability, less toxicity, and an acceptable mechanical strength (Mohite et al. 2018). Biodegradation is an extremely interesting feature when talking about hydrogels with biomedical applications, especially in tissue engineering and drug delivery applications, mainly because of their biological interaction with the body components (Jayaramudu et al. 2013).

Among biopolymers, polysaccharides have received special attention in forming hydrogels because they are abundant and readily available from renewable sources, such as plants, algae, or various microbial organisms, and also have a large variety of compositional and structural properties, making them facile to produce and versatile for gel formation as compared with synthetic polymers (Karoyo and Wilson 2017). Moreover, polysaccharides possess multiple functional groups, including acid, amine, hydroxyl, and aldehyde, which make them suitable for chemical conjugation of therapeutic agents (Thambi et al. 2016). Given this multitude of positive features, along with their structural and functional similarities with the extracellular matrix, it is quite easy to understand why polysaccharides have benefitted from a growing interest in the biomedical field, in recent decades (Tchobanian et al. 2019)

This chapter aims to provide an overview of the recent use of polysaccharides in obtaining stimuli-sensitive hydrogels and their applications, especially in the field of controlled drug delivery and tissue engineering. In addition, the main available sources of polysaccharides in nature, along with their structure and properties, important factors in subsequent biomedical applications, are presented.

9.2 POLYSACCHARIDES SOURCES AND PROPERTIES

Polysaccharides are high-molecular-weight carbohydrate polymers with sugar residues linked by glycosidic bonds. The most common constituent of polysaccharides is D-glucose, but D-fructose, D- or L-galactose, D-mannose, L-arabinose, and D-xylose are also frequent. The main functions played by polysaccharides in nature are either storage or structural functions, the properties of those two classes being dramatically different, although the compositions are quite similar (d'Ayala et al. 2008; Ozcan and Oner 2015).

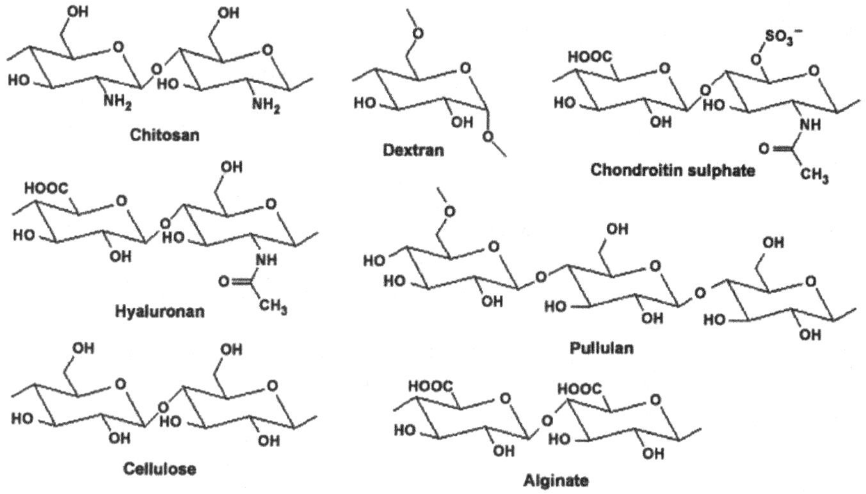

FIGURE 9.1 Chemical structures of naturally occurring polysaccharides. (Reprinted with permission from Oh et al. 2009).

Figure 9.1 shows the chemical structures of the most common polysaccharides in nature (Oh et al. 2009), and their sources of origin and the main structural features and functional groups are presented in Table 9.1.

9.2.1 POLYSACCHARIDES FROM HIGHER PLANTS

Cellulose is the most abundant biodegradable polymer, being the main constituent of plants and natural fibres (Sezer et al. 2018; Barman and Das 2018; Kayra and Aytekin 2018) and considered an almost inexhaustible source of raw material for the increasing demand of environmental-friendly and biocompatible products (Huq et al. 2012; Efthimiadou et al. 2014; Ciolacu 2018). Cellulose has properties that make it very promising and attractive material in many applications due to its low cost, biocompatibility, high mechanical and thermal stability. Its only limitation is low solubility, which restricts its use especially in biomedical and pharmaceutical domains, but this drawback is overcome by obtaining cellulose derivatives through various chemical modification procedures, such as esterification, etherification, or oxidation (Jeong et al. 2012; Marques-Marinho and Vianna-Soares 2013; Onofrei and Filimon 2016; Sezer et al. 2018). Cellulose hydrogels can be easily prepared either by physical interactions (chain entanglements, van der Waals forces, hydrogen bonds, hydrophobic, or electronic associations) or by chemical cross-linking (with cross-linking agents) (Ciolacu and Suflet 2018). Cellulose-based hydrogels have sustainable and biodegradable properties which have aroused interest in various biomedical applications (Fu et al. 2018), such as targeted drug delivery and smart sensors, or multiresponsive, injectable, and self-healing hydrogels with enhanced mechanical properties (Liu et al. 2017).

TABLE 9.1

The Source and Main Structural Features of Polysaccharides

Polysaccharide	Source	Structural Features	Functional Groups
Cellulose	Plant cell walls	Anhydro-D-glucopyranose units linked by β-1,4 linkages	Three hydroxyl groups per each glucose unit
Starch	Plant cell walls	Glucose molecules joined together with α-D-(1,4-) and(or α-D-(1,6-) linkages	Hydroxyl groups
Alginate	Seaweed, bacteria	(1,4) linked β-D-mannuronic and α-L-guluronic acids	Hydroxyl and carboxyl groups
Carrageenan	Seaweed (marine algae)	Linear, β(1,3)-sulfated-D-galactose and α(1,4)-3,6-anhydro-D-galactose (3-cross-linked-).	One (k-), two (i-), or three (l-) sulfate groups per disaccharide unit
Hyaluronan	Marine animals	Linear, D-glucuronic acid and N-acetyl-D-glucosamine linked by a β-1,3 linkage	Carboxylic functional groups and alcohols
Chitosan	Marine animals	Linear, two repeating units of (1,4)-2-acetamido-2-deoxy-β-D-glucan (N-acetyl D-glucosamine) and (1,4)-2-amino-2-deoxy-β-D-glucan (N-acetyl D-glucosamine)	Amino and hydroxyl groups
Dextran	Bacterial cultures	Branched, D-glucopyranose residues with 1,6- and some 1,3-glucosidic linkages	Hydroxyl groups
Bacterial cellulose	Aerobic bacteria	Anhydro-D-glucopyranose units linked by β-1,4 linkages	Three hydroxyl groups per each glucose unit

Starch is the most abundant storage polysaccharide in plants being considered a promising candidate for developing sustainable materials with various applications (Kunal et al. 2006; Lu et al. 2009; Ismail et al. 2013). The properties which make starch a suitable material, especially for biomedical applications, are its biodegradability, biocompatibility, non-toxic degradation products, high swelling capacity in water, and proper mechanical properties (Lu et al. 2009; Zhang 2015). One of its drawbacks is low surface area, but by chemical derivatization (e.g., grafting or etherification), its sorption capacities are improved (Kunal et al. 2006; Ismail et al. 2013; Zhang 2015). With a large number of hydroxyl groups, starch can be used to easily prepare hydrogels. It turns into a gel in a few thermally assisted steps: (i) the swelling occurs by water adsorption in the hydrophilic granules, (ii) the gelatinization takes place after starch is dissolved by heating, resulting in irreversible physical changes, (iii) the destruction of the granule structure (Ismail et al. 2013). Starch-based materials have received much attention in drug delivery and other biomedical applications, such as wound healing or tissue engineering, because of their extensive availability, low cost, and total compostability, without generating any hazardous residues (Waghmare et al. 2018).

9.2.2 ALGAL POLYSACCHARIDES

Alginate is extracted from various brown algal species, such as *Laminaria hyperborea*, *Macrocystis pyrifera*, and *Ascophyllum nodosum*, and is one of the most abundant and extensively characterized marine polysaccharide (Chan 2009). The proportion and arrangement of the two constituent segments (β-D-mannuronate and α-L-guluronate) vary with the source of algae species and are related to the physical strength or other physical properties of the alginate structure (Goh et al. 2012; Kaygusuz and Erim 2013; Dragan and Dinu 2019). Alginate forms gels by ionotropic gelation, at pH > 6, with divalent cations, such as Ca^{2+}, Ba^{2+}, or Zn^{2+} (Moller et al. 2007; Laurienzo 2010). Its hydrogels are pH-sensitive which shrinks in acidic conditions and are generally regarded as non-toxic, biocompatible, biodegradable, less expensive, and meets the important requirement of being suitable to sterilization and storage. All these advantages make alginates very useful materials for biomedical applications, especially for controlled delivery of drugs and other biologically active compounds (d'Ayala et al. 2008; Kaygusuz and Erim 2013).

Carrageenan is obtained from *Rhodophyceae* marine algae species, using dilute alkaline solution (Khalil et al. 2017). There are three main types of carrageenan, k-, i-, and l-carrageenan, based on the number and position of sulfate groups on the galactose/anhydrogalactose chain (Zarina and Ahmad 2015).

Carrageenan can be quite heterogeneous material depending on the species of algae, the seasonality, or even the extraction conditions. Thus, a wide variety of materials, including hydrogels, with a vast range of properties and applications, can be obtained (Silva et al. 2012). Due to its facile gelation, carrageenan is gaining interest for biomedical applications (Rocha et al. 2011). For instance, k-carrageenan can form thermally reversible gels, in the presence of potassium, and cooling below 10°C (Fouda et al. 2015; Feng et al. 2017). To improve their flexibility and attenuate the brittleness, which can limit their potential applications, sometimes plasticizers are added (Zarina and Ahmad 2015). The thermo-sensitive nature of k-carrageenan hydrogels, as well as their ability to form microparticles using in situ ionic gelation, makes these hydrogels interesting candidates for drug delivery applications (Hezaveh and Muhamad 2012).

9.2.3 POLYSACCHARIDES FROM MARINE ANIMALS

Chitosan is a polysaccharide derived from chitin, the second most widespread natural polymer after cellulose. The chitin from the exoskeleton of shrimp, lobster, and insects is converted to chitosan using alkali treatment (Dash et al. 2011; Croisier and Jérôme 2013; Ahmed and Ikram 2016; Chatterjee and Hui 2018). The presence of amino and hydroxyl groups in chitosan structure determines unique properties, of which the biological ones are especially distinguished including antibacterial, antifungal, analgesic, and hemostatic properties, along with biodegradability and biocompatibility (Abdelaal et al. 2007; Iacob et al. 2018; Vasiliu et al. 2018). The functional groups also allow covalent bonding or introduction of various species along the chitosan backbone, widening its possible applications spectrum (Cardoso et al. 2016; Domalik-Pyzik et al. 2018). Chitosan may present either reversible or irreversible gelation ability, depending on whether hydrogels were obtained by either physically associated or chemically cross-linked networks (Dash et al. 2011; Croisier and Jérôme 2013). Chitosans are a typical bio-based polymer sensitive to pH, as a result of the charges distributed on amino groups by protonation and deprotonation. Thus, chitosan hydrogel can be part in drug triggered release due to this special characteristic (Gao et al. 2019).

Hyaluronic acid, or hyaluronan, is a glucosaminoglycan found in extracellular tissue in many parts of the body and regulates numerous cellular processes (Selegard et al. 2017). It has a high solubility in water at room temperature and acidic medium and, due to its natural origin, has biocompatible, non-immunogenic, and non-inflammatory characteristics (Aravamudhan et al. 2014; Thambi et al. 2016). Its properties, both physical and biochemical, in solution or hydrogel form, are extremely attractive, finding applications in various fields, from tissue scaffolds to cosmetics (Collins and Birkinshaw 2013; Selegard et al. 2017). Hyaluronic acid is a pseudo-elastic material with shear thinning properties that is capable of gelation (Egbu et al. 2018). It contains functional groups (carboxylic acids and alcohols) along its backbone that can be used to introduce functional domains or to form a hydrogel by cross-linking (Collins and Birkinshaw 2013).

9.2.4 POLYSACCHARIDES FROM MICROORGANISMS

Bacterial cellulose, known as microbial cellulose or bio-cellulose, is biosynthesized by aerobic bacteria such as *Gluconacetobacter* (i.e., *Acetobacter xylinum*) (Khalil et al. 2017). It is a unique and interesting material, a purely cellulose polymer, with excellent mechanical strength and degradability, along with high water holding capacity, higher degree of polymerization and high crystallinity index compared to green plant cellulose. These excellent properties make bacterial cellulose a material with wide range of applications from food to functional materials such as electronic devices, wound dressings, additives in paper, membrane filters, and more. Thanks to its excellent properties, bacterial cellulose is an interesting candidate for hydrogel formation (Khalil et al. 2017; Mohite et al. 2018). It lacks stimuli-responsive property, but its functional composites with other polymers or inorganic materials have been prepared to enlarge its potential applications (Amin et al. 2014; de Sousa Moraes et al. 2016; Lin et al. 2017; Wang et al. 2018).

Dextran is a nontoxic and hydrophilic homopolysaccharide, secreted by lactic acid bacteria from Lactobacillacea family and, particularly, in the genera *Lactobacillus*, *Leuconostoc* and *Streptococcus* (Paulo et al. 2012; Sun and Mao 2012; Mollakhalili and Mohammadifar 2015). Unlike other polysaccharides, such as chitosan, alginate, or hyaluronic acid, which have various functional groups, dextran only has hydroxyl groups, which do not support cell attachment. New novel dextran derivatives can be made by incorporating different functionalities via chemical modification, without compromising primary properties of dextran (Van Tomme and Hennink 2007; Lima et al. 2011; Lin et al. 2017; O'Connor et al. 2018). Dextran can form hydrogel with three-dimensional structure capable of absorbing large amounts of water and biological fluids, using numerous different strategies that include either chemical or physical cross-linking methods (Ribeiro et al. 2013; Thambi et al. 2016; O'Connor et al. 2018). Modification of dextran backbone allows the development of hydrogels with specific characteristics, thus expanding the field of their utilization in various biomedical applications. A common approach is to incorporate vinyl groups via different types of acrylates (glycidyl acrylate, glycidyl methacrylate, methacrylate, acrylate, and hydroxyethyl methacrylate). These hydrogels were proven to be efficient as protein carriers (Sun et al. 2010; Sun and Mao 2012).

9.3 BIO-BASED STIMULI RESPONSIVE HYDROGELS

9.3.1 STIMULI-RESPONSIVE HYDROGELS: GENERAL CONSIDERATIONS

A hydrogel can be called *stimuli-responsive* or *smart material*, if at least one element of the hydrogel network can act as a receptor unit or is spectroscopic active. The effect of an external stimulus on hydrogel network can be measured in different forms at the macroscopic scale: shape and size change or changes in optical, electrical, or mechanical properties (Echeverria et al. 2018; Gao et al. 2019). During the last few decades, considerable attention has been given to environment-sensitive hydrogels, having in view that the most important substances in living systems are macromolecules which respond to their surroundings, in an intelligent or smart way (Kaith et al. 2010; Altomare et al. 2018; Chatterjee and Hui 2018). Generally, physical stimuli (temperature, electric or magnetic fields, mechanical stress) alter molecular interactions at critical onset points, and chemical stimuli (pH, ionic factors, and chemical agents) change the interactions between polymer chains or between polymer chains and solvent at the molecular level (Figure 9.2). Biochemical or biological stimuli are another category of stimuli which involves responses to enzymes, antigens, ligands, and other biochemical agents (Shang et al. 2019).

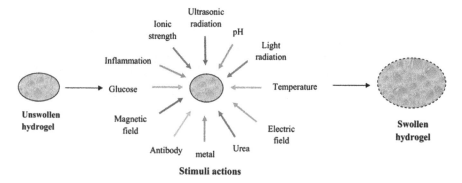

FIGURE 9.2 Stimuli responsible for swelling of hydrogels. (Reprinted with permission from Sood et al. 2016).

The development of stimuli-responsive hydrogels based on polysaccharides has been encouraged considering their versatile properties and advantages of these smart materials over the synthetic ones: non-toxicity, biodegradability, biocompatibility, similarity to biological environments, biological functionality at molecular level, and immune recognition. Furthermore, the use of polysaccharides is also encouraged in respect of environmental concerns, being the cheapest, most abundant, available, and renewable organic materials on earth (Das and Pal 2015; Onofrei and Filimon 2016). Polysaccharide-based hydrogels have been prepared by chemical (covalent bonds) or physical (electrostatic interactions, hydrogen bond, etc.) cross-linking approach (Fajardo et al. 2015). Apart from cross-linking, chemical modification of their functional groups (-OH, -COOH, -NH$_2$, -OSO$_3$H) has also been employed to provide required functionalities and responsiveness necessary for specific application (Camponeschi et al. 2015).

Cellulose, starch, alginate, chitosan, or hyaluronan are commonly polysaccharides used to prepare hydrogel matrices that provide some response to external stimuli. They are used in their natural or chemically modified form (Pasqui et al. 2012; Gao et al. 2014). Generally, the choice of the polysaccharide to form a hydrogel for a specific application has a critical role. For example, a polysaccharide that is easily degraded in stomach region (at acid conditions) is not a good candidate to form a colon-specific drug release system (Fajardo et al. 2015). Taking into account these polysaccharides, a great number of interesting studies about the polysaccharide-based stimuli-responsive hydrogels and their applications have been published.

9.3.2 Physical Stimuli-Responsive Hydrogels

Temperature-sensitive hydrogels. Thermo-responsive polymers are sensitive to environmental temperature changes (Gicquel 2017). Polymers which become insoluble upon heating present so-called LCST (low critical solution temperature), and those which become soluble upon heating have an UCST (upper critical solution temperature) (Figure 9.3). Characteristic of this type of polymers is that an increase or decrease in temperature results in gelation (Gandhi et al. 2015; Zhang and Khademhosseini 2017).

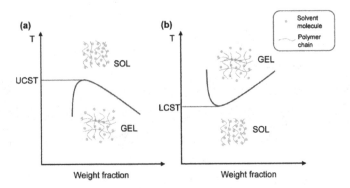

FIGURE 9.3 Thermo-responsive hydrogels: (a) UCST hydrogels; (b) LCST hydrogels. Reprinted with permission from Altomare et al. 2018.

Many natural polymers such as polysaccharides (cellulose, chitosan, xyloglucan, starch, xanthan gum, carrageenan, hyaluronic acid, dextran) and proteins (gelatin/collagen, and elastin) form thermally driven hydrogels (Jeong et al. 2012; Zhang and Khademhosseini 2017; Chatterjee and Hui 2018). Synthetic polymers are also known to lead to thermo-sensitive hydrogels presenting several advantages, such as facile modulation of the critical temperature, biodegradability, and mechanical properties. However, they lack biocompatibility and do not support cells or proteins adhesion. Thus, incorporation of natural polymers has been used to overcome these drawbacks for their use in biomedical applications (Kim et al. 2011). Among different possible stimuli, changes in temperature have proven to be easy and safe, especially for biomedical applications (Fajardo et al. 2015; Shang et al. 2019). Thus, temperature-responsive polymers have attracted high interest from researchers because this stimulus can easily be regulated and applied externally to the system (Echeverria et al. 2018). A large variety of applications which use temperature as stimulus in aqueous solvent have been developed, such as tissue engineering, drug delivery, catalysis, surface engineering, or information processing (Kim et al. 2011; Gandhi et al. 2015; Gicquel 2017).

Light-sensitive hydrogels. Light-sensitive polymers are sensitive to light exposure, at a specific wavelength and can be UV-sensitive or visible light-sensitive hydrogels (Shang et al. 2019). The development of light-responsive materials has many advantages: light is non-invasive and does not require direct contact, has no damage on human cells only a low thermal effect (biological-friendly applications), is highly accurate, the exposure time and intensity can be easily controlled (Gicquel 2017; Echeverria et al. 2018). Photo-responsive polymers contain light-sensitive chromophore groups such as, azobenzene, spiropyran, or nitrobenzyl (Figure 9.4) (Gicquel 2017; Shang et al. 2019). Azobenzenes are among the mostly used photo-switchable molecules, which undergo *trans*-to-*cis* or *cis*-to-*trans* isomerization when irradiated with UV- and blue-light, respectively (Echeverria et al. 2018; Su et al. 2018).

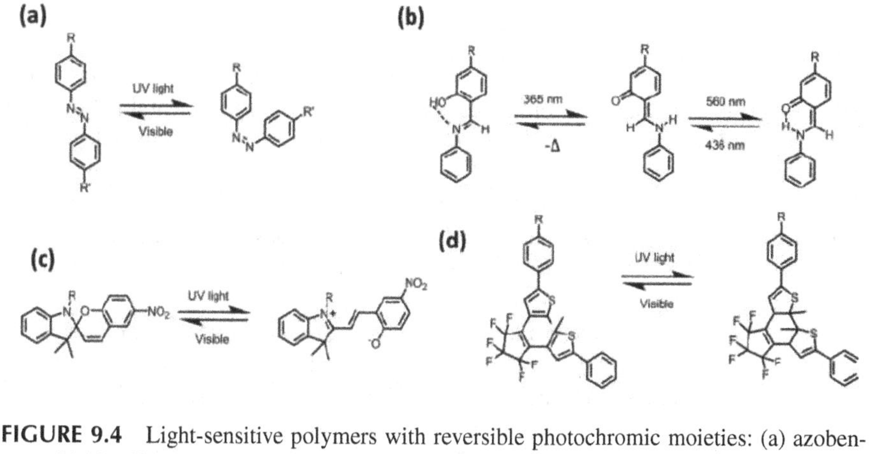

FIGURE 9.4 Light-sensitive polymers with reversible photochromic moieties: (a) azobenzene, (b) N-salicyliden aniline, (c) spiropyran, (d) dithienylethene. (Reprinted with permission from Cao et al. 2015).

(a)

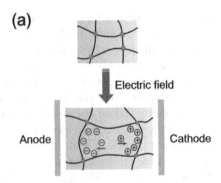

FIGURE 9.5 Mechanism of electric-responsive hydrogels. (Reprinted with permission from Shi et al. 2019).

Light-sensitive hydrogels found applications in the development of photo-responsive artificial muscle, in situ forming gels for cartilage tissue engineering or in local drug delivery to control the wound healing (Masteikova et al. 2003).

Electrical signal-sensitive hydrogels undergo shrinking or swelling in the presence of an applied electric field (Figure 9.5) (Shi et al. 2019). The use of electrical signal as environmental stimulus to induce responses in hydrogels has some advantages, such as reliable control of signal strength and direction, remote operation, and a precise "on/off" triggering switch (Shang et al. 2019).

Typical hydrogels sensitive to electrical signal are usually polyelectrolytes networks. Among polysaccharides, chitosan hydrogels were evaluated as matrices for electrically modulated drug delivery such as, hydrocortisone, benzoic acid, and lidocaine hydrochloride (Masteikova et al. 2003). Likewise, artificial muscles, robots and other smart devices were fabricated using polyelectrolyte polymers based on the electro-induced volume/shape changes (Shang et al. 2019).

9.3.3 Chemical Stimuli-Responsive Hydrogels

pH-sensitive hydrogels. A polymer is considered 'pH-responsive' if it contains ionizable groups able to donate/accept protons in response to environmental pH changes (Gicquel 2017; Echeverria et al. 2018). Depending on the nature of ionizable groups, acidic (carboxylic, sulfonic acids) or basic (ammonium salts), the hydrogels can be anionic or cationic, respectively. The ionizable groups of anionic hydrogels become ionized at a pH value greater than acid dissociation constant (pKa), and cationic hydrogel swell at pH lower than their pKa, respectively (Figure 9.6) (Chan 2009; Fajardo et al. 2015).

The process of swelling/shrinking in response to environmental pH changes is reversible, making this kind of hydrogels of interest especially for biomedical applications in developing a wide variety of drug delivery systems. These applications are based on the fact that, both in the healthy human body as well as being affected by some diseases, pH changes occur in different locations or organs (Chatterjee and Hui 2018). For instance, cationic hydrogels are suitable for drug release in the stomach,

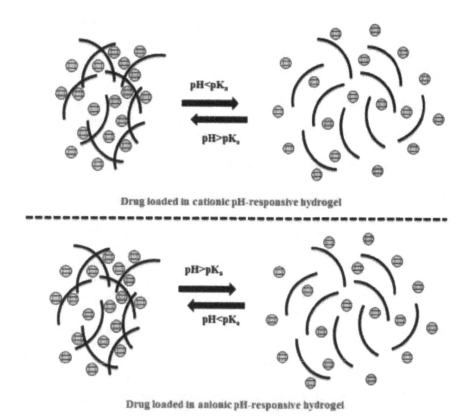

FIGURE 9.6 Schematic representation of swelling/shrinking of pH-responsive hydrogels. (Reprinted with permission from Chatterjee and Hui 2018).

providing protection of the drug in the oral cavity (pH 5.8–7.4), while releasing it in the stomach (pH 1–3.5) (Sharpe et al. 2014).

Ion-sensitive hydrogels. Some of the polysaccharides fall into the class of ion-sensitive ones, among which may be mentioned: (i) k-carrageenan which forms rigid and brittle gels in reply of small amount of K^+; (ii) i-carrageenan forms elastic gels mainly in the presence of Ca^{2+} (Masteikova et al. 2003); and (iii) alginate hydrogels are also formed in the presence of divalent/polyvalent cations (Ca^{2+}), their characteristics depending on guluronic (G) and mannuronic (M) acid blocks proportion: when G-segments predominate, alginate form stiffer and mechanically more stable gels with an egg box-like structure; in contrast, if alginate contains high proportion of M blocks units, soft and elastic gels were formed (Chowhan and Giri 2020). Ion-responsive hydrogels are widely applied in drug delivery systems for site-specific drug release, based on their abilities to exploit specific ion-gradients in the human body (Yoshida et al. 2013).

Redox-responsive hydrogels. Redox-responsive polymers are materials able to respond to reduction/oxidation of their constituent molecular components. The activation of such systems can be chemical or electrochemical (Echeverria et al. 2018).

Redox-responsive hydrogels can be regarded as promising biomaterials for biomedical applications, in tissue engineering or drug delivery (Shang et al. 2019). The redox-responsive system involved in a drug delivery system requires: (i) the use of disulfide cross-linkers in hydrogel's formation, the guest molecules being released after breaking disulfide bonds, (ii) the presence of ferrocene functions that can change the hydrophilic–hydrophobic balance of the gel depending on the redox state of the ferrocene moieties, or (iii) the use of conducting polymers, which respond to electrical current by releasing the encapsulated drug out of the hydrogel (Khodeir et al. 2019).

9.3.3.1 Biological Stimuli-Responsive Hydrogels

Enzyme-sensitive hydrogels. The sensitivity of an enzyme as a stimulus is unique due to its high selectivity. Moreover, enzyme-sensitive materials are sensitive only to specific enzymes, so by enzymatic regulation of the material properties an extremely selective sensitivity to biological signals is obtained (Abul-Haija and Ulijn 2014; Bawa et al. 2009). The enzyme-catalysed reactions generally require milder conditions compared to other stimuli, such as aqueous environments, relatively low temperatures, and neutral or slightly acidic/alkaline pH. These defining advantages make enzyme-catalysed reactions suitable for biomedical applications (Chandrawati 2016).

Enzyme-sensitive hydrogels are frequently used for targeted drugs administration, their potential to deliver enzymatically controlled drugs depending on the ability to allow the release of drugs when needed and only in the presence of specific enzymes (Billah et al. 2018). For example, a cancer-specific enzyme secreted by tumour cells can be used to trigger the release of a therapeutic drug to prevent or reduce metastasis, by immobilizing drug molecules to a polymeric backbone via enzyme-cleavable linkers. Ulijn and co-workers developed an enzyme-responsive hydrogel with physically entrapped guest molecules whose release is determined by charge-induced hydrogel swelling, which is controlled enzymatically (Ulijn et al. 2007). Liu et al. (2018) developed a dynamic gelatin-hyaluronic acid hybrid hydrogel system through integrating modular thiol-norbornene photopolymerization and enzyme-triggered on-demand matrix stiffening. In particular, gelatin was dually modified with norbornene and 4-hydroxyphenylacetic acid to render this bioactive protein photo-cross-linkable (through thiol-norbornene gelation) and responsive to tyrosinase-triggered on-demand stiffening (through HPA dimerization) (Figure 9.7).

Glucose-sensitive hydrogels. The glucose molecule is one of the most important stimuli because insulin-dependent diabetes mellitus, a major global public health epidemic, is characterized by increasing glucose concentration in the blood (Chang et al. 2018). Considerable research has been dedicated to develop self-regulated insulin delivery systems based on glucose-responsive hydrogels, which swell or shrink in response to changing glucose level, allowing the delivery of appropriate amounts of insulin, mimicking the natural response of the body (Masteikova et al. 2003; Ravaine et al. 2008; Peppas and Bures 2008; Roy et al. 2010).

9.3.3.2 Dual-Responsive Hydrogels

The systems which respond to a combination of two or more stimuli are called multi-responsive systems (Liu et al. 2017). Among all of the aforementioned stimuli, temperature and pH are the most common physical and chemical ones. Recently,

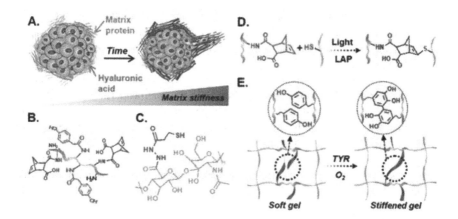

FIGURE 9.7 Enzyme-triggered on-demand stiffening of biomimetic hydrogels for *in vitro* cancer cell research. (A) Schematic of a tumour microenvironment with various matrix proteins and glycosaminoglycans (e.g., HA). Accumulation of these ECM leads to matrix stiffening and tumour progression. (B) Chemical structure of GelNB-HPA, (C) Chemical structure of THA. (D) Schematic of light and photoinitiator (LAP) induced thiol-norbornene cross-linking. (E) Schematic of tyrosinase-triggered di-HPA cross-linking and on-demand hydrogel stiffening. (Reprinted with permission from Liu et al. 2018).

numerous double-responsive materials, sensitive to both pH and temperature, have been developed (Zhang et al. 2011; Chatterjee and Hui 2018). Dual-stimuli or multi-stimuli-responsive hydrogels can be obtained by two or more kinds of single sensitive monomers, through free radical copolymerization or interpenetrating network structure. They are extremely applicable to drug release, being obvious that more stimuli responses would generate more flexibility and more precise control on the release spatiotemporal behaviour (Gao et al. 2019).

Poly(N-isopropylacrylamide) (PNIPAM) is a synthetic polymer very well known for its thermo-responsive character, its hydrogels being the most extensively exploited thermo-sensitive gels with a reversible phase transition near body temperature (LCST near 32°C) (Lee and Mooney 2012). However, the deswelling rate of PNIPAM hydrogel is very slow and this drawback is overcome by grafting PNIPAM into natural based polymers (Prabaharan and Mano 2006). A number of polysaccharides, including chitosan (Echeverria et al. 2018; Chatterjee and Hui 2018), alginate (Dumitriu et al. 2009; Lee and Mooney 2012), cellulose (Zhou et al. 2015; Masruchin et al. 2017), and dextran (Lima et al. 2011), were considered to be combined with these thermo-responsive materials, in order to develop novel hydrogels that simultaneously exhibit the properties of both components.

9.4 BIOMEDICAL APPLICATIONS OF POLYSACCHARIDE-BASED STIMULI-RESPONSIVE HYDROGELS

The biomedical field is constantly looking for new biomaterials with innovative properties, and natural polymers are good candidates for this matter having in view their unique properties (Silva et al. 2012). From this large group of natural polymers, polysaccharides in particular have received increasing attention in bio-medical and pharmaceutical fields due to natural abundance, their unique chemical structures and physicochemical/biological properties, and the ability to form hydro-gels (Chan 2009).

Hydrogels derived from natural occurring polysaccharides are high-water content polymeric materials that possess a number of favourable properties for biomedical applications (Salgueiro et al. 2013). The hydrogels' desired characteristics depend on the application: in wound healing, the ability to absorb wound exudates is very important, along with oxygen transport to the wound site; in tissue engineering, the ability to mimic the extracellular matrix and support cell growth is sought; in drug delivery, soft and rubbery consistency and low interfacial tension with water and biological fluids are considered as excellent characteristics (O'Connor et al. 2018; Fu et al. 2018).

9.4.1 STIMULI-RESPONSIVE HYDROGELS IN DRUG DELIVERY SYSTEMS

The primary goals in development of drug delivery systems are to protect an active therapeutic molecule from premature degradation, enhance drug efficiency by pro-longing in vivo drug actions and reducing its metabolism rate, and also minimize its side effects by reducing some possible drug toxicity (Sharpe et al. 2014; Fajardo et al. 2015). Oral drug delivery is the ideal administration route that involves low costs (Nokhodchi et al. 2012) and is desirable for a diversity of therapeutics for treatment of both systemic and local diseases. Unfortunately, this is limited only to conven-tional small-molecule drugs. For macro-molecular drugs, such as peptides, proteins, and chemotherapeutics, which suffer poor stability in the gastro-intestinal tract, being subject to enzymatic degradation, acidic denaturation, low solubility, and absorption, innovative drug delivery systems are required (Sharpe et al. 2014). Over the last few decades, extensive researches related to the hydrogels used in drug deliv-ery have been made, being encouraged by their biocompatible properties and easy control of solute transport (Jeong et al. 2012).

The hydrogel-based drug delivery systems are generally of two types: time-con-trolled systems and stimuli-induced release systems. The last category of responsive hydrogel systems is designed to deliver the drugs in response to a variable condition so that it can fulfill its requirements at the right time and place (Das and Pal 2015). Characteristics like the swelling behaviour in an aqueous medium, external stimuli sensitivity, and zero-order kinetics play important roles in the development of hydro-gel-based drug delivery systems (Dumitriu et al. 2009; Fu et al. 2018). In the last decades, huge progresses have been made in the use of polysaccharides to obtain stable and versatile hydrogels to be applied as carriers for drug delivery (Fajardo et al. 2015; Kumar et al. 2018) and the wide variety of polysaccharides response stimuli make them particularly attractive in this field (Alvarez-Lorenzo et al. 2013).

9.4.1.1 Thermo-Responsive Hydrogels in Drug Delivery

Temperature-sensitive hydrogels are probably the most commonly studied class of environment-sensitive polymer systems in drug delivery applications (Shen et al. 2016). Many natural polymers have been shown to exhibit gelation when temperature changes. They can be used alone or in combination with synthetic polymers to obtain thermally responsive hydrogels with desired properties (Klouda and Mikos 2008).

Cellulose-based thermo-responsive hydrogels. Cellulose-based materials have become very popular in recent years in obtaining stimuli-sensitive hydrogels due to their versatility, renewability, relatively low cost, abundance, and extensive study of these bio-polymers (Kumar et al. 2018). Cellulose derivatives have been extensively investigated for biomedical applications in drug delivery.

Methylcellulose (MC), the simplest cellulose derivative, exhibits thermo-reversible gelation properties in aqueous solutions: it gels on heating at temperature values between 60 and 80°C and turns into solution upon cooling. But this gelling temperature is found to be too high for biomedical applications, so methylcellulose is grafted with synthetic N-isopropylacrylamide (NIPAM) resulting in hydrogels which combine the thermo-gelling properties of both materials, in addition to improved mechanical characteristics due to methylcellulose (Klouda and Mikos 2008).

Carboxymethylcellulose (CMC), which is another functional derivative of cellulose, is capable of forming thermo-responsive hydrogels. The drug delivery system based on carboxymethyl cellulose and gelatin showed sol–gel transition near the body temperature and was used for transdermal drug therapy with lidocaine for treatment of atopic dermatitis. These hydrogels have provided both moisture and drug to the skin, protecting pathogenesis of atopic dermatitis (Chatterjee and Hui 2018).

Hydroxypropylcellulose (HPC) is a derivative of cellulose and a biodegradable, biocompatible macromolecule. HPC shows a well-defined LCST in water at about 41°C. According to its unique hydrophilic/hydrophobic change, thermo-responsive HPC-based hydrogels have been prepared for controlled delivery of hydrophilic drugs. One advantage of HPC over many other synthetic thermally responsive macromolecules is that HPC has been approved by the United States Food and Drug Administration for the use in food, drug, and cosmetics (Zhang et al. 2011).

Chitosan-based thermo-responsive hydrogels. The chitosan-based drug delivery systems, in various chemical and physical gel forms, have been developed and studied in the past decades (Wu et al. 2006). Chitosan-based materials lack intrinsic thermo-sensitive properties. Thus, other thermo-sensitive materials need to be combined with chitosan to make it applicable as a thermo-sensitive hydrogel (Domalik-Pyzik et al. 2018). Chitosan-based thermo-sensitive hydrogels have been prepared by grafting chitosan (CS) with poly(N-isopropylacrylamide) (PNIPAM) (Lee et al. 2004; Wang et al. 2009; Chen et al. 2018), with poly(ethylene glycol) (PEG) (Kim et al. 2011), or mixing chitosan with poly(vinyl alcohol) (PVA) and sodium bicarbonate (Tang et al. 2007; Ji et al. 2009).

Grafting PNIPAM side chains on chitosan backbone is an expanding research field, as it combines the most studied thermo-sensitive polymer with the most outstanding cationic polysaccharide to achieve materials with remarkable properties. The drug-loading characteristics of the chitosan-g-PNIPAM hydrogel, obtained by direct grafting via free radical polymerization, were tested by Lee and co-workers (Lee et al. 2004) using caffeine as the model drug. The study has proved the ability of this hydrogel to function as a temperature-controlled drug delivery system, the loading amount of the model drug at 20°C being affected by increasing the chitosan/

PNIPAM weight ratio and cross-linking agent amount, and the release rate is controlled by pore size and hydrogel swelling ratio at 37°C. The ability of the PNIPAM/chitosan semi-interpenetrating network hydrogel as controlled-release vehicles for pilocarpine hydrochloride, used in some ophthalmic diseases, was evaluated (Prabaharan and Mano 2006). The release process was found to be fast, with most of the drug being released in the first 40 min and the hydrogel loading/release properties were influenced by the chitosan proportion in the network.

A thermo-sensitive drug delivery vehicle using chitosan (CS), hyaluronic acid (HA), and synthetic PNIPAM was prepared and evaluated for controlled release of an analgesic drug called nalbuphine (Kim et al. 2011), proving the positive effect of CS/HA hydrogels incorporated in PNIPAM network by greater control of the drug release in vitro compared to PNIPAM hydrogels alone.

The efficiency of PEG-grafted chitosan thermo-sensitive hydrogels has been proven as protein delivery vehicles, using albumin as a model protein. The test findings showed that protein release was rapid in the first three days, followed by constant, controlled release in the next three days (Kim et al. 2011).

The combination of carboxymethyl-chitosan (CMCS) and PNIPAM has been investigated as localized drug delivery system for anticancer and anti-inflammatory drugs. The drug molecules release is controlled by the PNIPAM grafting proportion, pH and temperature of the release medium. The cytocompatibility of these hydrogels has also been confirmed by mesenchymal stem cell culturing, all these data indicating the great potentials of these hydrogels as localized drug delivery systems (Zhang et al. 2014).

Dextran-based thermo-responsive hydrogels. Dextran has no thermo-sensitivity; therefore, other thermo-sensitive materials are added to form thermo-sensitive hydrogels. Dextran, being part of a polymer block next to oligolactate, 2-hydroxyethyl methacrylate (HEMA), and PNIPAM, are capable of forming thermo-sensitive hydrogels with LCST around 32°C and exhibiting temperature-controlled release of incorporated albumin (Kim et al. 2011). The multifunctional and biodegradable thermo-responsive hydrogels made from PNIPAM, dextran, and poly(L-lactic acid) (PLLA) were used as drug delivery systems, and it showed gelation (LCST) around 32°C (Chatterjee and Hui 2018).

Injectable thermo-sensitive hydrogels in drug delivery systems. Injectable hydrogels are superior to preformed hydrogels, primarily due to their ability to fill and cover spaces of any shape and secondly because they do not require a surgical procedure for implantation (Upadhyay 2017). They have a very high affinity for body fluids and may be delivered into body through a catheter or by direct injection with a syringe (Figure 9.8) (Mathew et al. 2018). The high-water content, soft nature, pliability, and porous structure mimicking biological tissues of these injectable hydrogels are unique features minimizing mechanical irritation and damage to the surrounding tissues during subcutaneous administration. Stimuli-sensitive injectable hydrogels are able to switch sol-to-gel transitions, in aqueous solutions, in response to various stimuli including pH, temperature, light, enzyme, and magnetic field (Thambi et al. 2016). Polysaccharide-based injectable hydrogels are extremely advantageous and have a variety of biomedical applications in drug delivery and tissue engineering (Upadhyay 2017).

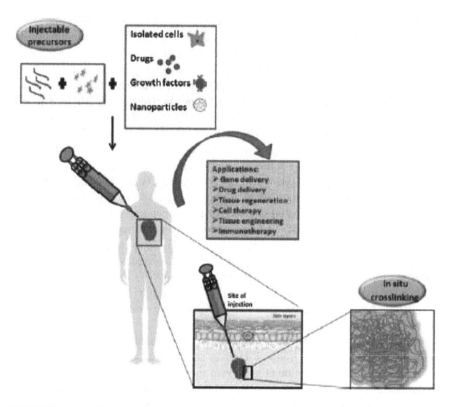

FIGURE 9.8 Application of injectable hydrogel systems in biomedical field.(Reprinted with permission from Mathew et al. 2018).

Injectable thermogel based on poly(ethylene glycol)-g-chitosan was synthesized, for drug release in vitro, using bovine serum albumin (BSA) as a model protein. In the release study, after an initial burst release in the first 5 h, a steady linear release of the protein from the hydrogel was achieved for a period of ~ 70 h. Prolonged quasi-linear release of protein up to 40 days was achieved by cross-linking the hydrogel with genipin in situ, in a fashion suitable for protein encapsulation, while maintaining the injectability of the hydrogel (Prabaharan and Mano 2006).

The thermo-sensitivity and rheological properties of injectable poly(vinyl alcohol)-chitosan hydrogels, formed by mixture of chitosan (CS), poly(vinyl alcohol) (PVA), and sodium bicarbonate was evaluated potentially as a drug release system using bovine serum albumin. PVA is frequently used as implant material for drug delivery systems and surgical repairs due to its excellent mechanical strength, biocompatibility, and nontoxicity. The hydrogel is liquid in aqueous solutions at low temperature (about 4°C) and forms gel implants in situ in response to increasing temperature to physiological value, when bioactive specie is safely and uniformly incorporated (Tang et al. 2007).

9.4.1.2 pH-Sensitive Hydrogels in Drug Delivery Systems

pH-sensitive hydrogels are another class of stimuli-sensitive hydrogels with considerable attractiveness for biomedical applications (Gao et al. 2014). The main feature of these hydrogels is their functional groups that undergo reversible ionization/deionization processes when the environmental pH changes and, implicitly, cause a swelling/collapse behaviour of the hydrogel network (Fajardo et al. 2015). The pH-responsive hydrogels are of particular interest for biomedical applications, especially for developing a wide variety of drug delivery systems, because substantial pH changes are found in various organs or locations in the body (gastrointestinal tract, blood vessels, intracellular vesicles) (Chatterjee and Hui 2018).

Chitosan-based pH-responsive hydrogels. Chitosan is an excellent example of pH-responsive natural polymer; its pH-responsiveness comes from primary amino groups which can be protonated or deprotonated depending on the pH of external environment (Lee et al. 2004). It forms cationic hydrogel network in water which swells in acidic pH and remains collapsed in basic pH (Sharpe et al. 2014). Chitosan-based systems are used for the delivery of proteins/peptides (Amidi et al. 2010), growth factors (Li et al. 2014), anti-inflammatory drugs (Badawi et al. 2008), antibiotics (Saha et al. 2010; Sevostyanov et al. 2015), as well as in gene therapy (Mao et al. 2010). Chitosan-based pH-sensitive hydrogels are performed via physical or chemical interactions (Delmar and Bianco-Peled 2016).

pH-sensitive xanthan–chitosan hydrogels, performed by ionic interactions between the free carboxyl group of xanthan and the free amino group of chitosan, were used for the encapsulation of neomycin sulfate drug. The hydrogels can sustain the release of neomycin sulfate for about 24 hours, as demonstrated by in vitro tests. The antimicrobial tests confirmed the prolonged drug release and effective bacterial inhibition when exposed to *Escherichia coli* and *Staphylococcus aureus* (Merlusca et al. 2018). Physically cross-linked pH-responsive hydrogel based on chitosan was developed via in situ free radical polymerization, using chitosan, acrylic acid (AA), and (2-dimethylaminoethyl) methacrylate (DMAEMA). These hydrogels proved to have very good mechanical strength and be suitable for controlled drug delivery of bovine serum albumin, and 5-fluorouracil (5-FU) in cancer therapy (Chatterjee and Hui 2018). Covalently cross-linked chitosan hydrogels with dialdehydes, such as glyoxal and glutaraldehyde (GA), lead to mechanical and chemical reinforced matrix, which are more stable and more suitable for intestinal protein delivery (Sharpe et al. 2014).

Interpenetrating polymeric network (IPN) gels based on chitosan and PVA were prepared and cross-linked either chemically with glutaraldehyde or by γ-irradiation. The obtained IPNs were characterized, and equilibrium swelling and the *in vitro* cumulated release of 5-fluorouracil (5-FU), as a model drug, in different buffer solutions were studied. The release percent of 5-FU depends directly on the content of PVA and more significantly on the pH of the releasing medium, where it was faster in an acidic medium than in a neutral or weakly alkaline one (Abdelaal et al. 2007).

Hollow silica nanoparticles (HNPs) covered with a chitosan layer have been investigated as drug delivery carriers to control the release of loaded anticancer drugs to the acidic local environment in tumour tissues. With mesoporous structures, silica nanoparticles are particularly attractive due to their large surface area, highly accessible pores, bio-inertness, and compatibility. The resulting nanocarriers exhibit good performance in gradually releasing tumour necrosis factor alpha (TNF-α) to breast cancer cells, both in vitro and in vivo, inducing apoptosis of targeted cells (Deng et al. 2011).

Cationic chitosan hydrogel N-(2-hydroxyl) propyl-3-trimethyl ammonium chitosan chloride (HTCC), prepared with different degree of substitution (DS) by a relatively easy chemical reaction of chitosan and epoxypropyl trimethyl ammonium chloride (EPTMAC), and the effect of DS on the pH- and glucose-sensitivity was discussed. With increasing DS, the cross-linking density decreased, and in the meantime the swelling ratio and pH-sensitivity of the hydrogel increased. HTCC hydrogels showed glucose concentration-dependent response release behaviour, and the step like release behaviour for incremental doses of glucose clearly showed the versatility of the system to respond to the sequential addition of glucose (Zou et al. 2015).

Starch-based pH-responsive hydrogels. Starch-based biodegradable polymers, in the form of microsphere or hydrogel, are suitable for drug delivery (Lu et al. 2009). Modified starch film can also act as a carrier for drug release. For instance, starch-based hydrogels have been reported for colon-targeted drug delivery. Starch-methacrylic acid (MAA) copolymer hydrogels was prepared and the release behaviour of ketoprofen as model drug was investigated. The starch-based hydrogels could retain the loaded drug (ketoprofen) at pH 1 and release it at pH 7. Thus, the hydrogel has good pH-sensitivity and can be a good candidate for colon drug delivery systems (Ismail et al. 2013).

Blending lignin with corn starch improves significantly the mechanical and water absorption properties of the starch films; and in addition, this film can also control the release of the load drug ciprofloxacin in response to the pH of the medium. Only 75% of ciprofloxacin was released in a medium with pH of 7.5, while almost 100% was released in a medium of pH 1 (Zhang 2015).

Alginate-based pH-responsive hydrogels. Alginate has been extensively used in preparing drug carriers due to its biocompatibility, good morphological and mechanical properties. It has the property of gelation in an aqueous solution with aid of divalent cations such as Ca^{2+} and Mg^{2+}. Park et al. studied the release profile of dual-stimuli-responsive hydrogel beads composed of calcium alginate and PNIPAM. The results showed that at 25°C the release rate from the beads was slower than that from the homo-PNIPAM beads due to the presence of the Ca-alginate chain network (Prabaharan and Mano 2006). Drug release from alginate hydrogels cross-linked with Ca^{2+} depends on the pH of the medium and the solubility of the drug. Generally, poorly water-soluble drugs are not released in acid pH due to the poor swelling by the hydrogels. In contrast, in phosphate buffer and simulated intestinal fluid the swelling and the release are promoted as phosphate ions extract the calcium from the hydrogels (Alvarez-Lorenzo et al. 2013).

Hyaluronic acid-based pH-responsive hydrogels. Another natural polymer commonly used in drug delivery formulations is hyaluronic acid (HA), which is an anionic glucosaminoglycan. The presence of one carboxylic group per repeat unit imparts a pH-responsiveness, and such behaviour is enhanced in cross-linked hydrogel network. Recent research found pH-responsive HA nanoparticles as a viable option for oral insulin delivery systems, showing enhanced delivery *via* transcellular pathway found in both in vitro and in vivo studies (Sharpe et al. 2014). Fiorica et al. developed a novel derivative of HA with increased carboxylic groups to optimize the system for delivery to the colon and demonstrated pH-sensitive release using α-chymotrypsin (Fiorica et al. 2013).

Cellulose-based pH-responsive hydrogels. Carboxymethyl cellulose (CMC), obtained by substitution of some of the hydroxyl groups of cellulose with carboxymethyl groups, displays pH-sensitivity due to the incorporation of carboxyl groups, making this material interesting for oral delivery systems (Gao et al. 2014).

9.4.1.3 Stimuli-Responsive Nanogels in Drug Delivery Systems

Hydrogels with dimensions in submicron range, known as nanogels, have received considerable attention as drug delivery systems, mainly because they combine the advantages resulting from their small dimensions (improved bioavailability, large surface area for multivalent bioconjugation, targeted delivery of drug, easier intracellular permeation, high stability, adjustable particle size) with the hydrogels properties like hydrophilicity and biocompatibility (Oh et al. 2008; Daniel-da-Silva et al. 2011; Hajebi et al. 2019).

Naturally derived nanogels can be prepared from polysaccharides, such as chitosan, hyaluronic acid, heparin, chondroitin sulfate, agarose, and alginate. For instance, hyaluronic acid-based nanogels (Yang et al. 2015; Pedrosa et al. 2017), chitosan-based nanogels (Duan et al. 2011; Wang et al. 2013; Feng et al. 2015), or hybrid microgels based on chitosan chemically cross-linked with hyaluronic acid (Schmitt et al. 2010) were prepared for delivery of different drugs. Different techniques have been applied to get small gel particles, as indicated in Figure 9.9.

Nanogels prepared from external stimuli-responsive polymers are of particular interest, being promising materials for the development of more effective disease treatments based on intelligent drug delivery systems, which can sense and respond directly to pathophysiological conditions (Daniel-da-Silva et al. 2011).

pH-responsive drug delivery systems based on nanogels are well explored by numerous researchers. Because of the different pH values observed in healthy tissues (pH 7.4), and tumour tissues (pH 6.5–7.0), nanogels have been designed to be sensitive to the particular pH range of interest allowing drug release only in the tissue to be targeted (Hajebi et al. 2019). For example, nanogels containing pH-responsive functional

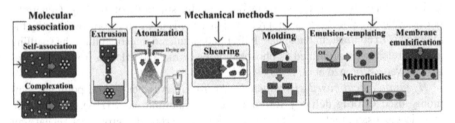

FIGURE 9.9 Different preparation methods for obtaining micro/nanogel particles. (Reprinted with permission from del Valle et al. 2017).

linkages in structures have been developed for the attachment of pH-triggered drug delivery for cancer therapy using the facts that tumour extracellular environment is more acidic than blood (Debele et al. 2016).

Nanogels composed of *chitosan* (CS) and carboxymethyl chitosan (CMCS) were generated *via* electrostatic interaction between positively charged CS and negatively charged CMCS in the presence of sodium tripolyphosphate (TPP) as ionic cross-linker. The doxorubicin hydrochloride (DOX) loading capacity of these nanogels has been greatly improved by incorporation of CMCS compared with CS only, mainly due to better interaction between positive charged DOX and negatively charged CMCS. The drug release rate of DOX:CS/CMCS nanogels was much lower in simulating gastric acid environment (< 18%) than that in simulating intestinal environment, which was in favour of targeted delivery of drug to the colon (Debele et al. 2016)

Duan and co-workers synthesized chitosan grafted poly(N-isopropyl acrylamide) (PNIPAM) nanogels using N,N'-methylene-bis-acrylamide (MBA) as the cross-linking agent, and a potent anticancer agent for variety of tumours and cancer cell types (Oridonin, ORI) was successfully encapsulated. In vitro drug release result indicated that, ORI-loaded nanogels could enhance the antitumor activity under an acidic environment, because these nanogels exhibited a pH-triggered fast drug release under a slightly acidic condition: the release rates drastically increased from about 50% at pH 7.4 to more than 80% at pH 6.5, 6.0 and 5.0, owing in part to the ionization of chitosan (Duan et al. 2011).

Hyaluronic acid nanogels with enzyme-sensitiveness were prepared for doxorubicin (DOX) delivery (Yang et al. 2015) The DOX-loaded nanogels showed superior antitumour efficacy compared with the free DOX, demonstrated by great tumour penetrating capacity, enhanced DOX accumulation in the tumour site, and prolonged DOX circulation time. Therefore, these targeting, biocompatible, and multi-enzymatic degradable HA nanogels have great potential in drug delivery for cancer therapy.

Redox-sensitive dextran nanogels, for intracellular delivery of antigens (ovalbumin, OVA) were synthesized (Li et al. 2015). Disulfide bonds of dextran nanogels are stable in the extracellular environment, but are reduced in the cytosol of dendritic cells (DCs), due to the presence of glutathione (GSH). OVA-conjugated nanogels show intracellular release of the antigen in DCs and boost the major histocompatibility complex class I (MHC-I) antigen presentation, demonstrating the feasibility of this concept for the aimed intracellular antigen delivery.

Redox-sensitive nanogels based on carboxymethyl cellulose with a diameter of about 192 ± 2 nm were synthesized from methacrylated carboxymethyl cellulose (MACMC) and disulfide containing cystamine bisacrylamide (CBA). In vitro cytotoxicity, indicated that the relatively faster release of DOX from MACMC–CBA nanogels in the reductive environment of the cytoplasm can induce more inhibition of cell proliferation, leading to an improved antitumour efficiency. In vivo antitumour evaluation also showed that DOX loaded nanogels exhibited a significantly superior antitumour effect than the free DOX by combining the tumor volume measurement and the examination of cell apoptosis and proliferation in tumour tissues (Qian et al. 2014).

9.4.2 STIMULI-RESPONSIVE HYDROGELS BASED ON POLYSACCHARIDES IN TISSUE ENGINEERING

In tissue engineering, scaffolds play a very important role, developed as temporary and artificial matrix to support the cells attachment and the three-dimensional (3D) tissue formation (Barman and Das 2018). The principle of tissue engineering is conceptually very simple: cells are cultivated in vitro on scaffolds with 3D network structure, which offer support for cells to attach, migrate, proliferate, and differentiate (Kim et al. 2011).

An ideal scaffold must mimic the native extracellular matrix. Hydrogels, with 3D network structure, with ability to take on large amounts of water and physiological fluids, and with physical properties similar to those of living tissues can be considered ideal scaffolds (Thambi et al. 2016; Onofrei and Filimon 2016). Polysaccharide-based stimuli-responsive hydrogels as scaffolds for cell cultures is a promising field; they can provide tailored biofunctions at molecular level, adjustable properties, and morphologies according to external/internal stimuli, as well as biocompatibility and biodegradability. Thus, the polysaccharides hydrogels represent a promising microenvironment for cell growth and tissue formation (Fajardo et al. 2015).

9.4.2.1 Thermo-Responsive Hydrogels in Tissue Engineering

In tissue engineering, thermo-responsive polymers are commonly used in two situations: (i) as substrates that enable the cells growth and proliferation and (ii) as injectable gels, for in situ scaffold generation. In case of thermo-responsive polymer used as a substrate, its ability to be sensitive to temperature is important to adjust the cells' attachment/detachment from the surface. If injectable gels are used, the procedure involves the mixing of cells with the thermo-responsive polymer at room temperature, followed by the injection into the body; due to the temperature increase to 37°C, the polymer forms a physical gel, the cells being encapsulated within the gel 3D structure (Gandhi et al. 2015).

Cellulose-based hydrogels usually couple their biodegradability with a smart stimuli-sensitive behaviour. These features, together with the large availability of cellulose in nature and the low cost of cellulose derivatives, make cellulose-based hydrogels particularly attractive as tissue engineering scaffolds (Sannino et al. 2009). The degradation of cellulose results in glucose, which serves as a nutrient for cell; thus, cellulose hydrogels as scaffolds can provide structural integrity to tissue constructions and, additionally, serve as nutrient for cell proliferation (Barman and Das 2018).

Cellulose derivatives, such as MC, CMC, and HPMC, due to their excellent biocompatibility, biodegradability, and environmental friendliness, are extensively used in the regeneration of different tissues, such as bone, heart, cartilage, blood vessel, and nerve (Rusu et al. 2019). Methylcellulose (MC) has the transition sol–gel temperature at about 60~80°C, so in order to be used as temperature-sensitive hydrogel for tissue engineering applications, some modifications of polymer are necessary. Grafting methylcellulose with synthetic polymer N-isopropylacrylamide (NIPAM) combines the thermo-sensitive properties of both materials, and in addition it is possible to modulate LCST of the resulting hydrogel by adjusting the ratios of the two components (Kim et al. 2011; Contessi et al. 2017; Chatterjee and Hui 2018). Thermo-responsive hydrogels MC-PNIPAM have been evaluated for proliferation of human adipose derived stromal/stem cells (hASC). Cell sheets have maintained cell viability for up to 21 days, and the spontaneously detached cell sheets after incubation at room temperature, kept sheet morphology, extracellular matrix structure, and cell–cell interactions largely intact. It is clear that MC-PNIPAM have potential clinical use in tissue engineering applications (Forghani et al. 2017).

Chitosan is an attractive material both as tissue engineering scaffold and as injectable hydrogel, due to structural similarities to glucosaminoglycans and its hydrophilicity, but its hydrogels have low mechanical strength and slow temperature response, so further modifications or combination with other material are needed (Ahmed and Ikram 2016; Liu et al. 2017; Huang et al. 2019).

Thermo-sensitive formulations based on chitosan-β-glycerophosphate (CGP) with different concentrations of starch were successfully used as injectable vehicles for cell delivery and support, aiming cartilage regeneration through engineered strategies. The addition of starch did not change the transition temperature of CGP and, in addition, improved the degradation profile. The heating inducing gelation was confirmed for all solutions, which remain liquid at room temperature and solidify into hydrogels near body temperature, supporting the ability of these novel formulations to be applied as minimal invasive systems. All these data suggest the potential of these novel thermo-responsive CGP hydrogels to be used as injectable scaffolds for cartilage engineered applications (Sa-Lima et al. 2010).

Hydrogels with different ratios of chitosan and collagen have been fabricated by initiating gelation using β-glycerophosphate (β-GP) and temperature, in order to be used in delivery of human bone marrow-derived stem cells. The presence of collagen was associated with high cell proliferation, as well as increased gel compaction and a resulting stiffer matrix. These chitosan–collagen composite hydrogel materials have potential applications in regenerative medicine, particularly in applications where injectable cell carriers are advantageous. Such materials can be used for cell encapsulation and delivery, or as in situ gel-forming materials for tissue repair (Wang and Stegemann 2010).

An injectable hydrogel was formed *via* ionic and hydrophobic interactions between chitosan and methylcellulose chains, in the presence of various salts (NaCl, Na_3PO_4, $NaHCO_3$, and glycerophosphate), under mild conditions without organic solvent, high temperature, or harsh pH. Such blends were liquid at low temperature, but gel under physiological conditions (37°C). Spectroscopic characterization showed that the hydrogels had good miscibility/uniformity/homogeneity and different pore structures (ranging from nonporous to microporous). When used as a scaffold for chondrocytes, hydrogel resulted in good cell viability and proliferation, indicating potential use as a three-dimensional synthetic matrix for tissue engineering (Tang et al. 2010).

Hyaluronic acid (HA) is an important component of cellular microenvironments, because it can directly affect tissue organization *via* the interaction with cell-surface receptors, which promotes the migration of cells and stimulates the production of collagen II as well as cell proliferation (Thambi et al. 2016). Due to its properties, HA has been considered an attractive starting material for the construction of hydrogels with desired morphology, stiffness, and bioactivity to be applied as scaffolds (Fajardo et al. 2015). HA offers many advantages as a tissue scaffold which include the following: biodegradability, biocompatibility, and bio-resorbability. This contains functional groups (carboxylic acids and alcohols) along its backbone that can be used to introduce functional domains or to form a hydrogel by cross-linking. Thus, it can be part of a new kind of tissue engineering scaffold that is bioactive both in its full length and in the degraded form (Collins and Birkinshaw 2013). The use of HA in its native form in tissue engineering appli-cations is pretty limited, due to its low stability in such applications. Several cross-linking methods are used to increase the stability of HA: water-soluble carbodiimide, divinyl sulfone, disulfide, and photo-cross-linking through glycidyl methacrylate-hyaluronic acid conjugates. Hence, these techniques of covalent cross-linking pro-vide the opportunity to combine hyaluronic acid with more mechanically stronger polymers (Aravamudhan et al. 2014).

Hyaluronic acid was used as a component along with thermo-responsive poly(N-isopropylacrylamide) (PNIPAM) to prepare injectable gels with ophthalmic applica-tions. The gel systems were prepared in the presence of the antibody infliximab (INF): on one side, tyramine-substituted HA (HA-Tyr) was enzymatically cross-linked in the presence of INF to form HA-Tyr-INF, and on the other hand, PNIPAM was chemically cross-linked in the presence of HA and INF with poly(ethylene gly-col) diacrylate (PEGDA) to form PEGDA-PNIPAM-HA-INF. All hydrogels demon-strated sustained release of INF in vitro using human eye model called the PK-Eye. The 1% PEGDA-PNIPAM-HA-INF hydrogel displayed the slowest release (24.9 ± 0.4% INF release by day 9) in phosphate buffered saline (PBS, pH 7.4), which is a better release profile than the free drug alone. These results suggest that PEGDA-PNIPAM-HA has potential for development of formulations to prolong the intraocu-lar release of proteins (Egbu et al. 2018).

The use of *carrageenans* in the biomedical field is being explored, mainly taking advantage of their biological activities mostly related to the sulfate content. From this perspective, carrageenan has been considered not only for growth factor or enzyme immobilization but also in encapsulation of several cell types for their in vivo deliv-ery, envisioning cartilage regeneration (Silva et al. 2012).

k-Carrageenan can be a suitable material for cartilage tissue engineering as it is non-cytotoxic and an efficient cell entrapment. Human adipose-derived stem cells (hASCs) were encapsulated with transforming growth factor-β1 (TGF-β1) in the k-carrageenan hydrogel networks. The k-carrageenan was found to be a suitable bio-material for cell and growth factor encapsulation. The incorporation of TGF-β1 within the carrageenan-based hydrogel enhanced the cartilage differentiation of hASCs. These findings indicate that this new system for cartilage TE is very promis-ing for injectable thermo-responsive formulation applications (Rocha et al. 2011).

Ternary composite collagen-hydroxyapatite-k-carrageenan (COL-HAP/KCAR) was designed as a substitute material for bone tissue and fabricated *via* in situ syn-thetic reaction, by introduction of k-carrageenan to the collagen-hydroxyapatite (COL-HAP) system. The studies on the mechanical properties, hygroscopicity, bio-degradability, cytotoxicity, and hemolytic phenotype indicated that the material is a potential bone repair material (Feng et al. 2017).

9.5 CONCLUSIONS

In the past few decades, a considerable attention has been given to biopolymers, due to their sustainability, eco-efficiency, and renewable nature. One of the most abundant and widely available renewable resources are polysaccharides, which can be found as main constituents of plants and algae or can be synthesized by microbial organisms. These exhibit a large variety of structural features and physicochemical properties, which allows them to produce versatile hydrogels with tunable properties. Thus, the synthesis of polysaccharide-based stimuli-responsive hydrogels is of great interest, due to their versatility and the possibilities to be used in various application fields.

The current trend in the eco-friendly design of these hydrogels directed the researches to find new non-toxic cross-linking agents, in order to improve the safety of final product, as well as of the manufacturing process. In addition, to help the patient, injectable hydrogel formulations have been developed that have an easy administration and minimally invasive treatments, consequently, determining an improvement in patient comfort. Furthermore, to improve clinical administration, new possibilities to modulate the drug release, to allow the controlled release of more drugs from the same matrices and the degradation profiles were found by varying the composition and gelation conditions of the hydrogels.

Even if the significant advantages and versatility of polysaccharide-based stimuli-responsive hydrogels have been highlighted in this chapter, the suitable polymers for biological media in a living system, as well as the hydrogels for multiple applications and specific treatments per patient, are still a challenge.

ACKNOWLEDGEMENTS

This work was supported by a grant of the Romanian Ministry of Research and Innovation, CCCDI-UEFISCDI, project number PN-III-P1-1.2-PCCDI-2017-0697/13PCCDI/2018, within PNCDI III.

LIST OF ABBREVIATIONS

AA	acrylic acid
BSA	bovine serum albumin
CMC	carboxymethylcellulose
CMCS	carboxymethyl-chitosan
CS	chitosan
COL	collagen
CBA	cystamine bisacrylamide
DS	degree of substitution
DC	dendritic cells
DMAEMA	(2-dimethylaminoethyl) methacrylate
DOX	doxorubicin hydrochloride
EPTMAC	epoxypropyl trimethyl ammonium chloride

5-FU	5-fluorouracil
β-GP	β-glycerophosphate
GA	Glutaraldehyde
GSH	glutathione
HNP	hollow silica nanoparticles
hASC	human adipose derived stem cells
HA	hyaluronic acid
HAP	hydroxyapatite
HEMA	2-hydroxyethyl methacrylate
HPC	hydroxypropylcellulose
INF	infliximab
IPN	interpenetrating polymeric network
NIPAM	N-isopropylacrylamide
KCAR	k-carrageenan
LCST	low critical solution temperature
MHC-I	major histocompatibility complex class I antigen
MACMC	methacrylated carboxymethyl cellulose
MAA	methacrylic acid
MC	methylcellulose
HTCC	N-(2-hydroxyl) propyl-3-trimethyl ammonium chitosan chloride
MBA	N,N-methylene-bis-acrylamide
ORI	Oridonin
PEG	poly(ethylene glycol)
PEGDA	poly(ethylene glycol) diacrylate
PNIPAM	poly(N-isopropylacrylamide)
PLLA	poly(L-lactic acid)
PVA	poly(vinyl alcohol)
TGF-β1	transforming growth factor-β1
TPP	tripolyphosphate
Tyr	tyramine
TNF-α	tumour necrosis factor alpha
UCST	upper critical solution temperature

REFERENCES

Abdelaal, M. Y., Abdel-Razik, E. A., Abdel-Bary, E. M., et al. 2007. Chitosan-based inter-polymeric pH-responsive hydrogels for in vitro drug release. *J Appl Polym Sci* 103:2864–2874.

Abul-Haija, Y. M., and R. V. Ulijn. 2014. Enzyme-responsive hydrogels for biomedical applications. In: *Hydrogels in cell-based therapies*, ed. C. J. Connon and I. W. Hamley, 112–134. The Royal Society of Chemistry.

Ahmed, S., and S. Ikram. 2016. Chitosan based scaffolds and their applications in wound healing. *Achiev Life Sci* 10:27–37.

Altomare, L., Bonetti, L., Campiglio, C. E. et al. 2018. Biopolymer-based strategies in the design of smart medical devices and artificial organs. *Int J Artif Organs* 41(6):337–359.

Alvarez-Lorenzo, C., Blanco-Fernandez, B., Puga, A. M., et al. 2013. Crosslinked ionic polysaccharides for stimuli-sensitive drug delivery. *Adv Drug Deliver Rev* 65:1148–1171.

Amidi, M., Mastrobattista, E., Jiskoot, W., et al. 2010. Chitosan-based delivery systems for protein therapeutics and antigens. *Adv Drug Deliver Rev* 62:59–82.

Amin, M. C. I. M., Ahmad, N., Pandey, M., et al. 2014. Stimuli-responsive bacterial cellulose-g-poly(acrylic acid-co-acrylamide) hydrogels for oral controlled release drug delivery. *Drug Dev Ind Pharm* 40(10):1340–1349.

Aravamudhan, A., Ramos, D. M., Nada, A. A., et al. 2014. Natural polymers: Polysaccharides and their derivatives for biomedical applications. In: *Natural and synthetic biomedical polymers*, ed. S. G. Kumbar, C. T. Laurencin and M. Deng, 67–89. Elsevier.

Babu, R. P., O'Connor, K., and R. Seeram. 2013. Current progress on bio-based polymers and their future trends. *Prog Biomater* 2:8–24.

Badawi, A. A., El-Laithy, H. M., El-Qidra, R. K., et al. 2008. Chitosan based nanocarriers for indomethacin ocular delivery. *Arch Pharm Res* 31(8):1040–1049.

Barman, A., and M. Das. 2018. Cellulose-based hydrogels for pharmaceutical and biomedical applications. In: *Cellulose-based superabsorbent hydrogels*, ed. Md. I. H. Mondal, chapter 36, 1–28. Springer International Publishing.

Bawa, P., Pillay, V., Choonara, Y. E., et al. 2009. Stimuli-responsive polymers and their applications in drug delivery. *Biomed Mater* 4:022001.

Billah, S. M. R., Mondal, Md. I. H., Somoal, S. H., et al. 2018. Enzyme-responsive hydrogels. In: *Cellulose-based superabsorbent hydrogels*, ed. Md. I. H. Mondal, chapter 10, 309–330. Springer International Publishing.

Camponeschi, F., Atrei, A., Rocchigiani, G., et al. 2015. New formulations of polysaccharide-based hydrogels for drug release and tissue engineering. *Gels* 1:3–23.

Cao, P. F., Mangadlao, J. D., and R. C. Advincula. 2015. Stimuli-responsive polymers and their potential applications in oil-gas industry. *Polym Rev* 55(4):706–733.

Cardoso, M. J., Costa, R. R., and J. F. Mano. 2016. Marine origin polysaccharides in drug delivery systems. *Mar Drugs* 14:34–61.

Chan, A. W. J. 2009. Controlled synthesis of stimuli-responsive network alginate. PhD diss., Queen's University Kingston, Ontario, Canada.

Chandrawati, R. 2016. Enzyme-responsive polymer hydrogels for therapeutic delivery. *Exp Biol Med* 241:972–979.

Chang, R., Li, M., Ge, S., et al. 2018. Glucose-responsive biopolymer nanoparticles prepared by co-assembly of concanavalin A and amylopectin for insulin delivery. *Ind Crop Prod* 112:98–104.

Chatterjee, S., and P. C. Hui. 2018. Stimuli-responsive hydrogels: An interdisciplinary overview. In: *Hydrogels - smart materials for biomedical applications*, ed. L. Popa, M. V. Ghica, C. Dinu-Pirvu, chapter 3, 1–23. Intechopen.

Chen, W., He, H., Zhu, H., et al. 2018. Thermo-responsive cellulose-based material with switchable wettability for controllable oil/water separation. *Polymers* 10:592–607.

Chowhan, A., and T. K. Giri. 2020. Polysaccharide as renewable responsive biopolymer for in situ gel in the delivery of drug through ocular route. *Int J Biol Macromol* 150:559–572.

Ciolacu, D. E. 2018. Structure-property relationships in cellulose-based hydrogels. In: *Cellulose-based superabsorbent hydrogels*, ed. Md. I. H. Mondal, chapter 3, 65–95. Springer International Publishing.

Ciolacu, D. E., and D. M. Suflet. 2018. Cellulose-based hydrogels for medical/pharmaceutical applications. In: *Biomass as renewable raw material to obtain bioproducts of high-tech value*, ed. V. I. Popa and I. Volf, chapter 11, 401–439. Elsevier.

Collins, M. N., and C. Birkinshaw. 2013. Hyaluronic acid based scaffolds for tissue engineering – A review. *Carbohyd Polym* 92:1262–1279.

Contessi, N., Altomare, L., Filipponi, A., et al. 2017. Thermo-responsive properties of methylcellulose hydrogels for cell sheet engineering. *Mater Lett* 207:157–160.

Croisier, F., and C. Jérôme. 2013. Chitosan-based biomaterials for tissue engineering. *Eur Polym J* 49:780–792.

d'Ayala, G. G., Malinconico, M., and P. Laurienzo. 2008. Marine derived polysaccharides for biomedical applications: Chemical modification approaches. *Molecules* 13:2069–2106.

Daniel-da-Silva, A. L., Ferreira, L., Gil, A. M., et al. 2011. Synthesis and swelling behavior of temperature responsive k-carrageenan nanogels. *J Colloid Interf Sci* 355:512–517.

Das, D., and S. Pal. 2015. Modified biopolymer-dextrin based crosslinked hydrogels: Application in controlled drug delivery. *RSC Adv* 5:25014–25050.

Dash, M., Chiellini, F., Ottenbrite, R.M., et al. 2011. Chitosan – A versatile semi-synthetic polymer in biomedical applications. *Prog Polym Sci* 36:981–1014.

de Sousa Moraes, P. R. F., Saska, S., Barud, H. et al. 2016. Bacterial cellulose/collagen hydrogel for wound healing. *Mater Res* 19(1):106–116.

Debele, T. A., Mekuria, S. L., and H. C. Tsai. 2016. Polysaccharide based nanogels in the drug delivery system: Application as the carrier of pharmaceutical agents. *Mater Sci Eng C* 68:964–981.

del Valle, L. J., Díaz, A., and J. Puiggalí. 2017. Hydrogels for biomedical applications: Cellulose, chitosan, and protein/peptide derivatives. *Gels* 3:27–55.

Delmar, K., and H. Bianco-Peled. 2016. Composite chitosan hydrogels for extended release of hydrophobic drugs. *Carbohyd Polym* 136:570–580.

Deng, Z., Zhen, Z., Hu, X., et al. 2011. Hollow chitosan-silica nanospheres as pH-sensitive targeted delivery carriers in breast cancer therapy. *Biomaterials* 32:4976–4986.

Domalik-Pyzik, P., Chłopek, J., and K. Pielichowska. 2018. Chitosan-based hydrogels: Preparation, properties, and applications. In: *Cellulose-based superabsorbent hydrogels*, ed. Md. I. H. Mondal, chapter 56, 1665–1693. Springer International Publishing.

Dragan, E. S., and M. V. Dinu. 2019. Polysaccharides constructed hydrogels as vehicles for proteins and peptides. A review. *Carbohyd Polym* 225:115210.

Duan, C., Zhang, D., Wang, F. et al. 2011. Chitosan-g-poly(N-isopropylacrylamide) based nanogels for tumor extracellular targeting. *Int J Pharm* 409:252–259.

Dumitriu, R. P., Oprea, A. M., and C. Vasile. 2009. A drug delivery system based on stimuli-responsive alginate/N-isopropylacryl amide hydrogel. *Cell Chem Technol* 43(7–8):251–262.

Echeverria, C., Fernandes, S. N., Godinho, M. H., et al. 2018. Functional stimuli-responsive gels: Hydrogels and microgels. *Gels* 4:54–91.

Efthimiadou, E. K., Metaxa, A. F., and G. Kordas. 2014. Modified polysaccharides as drug delivery. In: *Polysaccharides: Bioactivity and Biotechnology*, ed. K. Ramawat, J. M. Mérillon, chapter 57, 1805–1835. Springer International Publishing.

Egbu, R., Brocchini, S., Khaw, P. T., et al. 2018. Antibody loaded collapsible hyaluronic acid hydrogels for intraocular delivery. *Eur J Pharm Biopharm* 124:95–103.

Emas, R. 2015. The concept of sustainable development: Definition and defining principles. Brief for GSDR. United Nations' 2015 Global Sustainable Development Report.

Fajardo, A. R., Pereira, A. G. B., Rubira, A. F., et al. 2015. Stimuli-responsive polysaccharide-based hydrogels. In: *Polysaccharide hydrogels: Characterization and biomedical applications*, ed. P. Matricardi, F. Alhaique, and T. Coviello, chapter 9, 319–360. Jenny Stanford Publishing.

Feng, C., Li, J., Kong, M. et al. 2015. Surface charge effect on mucoadhesion of chitosan based nanogels forlocal anti-colorectal cancer drug delivery. *Colloid Surface B: Biointerfaces* 128:439–447.

Feng, W., Feng, S., Tang, K., et al. 2017. A novel composite of collagen-hydroxyapatite/kappa-carrageenan. *J Alloy Compd* 693: 482–489.

Fiorica, C., Pitarresi, G., Palumbo, F. S., et al. 2013. A new hyaluronic acid pH sensitive derivative obtained by ATRP for potential oral administration of proteins. *Int J Pharmaceut* 457:150–157.

Forghani, A., Kriegh, L., Hogan, K., et al. 2017. Fabrication and characterization of cell sheets using methylcellulose and PNIPAAm thermoresponsive polymers: A comparison Study. *J Biomed Mater Res Part A* 105A:1346–1354.

Fouda, M. M. G., El-Aassar, M. R., El-Fawal, G. F., et al 2015. k-Carrageenan/poly vinyl pyrollidone/polyethylene glycol/silver nanoparticles film for biomedical application. *Int J Biol Macromol* 74:179–184.

Fu, L. H., Qi, C., Ma, M. G., et al. 2018. Multifunctional cellulose-based hydrogels for biomedical applications. *J Mater Chem B* 7:1541–1562.

Gandhi, A., Paul, A., Sen, S. O., et al. 2015. Studies on thermoresponsive polymers: Phase behaviour, drug delivery and biomedical applications. *Asian J Pharm Sci* 10:99–107.

Gao, S., Tang, G., Hua, D. et al. 2019. Stimuli-responsive bio-based polymeric systems and their applications. *J Mater Chem B.* 7:709–729.

Gao, X., Cao, Y., Song, X. et al. 2014. Biodegradable, pH-responsive carboxymethyl cellulose/poly(acrylic acid) hydrogels for oral insulin delivery. *Macromol Biosci* 14:565–575.

Gicquel, E. 2017. Development of stimuli-responsive cellulose nanocrystals hydrogels for smart applications. PhD diss., Grenoble Alpes Univ.

Goh, C. H., Heng, P. W. S., and L. W. Chan. 2012. Alginates as a useful natural polymer for microencapsulation and therapeutic applications. *Carbohyd Polym* 88:1–12.

Hajebi, S., Rabiee, N., Bagherzadeh, M. et al. 2019. Stimulus-responsive polymeric nanogels as smart drug delivery systems. *Acta Biomater* 92:1–18.

Hezaveh, H., and I. I. Muhamad. 2012. Effect of natural cross-linker on swelling and structural stability of kappa-carrageenan/hydroxyethyl cellulose pH-sensitive hydrogels. *Korean J Chem Eng* 29(11):1647–1655.

Huang, H., Qi, X., Chen, Y., et al. 2019. Thermo-sensitive hydrogels for delivering biotherapeutic molecules: A review. *Saudi Pharm J* 27:990–999.

Huq, T., Salmieri, S., Khan, A. et al. 2012. Nanocrystalline cellulose (NCC) reinforced alginate based biodegradable nanocomposite film. *Carbohyd Polym* 90:1757–1763.

Iacob, A. T., Dragan, M., Ghetu, N. et al. 2018. Preparation, characterization and wound healing effects of new membranes based on chitosan, hyaluronic acid and arginine derivatives. *Polymers* 10:607–626.

Ismail, H., Irani, M., and Z. Ahmad. 2013. Starch-based hydrogels: Present status and applications. *Int J Polym Maters Po* 62:411–420.

Jayaramudu, T., Raghavendra, G. M., Varaprasad, K., et al. 2013. Iota-Carrageenan-based biodegradable Ag⁰ nanocomposite hydrogels for the inactivation of bacteria. *Carbohyd Polym* 95:188–194.

Jeong, B., Kim, S. W., and Y. H. Bae. 2012. Thermosensitive sol-gel reversible hydrogels. *Adv Drug Deliver Rev* 64:154–162.

Ji, Q. X., Chen, X. G., Zhao, Q. S., et al. 2009. Injectable thermosensitive hydrogel based on chitosan and quaternized chitosan and the biomedical properties. *J Mater Sci: Mater Med* 20:1603–1610.

Kaith, B. S., Jindal, R., Mittal, H., et al. 2010. Temperature, pH and electric stimulus responsive hydrogels from Gum ghatti and polyacrylamide-synthesis, characterization and swelling studies. *Der Chemica Sinica* 1(2):44–54.

Karoyo, A. H., and L. D. Wilson. 2017. Physicochemical properties and the gelation process of supramolecular hydrogels: A review. *Gels* 3:1–19.

Kaygusuz, H., and F. B. Erim. 2013. Alginate/BSA/montmorillonite composites with enhanced protein entrapment and controlled release efficiency. *React Funct Polym* 73:1420–1425.

Kayra, N., and A. Ö. Aytekin. 2018. Synthesis of cellulose-based hydrogels: Preparation, formation, mixture, and modification, In: *Cellulose-based superabsorbent hydrogels*, ed. Md. I. H. Mondal, chapter 14, 1–28, Springer International Publishing.

Khalil, H. P. S. A., Tye, Y. Y., Saurabh, C. K. et al. 2017. Biodegradable polymer films from seaweed polysaccharides: A review on cellulose as a reinforcement material. *Express Polym Lett* 11(4):244–265.

Khodeir, M., Ernould, B., Brassinne, J. et al. 2019. Synthesis and characterisation of redox-responsive hydrogels based on stable nitroxide radicals. *Soft Matter* 15:6418–6426.

Kim, M. S., Park, S. J., Chun, H. J., et al. 2011. Thermosensitive hydrogels for tissue engineering. *Tissue Eng Regen Med* 8(2):117–123.

Kimura, Y. 2009. Molecular, structural, and material design of bio-based polymers. *Polym J* 41(10):797–807.

Klouda, L., and A. G. Mikos. 2008. Thermoresponsive hydrogels in biomedical applications. *Eur J Pharm Biopharm* 68:34–45.

Kumar, R., Sharma, R. Kr., and A. P. Singh. 2018. Grafted cellulose: A bio-based polymer for durable applications. *Polym Bull* 75:2213–2242.

Kunal, P., Banthia, A. K., and D. K. Majumdar. 2006. Starch based hydrogel with potential biomedical application as artificial skin. *Afr J Biomed Res* 9:23–29.

Laurienzo, P. 2010. Marine polysaccharides in pharmaceutical applications: An overview. *Mar Drugs* 8:2435–2465.

Lee, K. Y., and D. J. Mooney. 2012. Alginate: Properties and biomedical applications. *Prog Polym Sci* 37:106–126.

Lee, S. B., Ha, D. I., Cho, S. K., et al. 2004. Temperature/pH-sensitive comb-type graft hydrogels composed of chitosan and poly(N-isopropylacrylamide). *J Appl Polym Sci* 92:2612–2620.

Li, D., Kordalivand, N., Fransen, M. F. et al. 2015. Reduction-sensitive dextran nanogels aimed for intracellular delivery of antigens. *Adv Funct Mater* 15(25):2993–3003.

Li, H., Koenig, A. M., Sloan, P., et al. 2014. In vivo assessment of guided neural stem cell differentiation in growth factor immobilized chitosan-based hydrogel scaffolds. *Biomaterials* 35:9049–9057.

Lima, A. C., Song, W., Blanco-Fernandez, B., et al. 2011. Synthesis of temperature-responsive dextran-MA/PNIPAAm particles for controlled drug delivery using superhydrophobic surfaces. *Pharm Res* 28:1294–1305.

Lin, S. P., Kung, H. N., Tsai, Y. S., et al. 2017. Novel dextran modified bacterial cellulose hydrogel accelerating cutaneous wound healing. *Cellulose* 24:4927–4937.

Liu, H. Y., Korc, M., and C. C. Lin. 2018. Biomimetic and enzyme-responsive dynamic hydrogels for studying cell-matrix interactions in pancreatic ductal adenocarcinoma. *Biomaterials* 160:24–36.

Liu, H., Rong, L., Wang, B. et al. 2017. Facile fabrication of redox/pH dual stimuli responsive cellulose hydrogel. *Carbohyd Polym* 176:299–306.

Lu, D. R., Xiao, C. M., and S. J. Xu. 2009. Starch-based completely biodegradable polymer materials. *Express Polym Lett* 3(6):366–375.

Mao, S., Sun, W., and T. Kissel. 2010. Chitosan-based formulations for delivery of DNA and siRNA. *Adv Drug Deliver Rev* 62:12–27.

Marques-Marinho, F. D., and C. D. Vianna-Soares. 2013. Cellulose and its derivatives use in the pharmaceutical compounding practice. In: *Cellulose – medical, pharmaceutical and electronic applications*, ed. T. G.M. Van De Ven, chapter 8, 141–162. Intechopen.

Masruchin, N., Park, B. D., and V. Causin. 2017. Characterisation of cellulose nanofibrils-PNIPAAm composite hydrogels at different carboxyl contents. *Cell Chem Technol* 51(5–6):497–506.

Masteiková, R., Chalupová, Z., and Z. Šklubalová. 2003. Stimuli-sensitive hydrogels in controlled and sustained drug delivery. *Medicina* 39:19–24.

Mathew, A. P., Uthaman, S., Cho, K. H., et al. 2018. Injectable hydrogels for delivering biotherapeutic molecules. *Int J Biol Macromol* 110:17–29.

Merlusca, I. P., Ibanescu, C., Tuchilus, C., et al. 2018. Characterization of neomycin-loaded xanthan-chitosan hydrogels for topical applications. *Cell Chem Technol* 53(7–8):709–719.

Mohite, B. V., Koli, S. H., and S. V. Patil. 2018. Bacterial cellulose-based hydrogels: Synthesis, properties, and applications. In: *Cellulose-based superabsorbent*, ed. Md. I. H. Mondal, chapter 42, 1255–1276, Springer International Publishing.

Mollakhalili Meybodi, N., and M. A. Mohammadifar. 2015. Microbial exopolysaccharides: A review of their function and application in food sciences. *J Food Qual Hazards Control* 2:112–117.

Moller, S., Weisser, J., Bischoff, S., et al. 2007. Dextran and hyaluronan methacrylate based hydrogels as matrices for soft tissue reconstruction. *Biomol Eng* 24:496–504.

Nakajima, H., Dijkstra, P., and K. Loos. 2017. The recent developments in biobased polymers toward general and engineering applications: Polymers that are upgraded from biodegradable polymers, analogous to petroleum-derived polymers, and newly developed. *Polymers* 9:523–549.

Nokhodchi, A., Raja, S., Patel, P., et al. 2012. The role of oral controlled release matrix tablets in drug delivery systems. *BioImpacts* 2(4):175–187.

O'Connor, N. A., Jitianu, M., Nunez, G. et al. 2018. Dextran hydrogels by crosslinking with amino acid diamines and their viscoelastic properties. *Int J Biol Macromol* 111:370–378.

Oh, J. K., Drumright, R., Siegwart, D. J., et al. 2008. The development of microgels/nanogels for drug delivery applications. *Prog Polym Sci* 33:448–477.

Oh, J. K., Lee, D. I., and J. M. Park. 2009. Biopolymer-based microgels/nanogels for drug delivery applications. *Prog Polym Sci* 34(12):1261–1282.

Onofrei, M. D., and A. Filimon. 2016. Cellulose-based hydrogels: Designing concepts, properties, and perspectives for biomedical and environmental applications. In: *Polymer science: Research advances, practical applications and educational aspects*, ed. A. Mendez-Vilas, and A. Solano, 108–120. Formatex Research Center S.L.

Özcan, E., and E. T. Öner. 2015. Microbial production of extracellular polysaccharides from biomass sources. In: *Polysaccharides: Bioactivity and Biotechnology*, ed. K.G. Ramawat, and J. M. Merillon, chapter 6, 161–184. Springer International Publishing.

Pasqui, D., De Cagna, M., and R. Barbucci. 2012. Polysaccharide-based hydrogels: The key role of water in affecting mechanical properties. *Polymers* 4:1517–1534.

Paulo, E. M., Boffo, E. F., Branco, A. et al. 2012. Production, extraction and characterization of exopolysaccharides produced by the native *Leuconostoc pseudomesenteroides* R2 strain. *An Acad Bras Ciênc* 84(2):495–507.

Pedrosa, S. S., Pereira, P., Correia, A., et al. 2017. Targetability of hyaluronic acid nanogel to cancer cells: In vitro and in vivo studies. *Eur J Pharm Sci* 104:102–113.

Peppas, N. A., and C. D. Bures. 2008. Glucose-responsive hydrogels. In: *Encyclopedia of biomaterials and biomedical engineering*, ed. G. E. Wnek and G. L. Bowlin, 1163–1174. CRC Press.

Prabaharan, M., and J. F. Mano. 2006. Stimuli-responsive hydrogels based on polysaccharides incorporated with thermo-responsive polymers as novel biomaterials. *Macromol Biosci* 6:991–1008.

Qian, H., Wang, X., Yuan, K. et al. 2014. Delivery of doxorubicin in vitro and in vivo using bio-reductive cellulose nanogels. *Biomater Sci* 2:220–232.

Ravaine, V., Ancla, C., and B. Catargi. 2008. Chemically controlled closed-loop insulin delivery. *J Control Release* 132:2–11.

Ribeiro, M. P., Morgado, P. I., Miguel, S. P., et al. 2013. Dextran-based hydrogel containing chitosan microparticles loaded with growth factors to be used in wound healing. *Mater Sci Eng C* 33:2958–2966.

Rocha, P. M., Santo, V. E., Gomes, M. E., et al. 2011. Encapsulation of adipose derived stem cells and transforming growth factor-β1 in carrageenan-based hydrogels for cartilage tissue engineering. *J Bioact Compat Pol* 26(5):493–507.

Roy, D., Cambre, J. N., and B. S. Sumerlin. 2010. Future perspectives and recent advances in stimuli-responsive materials. *Prog Polym Sci* 35:278–301.

Rusu, D., Ciolacu, D., and B. C. Simionescu. 2019. Cellulose-based hydrogels in tissue engineering applications. *Cell Chem Technol* 53(9–10):907–923.

Saha, P., Goyal, A. K., and G. Rath. 2010. Formulation and evaluation of chitosan-based ampicillin trihydrate nanoparticles. *Trop J Pharm Res* 9(5):483–488.

Salerno, A., and C. D. Pascual. 2015. Bio-based polymers, supercritical fluids and tissue engineering. *Process Biochem* 50:826–838.

Salgueiro, A. M., Daniel-da-Silva, A. L., Fateixa, S., et al. 2013. k-Carrageenan hydrogel nanocomposites with release behavior mediated by morphological distinct Au nanofillers. *Carbohyd Polym* 91:100–109.

Sa-Lima, H., Caridade, S. G., Mano, J. F., et al. 2010. Stimuli-responsive chitosan-starch injectable hydrogels combined with encapsulated adipose-derived stromal cells for articular cartilage regeneration. *Soft Matter* 6:5184–5195.

Sannino, A., Demitri, C., and M. Madaghiel. 2009. Biodegradable cellulose-based hydrogels: Design and applications. *Materials* 2:353–373.

Schmitt, F., Lagopoulos, L., Käuper, P. et al. 2010. Chitosan-based nanogels for selective delivery of photosensitizers to macrophages and improved retention in and therapy of articular joints. *J Control Release* 144:242–250.

Selegård, R., Aronsson, C., Brommesson, C., et al. 2017. Folding driven self-assembly of a stimuli-responsive peptide-hyaluronan hybrid hydrogel. *Sci Rep* 7:7013.

Sevostyanov, M. A., Fedotov, A. Y., Nasakina, E. O. et al. 2015. Kinetics of the release of antibiotics from chitosan-based biodegradable biopolymer membranes. *Dok Chem* 465(1):278–280.

Sezer, S., Şahin, İ., Öztürk, K., et al. 2018. Cellulose-based hydrogels as biomaterials. In: *Cellulose-based superabsorbent hydrogels*, ed. Md. I. H. Mondal, chapter 39, 1–27. Springer International Publishing.

Shang, J., Le, X, Zhang, J., et al. 2019. Trends in polymeric shape memory hydrogels and hydrogel actuators. *Polym Chem.* 10:1036–1055.

Sharpe, L. A., Daily, A. M., Horava, S. D., et al. 2014. Therapeutic applications of hydrogels in oral drug delivery. *Expert Opin Drug Deliv* 11(6):901–915.

Shen, X., Shamshina, J. L., Berton, P., et al. 2016. Hydrogels based on cellulose and chitin: Fabrication, properties, and applications. *Green Chem* 18:53–75.

Shi, Q., Liu, H., Tang, D., et al. 2019. Bioactuators based on stimulus-responsive hydrogels and their emerging biomedical applications. *NPG Asia Mater* 11:64–85.

Silva, T. H., Alves, A., Popa, E. G. et al. 2012. Marine algae sulfated polysaccharides for tissue engineering and drug delivery approaches. *Biomatter* 2(4):278–289.

Sood, N., Bhardwaj, A., Mehta, S., et al. 2016. Stimuli-responsive hydrogels in drug delivery and tissue engineering. *Drug Deliv* 23(3):748–770.

Su, X., Xiao, C., and C. Hu. 2018. Facile preparation and dual responsive behaviors of starch-based hydrogel containing azo and carboxylic groups. *Int J Biol Macromol* 115:1189–1193.

Sun, G., and J. J. Mao. 2012. Engineering dextran-based scaffolds for drug delivery and tissue repair. *Nanomedicine (Lond)* 7(11):1771–1784.

Sun, G., Shen, Y. I., Ho, C. C., et al. 2010. Functional groups affect physical and biological properties of dextran-based hydrogels. *J Biomed Mater Res A* 93(3):1080–1090.

Tang, Y. F., Du, Y. M., Hu, X. W., et al. 2007. Rheological characterization of a novel thermosensitive chitosan/poly(vinyl alcohol) blend hydrogel. *Carbohydr Polym* 67:491–499.

Tang, Y., Wang, X., Li, Y. et al. 2010. Production and characterization of novel injectable chitosan/methylcellulose/salt blend hydrogels with potential application as tissue engineering scaffolds. *Carbohydr Polym* 82:833–841.

Tchobanian, A., Van Oosterwyck, H., and P. Fardim. 2019. Polysaccharides for tissue engineering: Current landscape and future prospects. *Carbohyd Polym* 205:601–625.

Thambi, T., Phan, V. H. G., and D. S. Lee. 2016. Stimuli-sensitive injectable hydrogels based on polysaccharides and their biomedical applications. *Macromol Rapid Commun* 37:1881–1896.

Ulijn, R. V., Bibi, N., Jayawarna, V. et al. 2007. Bioresponsive hydrogels. *Mater Today* 10(4):40–48.

Upadhyay, R. K. 2017. Use of polysaccharide hydrogels in drug delivery and tissue engineering. *Adv Tissue Eng Regen Med* 2(2):145–151.

Van Tomme, S. R., and W. E. Hennink. 2007. Biodegradable dextran hydrogels for protein delivery applications. *Expert Rev Med Devic* 4:147–164.

Vasiliu, S., Racovita, S., Popa, M., et al. 2018. Chitosan-based polyelectrolyte complex hydrogels for biomedical applications. In: *Cellulose-based superabsorbent hydrogels*, ed. Md. I. H. Mondal, chapter 57, 1695–1725. Springer International Publishing.

Voevodina, I., and A. Kržan. 2013. Innovative value chain development for sustainable plastics in Central Europe. Project presentation. www.plastice.org

Waghmare, V. S., Wadke, P. R., Dyawanapelly, S., et al. 2018. Starch based nanofibrous scaffolds for wound healing applications. *Bioact Mater* 3:255–266.

Wang, C., Mallela, J., Garapati, U. S. et al. 2013. A chitosan-modified graphene nanogel for noninvasive controlled drug release. *Nanomedicine* 9:903–911.

Wang, L., and J. P. Stegemann. 2010. Thermogelling chitosan and collagen composite hydrogels initiated with β-glycerophosphate for bone tissue engineering. *Biomaterials* 31:3976–3985.

Wang, Q., Asoh, T. A., and H. Uyama. 2018. Rapid uniaxial actuation of layered bacterial cellulose/poly(N-isopropylacrylamide) composite hydrogel with high mechanical strength. *RSC Adv.* 8:12608–12613.

Wang, T., Wu, D. Q., Jiang, X. J. et al. 2009. Novel thermosensitive hydrogel injection inhibits post-infarct ventricle remodeling. *Eur J Heart Fail* 11:14–19.

Wu, J., Su, Z. G., and G. H. Ma. 2006. A thermo- and pH-sensitive hydrogel composed of quaternized chitosan/glycerophosphate. *Int J Pharm* 315:1–11.

Yang, C., Wang, X., Yao, X., et al. 2015. Hyaluronic acid nanogels with enzyme-sensitive cross-linking group for drug delivery. *J Control Release* 205:206–217.

Yoshida, T., Lai, T. C., Kwon, G. S., et al. 2013. pH- and ion-sensitive polymers for drug delivery. *Expert Opin Drug Deliv* 10(11):1497–1513.

Zarina, S., and I. Ahmad. 2015. Biodegradable composite films based on k-carrageenan reinforced by cellulose nanocrystal from Kenaf fibres. *BioResources* 10(1):256–271.

Zhang, J. 2015. Bioactive films and hydrogels based on potato starch and phenolic acids using subcritical water technology. MSc diss., Alberta Univ.

Zhang, L., Wang, L., Guo, B., et al. 2014. Cytocompatible injectable carboxymethylchitosan/N-isopropylacrylamide hydrogels for localized drug delivery. *Carbohyd Polym* 103:110–118.

Zhang, Y. S., and A. Khademhosseini. 2017. Advances in engineering hydrogels. *Science* 05(356):6337–6374.

Zhang, Z., Chen, L., Deng, M., et al. 2011. Biodegradable thermo- and pH-responsive hydrogels for oral drug delivery. *J Polym Sci Pol Chem* 49:2941–2951.

Zhou, H., Zhu, H., Yang, X. et al. 2015. Temperature/pH sensitive cellulose-based hydrogel: Synthesis, characterization, loading, and release of model drugs for potential oral drug delivery. *BioResources* 10(1):760–771.

Zou, X., Zhao, X., and L. Ye. 2015. Synthesis of cationic chitosan hydrogel and its controlled glucose-responsive drug release behavior. *Chem Eng J* 273:92–100.

10 Curdlan Derivatives
New Approaches in Synthesis and Their Applications

Dana M. Suflet

CONTENTS

10.1 INTRODUCTION

Since its discovery by Harada et al. 1966, curdlan production has drawn considerable interest due to its unique rheological and thermal gelling properties. Curdlan was approved in 1996 by the U.S. Food and Drug Administration (FDA) (Federal Register 1996) as biopolymer used in food industries as food additive (gelling agent) in manufacturing process of jelly, noodles, and dietary fibres. Published studies on the immunostimulatory or antitumour effects of β-glucans have exponentially increased interest in the study of these polysaccharides and their application in the biomedical field. Some studies report that only glucans with $(1{\rightarrow}3)$-linked β-glucan backbone and with $(1{\rightarrow}6)$-linked β-D-glucopyranosyl units as branches produce a complete inhibition of tumour growth, while glucans with $(1{\rightarrow}3)$-β-glucosidic bonds present a percentage of 98–99% inhibition of tumour, other polysaccharides exhibit only 10–40% inhibition. However, the water insolubility of curdlan, generally attributed to the existence of extensive intra/intermolecular hydrogen bonds and the triple helix structure, limits its applications especially in the medical field. This disadvantage was eliminated by various substitution reactions of hydroxyl groups.

From this point of view, this chapter presents an overview of the chemical changes of curdlan and the most important applications of these derivatives.

10.2 STRUCTURE

Curdlan is an extracellular homopolysaccharide, produced by microorganisms such as *Alcaligenes faecalis var. myxogenes 10C3K, Agrobacterium (sp., radiobacter IFO12607, rhizogenes IFO13259)*, or *Streptococcus mutants*, which belongs to the β-glucans class (Harada et al. 1968; Hisamatsu et al. 1977; McIntosh et al. 2005; Kalyanasundaram et al. 2012). Curdlan is a linear β-glucan composed entirely of D-glucose units linked by (1-3)-β-glucosidic bonds (Figure 10.1). It is a neutral polysaccharide, insoluble in water, alcohols, and most organic solvents but dissolves in alkaline diluted bases (0.25M NaOH), dimethylsulfoxide (DMSO), formic acid, and some aprotic solvents such as N-methylmorpholine-N-oxide (NMMO), dimethylacetamide/lithium chloride (DMAc/LiCl), or dimethylphormanide/lithium chloride (DMF/LiCl) (Ogawa et al. 1973; Futatsuyama et al. 1999).

Curdlan can form in water two thermogels with different structures depending on its gelation temperature. When the temperature is raised to 55–70°C, aqueous suspensions of curdlan form low-set gels, whereas high-set gels are obtained when the gelation temperature is high between 80 and 130°C. In both cases, the thermogels retain their structure once cooled down at room temperature. In addition the studies have shown that both types of gels are different, both at the macromolecular and nanoscopic scales. Thus analysing the results obtained by X-ray diffraction (XRD), Atomic Force Microscopy (AFM), and Transmission Electron Microscopy (TEM) three distinct forms of the curdlan structure *Form I, II*, and *III* were highlighted (Figure 10.2) (McIntosh et al. 2005; Miyoshi et al. 2004).

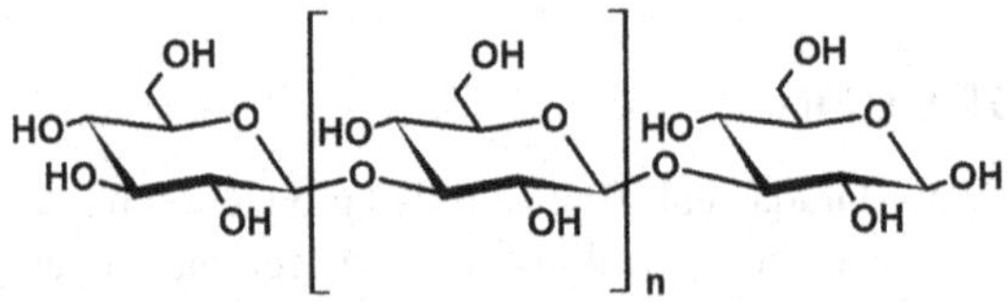

FIGURE 10.1   Molecular structure of curdlan.

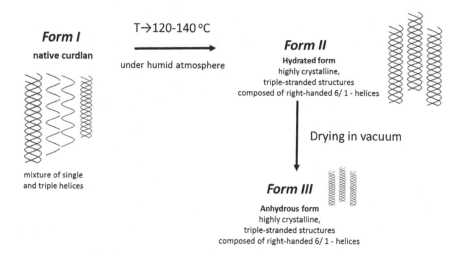

FIGURE 10.2 Schematic representation of structural changes between the three forms of curdlan.

Some natural organisms can produce curdlan as a single polysaccharide chain, but it has been shown that a humid atmosphere normally leads to a rearrangement in triple helix structures, in which the three chains are linked together by triads of strong hydrogen bonds through between the hydroxyl groups (Miyoshi et al. 2004; Deslandes et al. 1980).

The weight average molecular mass of curdlan was found to be about 2.0–0.05×10^6 Da being determined by various methods such as Dynamic Light scattering (DLS), viscosity measurements made on solutions of curdlan in 0.3M NaOH (Nakata et al. 1998; Zhang et al. 1997; Futatsuyama et al. 1999), size exclusion chromatography (SEC) (Zhang et al. 1999), MALDI–TOF mass spectrometry (Chan et al. 2006).

In 1978, Sasaki and co-workers (Sasaki et al. 1978) published their studies regarding the inhibitory effect of curdlan on Sarcoma 180 by subcutaneous administration of 5–50 mg kg^{-1} for 10 days and also very high activities at doses of 50 and 100 mg kg^{-1} administered intraperitoneally in a single injection. Since the publication of these studies, the interest in this polysaccharide has been steadily growing.

The nondestructive structure of curdlan was studied by FTIR (Fourier-Transform Infrared) (Jin et al. 2006) and NMR (Nuclear Magnetic Resonance) spectroscopies (Saito et al. 1989; Kim et al. 2000; Jin et al. 2006) and by X-ray diffraction analysis (Pelosi et al. 2006). Also, the differential scanning calorimetry (DSC) and dynamic viscoelastic measurements of aqueous dispersions of curdlan were used in order to clarify the gelation mechanism (Harashima et al. 1997; Jin et al. 2006).

10.3 CHEMICAL MODIFICATION

In recent years, several reviews have focused on the biosynthesis, structural analysis, and general biomedical/food applications of curdlan (Cai and Zhang 2017). However, its property of being insoluble in water is unfavourable for many applications,

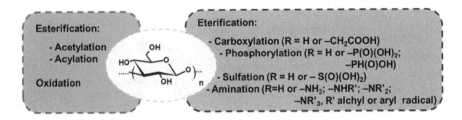

FIGURE 10.3 Chemical approaches for modification of curdlan.

especially for biomedical applications. This prompted the researchers to explore another way of using this polysaccharide, synthesizing curdlan derivatives with good water solubility as well as biological activity. In this sense, the chapter presents the latest important chemical approaches described in the literature for modifying the curdlan with potential industrial or medicinal importance (Figure 10.3).

10.3.1 CARBOXYLATION

A chemical modification procedure of polysaccharides by substitution reactions is the carboxymethylation, which gives products in which the primary and/or secondary alcohol groups are etherified with carboxymethyl groups.

Curdlan was carboxymethylated in an aqueous alkaline medium using monochloroacetic acid as the etherifying agent (Sasaki et al. 1979). The yield of this method was about 95%, and the carboxymethylated curdlan (CMCurd) was a water-soluble derivative (70 mg mL^{-1}) with degree of substitution about 0.47–0.65 (Jin et al. 2006). Table 10.1 includes some key features of CMCurd.

TABLE 10.1
The Main Characteristic of Carboxymethylated Curdlan

Sample	Reaction Conditions	DS	[η], dL g^{-1}	M_w	M_w/M_n	References
CMCurd	Monochloroacetic acid	–	0.83	1.62	4.78×10^6	2.99	Jin et al. (2006)
		60–70°C; 5h	0.47–0.65	–	–	–	Sasaki et al. (1979)
		50°C; 2h	0.98–1.03	–	–	–	Sasaki et al. (1979)
		60°C; 1.5h	1.27	2.41	3.89×10^4	1.3	Wang and Zhang (2006)
		–	0.81	–	6.30×10^5	1.55	Wu et al. (2012)
		55°C; 5h	0.49	–	–	–	Gao et al. (2008); Rafigh et al. (2016)

The structure of carboxymethylated curdlan was analysed by FTIR and NMR spectroscopy, which revealed that the carboxymethyl group was introduced mainly at the C-6 position as well as at the C-2 and C-4 positions. Figure 10.4 shows the comparative FTIR spectra of native curdlan and CMCurd, and the specific absorption bands are summarized in Table 10.2.

The NMR spectroscopy also allowed the confirmation of the carboxylation reaction of curdlan, highlighting the specific signals for each atom: in spectrum of the ^{13}C

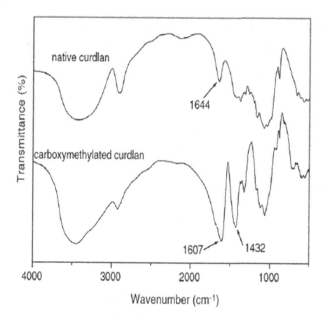

FIGURE 10.4 FTIR spectra of native curdlan and the carboxymethylated derivative (Jin et al. 2006).

TABLE 10.2
Specific Absorption Bands of Curdlan and Carboxymethylated Derivative

Wavenumber (cm^{-1})	Fragment	References
3370	OH	
2917	CHCH$_2$	
1644	H–O–H (the existence of water)	
1373	CH	
1317	CH$_2$	
1261	C–OH	Jin et al. (2006);
1234	C–O	Pretsch et al. (2010);
1160	C$_1$–O–C$_3$	Gao et al. (2008)
1080	C–O	
1607–1601	asymmetrical COO$^-$ stretching vibration	
1718	carboxyl group in protonated form	
1432–1428	symmetrical COO$^-$stretching vibration	

NMR the signals 103.7–96.7 (C-1), 75.1–73.9 (C-2), 86.7–76.7 (C-3), 70.6–68.6 (C-4), 76.6–76.8 (C-5), and 61.3–61.7 (C-6) ppm are attributed to the backbone chain of curdlan. In the spectrum of CMCurd, a new signal appears due to the carboxyl group at about 178 ppm. The distinct increase in signal intensities around 70 ppm is attributed to the shift of primary carbon (C-6) from the region around 60 ppm after substitution with –CH₂COOH group on the primary carbon. CMCurd was also used as parent polymer to synthesized new derivatives such as deoxycholic acid modified CMCurd (Gao et al. 2008).

Shibakami et al. synthesized a carboxylate derivative of curdlan by reaction with succinic anhydride in DMAc/LiCl medium (Shibakami et al. 2013). The reaction took place in a homogeneous medium leading to a white, water-soluble product with a DS value of 0.46.

The carboxylic derivative of curdlan with vinyl groups was synthesized by esterification reaction with maleic anhydride (Popescu et al. 2018). The opening of the anhydride cycle and the free carboxylic group obtained ensured the pH sensitivity of the curdlan derivative. The reaction was performed with maleic anhydride and TEA as catalyst, in DMF/LiCl medium, at 60°C. The maleilated curdlan (DS = 0.8 calculated from ^{1}H NMR spectrum) was used as macromolecular cross-linker in the polymerization of N-isopropylacrylamide (NIPAM) for obtaining thermo and pH-sensitive networks with high porosity.

10.3.2 SULFATION

The chemical modification of curdlan by the introduction of sulfate groups has been extensively studied due to the new properties conferred by these derivatives, such as anticoagulant, antithrombotic activities. The protocol of the sulfation reaction developed for some insoluble polysaccharides, such as cellulose or starch, was also applied to curdlan. The most used sulfating reagents are summarized in Table 10.3.

Chemical structure of curdlan sulfate (CurdS) and positions of sulfate groups in anhydrogluco pyranose units (AGUs) were extensively studied by FTIR and NMR spectroscopy (^{1}H NMR, ^{13}C NMR, 2-D correlation experiments). The sulfation reactions take place almost exclusively at primary OH (C-6) when lower values of DS are obtained and then at secondary OH (C-2) when DS is increased. A very little substitution at OH (C-4) was reported in case when DS > 1.6. Compared with the FTIR spectra of curdlan (see Figure 10.4), new adsorption peaks appeared at 870–690 cm⁻¹ attributed to symmetrical C-S-O stretching, and 1260–1200 cm⁻¹, 1070–1030 cm⁻¹ represent the asymmetrical and symmetrical S=O stretching vibrations from R-SO₃⁻ groups in CurdS (Pretsch et al. 2010; Li et al. 2014; Jin et al. 2020).

The ^{13}C-NMR spectrum of curdlan sulfate recorded in DMSO-d₆ at 60°C, evidenced three new signals at C-6s (67.9 ppm), C-2s (86.0 ppm), and C-4s (79.0 ppm) which were attributed at sulfated groups from C-6, C-4, and C-2 positions of AGU.

The molecular weight (M_w) and intrinsic viscosity [η] of sulfated derivatives were determined by multi-angle light scattering and viscosimetry. The M_w value (2.4×10^4) of sulfated derivatives was much lower than parent curdlan (44.5×10^4) which indicates a degradation of the polymer chain during the reaction. The Mark–Houwink equation and average value of characteristic ratio C_∞ ($C_\infty = 16$) in 0.2 M NaCl

TABLE 10.3

The Main Sulfating Agents of Curdlan and the Characteristics of the Derivatives

Sample	Sulfating Reagent	Main Characteristics	References
Curdlan sulfate	Piperidine-N-sulfonic acid/DMSO	DS = 0.4–1.6; water soluble; $M_n = 2 \times 10^4$ Da	Yoshida et al. (1995)
	SO_3-pyridine/ pyridine	DS = 2.2–2.6; water soluble; $M_n = 6.3 \times 10^4$	Yoshida et al. (1995) Lee et al. (2005)
	SO_3-pyridine/DMF	DS = 0.93–2.1; water soluble	Osawa et al. (1993)
	SO_3-pyridine/DMSO	DS = 0.17 - 1.13; water soluble; $M_w = 2.5 \times 10^4 - 1.2 \times 10^5$	Li et al. (2014); Jin et al. (2020); Sun et al. (2009)
	SO_3–pyridine/ DMSO/LiCl	DS = 1.2–1.6; water soluble; $M_w = 2.4 \times 10^4$	Zang et al. (2000)
	Chlorosulfonic acid/ pyridine	DS = 2.6–3.0; water soluble; $M_n = 1.3 \times 10^4$	Yoshida et al. (1995)
	Chlorosulfonic acid/ DMSO/pyridine	DS = 0.28–0.54; water soluble	Tao et al. (2006)
	Chlorosulfonic acid/ DMF/pyridine	Water soluble; DS = 0.86; $M_n = 14.5 \times 10^4$; DS = 1.02; $M_n = 2.65 \times 10^4$	Huang et al. (2006)
	H_2SO_4	DS = 0.03; water soluble; $M_n = 7.5 \times 10^4$	Wong et al. (2010)
	H_2SO_4/ ultrasonication	DS = 0.115; water soluble; $M_n = 4.9 \times 10^4$	Wong et al. (2010)
Sulfopropyl curdlan (SPrCurd)	Propane-1,3-sultone/ DMSO/NaH	DS = 0.6–2.4; water soluble	Osawa et al. (1993)
	Propane-1,3-sultone/ isopropyl alcohol	Water soluble	Cremer et al. (2010)

aqueous solution at 25°C was found to be $[\eta] = 1.32 \times 10^{-3} M_w^{1.06}$ (cm^3 g^{-1}) (Zang et al. 2000). The 1.06 value of exponent of the equation indicates that the sulfated curdlan has more extended polymeric chains than original curdlan, where the $[\eta] - M_w$ relationship was given to be $[\eta] = 71 \times 10^{-3} M_w^{0.06}$ cm^3 g^{-1}, due to the electrostatic repulsion of sulfate groups.

Curdlan sulfate has also been used as parent polymer in different derivatization syntheses such as acetyl curdlan sulfate (Osawa et al. 1993), azidothymidine curdlan sulfate (Gao et al. 1998), *O*-palmitoyl curdlan sulfate (Lee et al. 2005; Cremer et al. 2010), 6-amino-6-deoxy curdlan sulfate (Borjihan et al. 2003).

10.3.3 PHOSPHORYLATION

The derivatives curdlan with phosphoric groups (mono- or dibasic groups) were synthesized by reaction of curdlan with phosphorous acid in molten urea (Suflet et al. 2011a) or phosphoric acid in presence of urea/DMSO (Chen et al. 2009).

The structure of monobasic curdlan phosphate (PCurd) was investigated by FTIR and NMR spectroscopy and the substitution degree was calculated from

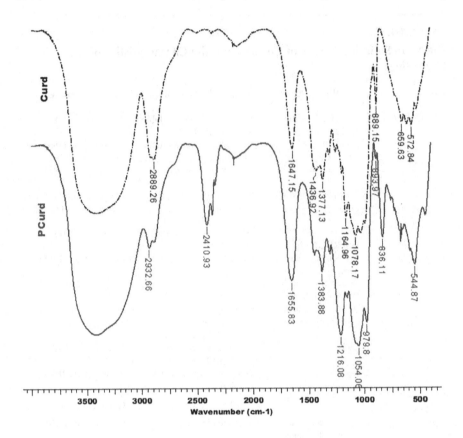

FIGURE 10.5 FTIR spectra of monobasic curdlan phosphate compared with native curdlan.

electrochemical titration. In Figure 10.5 the PCurd FTIR spectrum are presented, compared with native curdlan. In the FTIR spectrum of PCurd several new bands appeared: a band at 2410 cm⁻¹ corresponding to the P–H bond, another one at 1216 cm⁻¹ corresponding to the P–O bond, a shoulder at 1054 cm⁻¹ attributable to the P–OH bond, and a band at 836 cm⁻¹ corresponding to the P–O–C bonds (Pretsch et al. 2010; Suflet et al. 2011). The NMR data provide evidence for the formation of phosphorylated curdlan. In the ^{1}H NMR spectrum (Figure 10.6) of PCurd, two peaks at 6.17 and 7.84 ppm are observed which correspond to a doublet with a coupling constant $J_{H–P}$ of 668 Hz. This value is characteristic for the P–H bond.

Fully water-soluble PCurd with M_w of 178,000 g mol⁻¹ was obtained with a DS up to 1, with a dissociation constant $pK_0 = 2.79–2.81$ and intrinsic viscosity [η] = 81.93 and 0.69 dL g⁻¹ in pure water, and at infinite ionic strength, respectively (Suflet et al. 2011).

The dibasic curdlan phosphate was synthetized by reaction with phosphorus acid in DMSO and urea. This derivative with only 0.056–0.075 substitution degree exhibit antitumour activities in vitro and relatively strong inhibition ratios against *Sarcoma 180* (S-180) tumour cells in vivo (Chen et al. 2009).

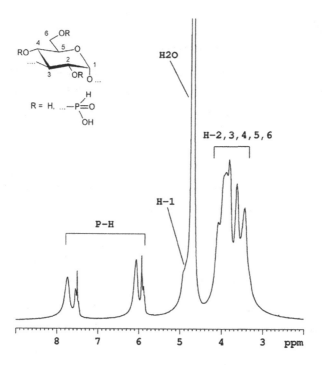

FIGURE 10.6   ¹H NMR spectrum of monobasic curdlan phosphate in D2O (Suflet et al. 2011).

10.3.4 OXIDATION

Regioselective oxidation of C6 primary hydroxyls from AGU units of polysaccharides was performed using the 2,2,6,6-tetramethylpiperidine-1-oxyl radical (TEMPO) system to obtain 1,3-β-polyglucoglucuronic acid. This reaction opened a new horizon for polysaccharide chemistry by improving their viscosity and solubility in water, thus expanding applications in food, cosmetic, textile, and even medical and pharmaceutical fields. Any water-insoluble polysaccharide like chitins, regenerated celluloses become water-soluble by the TEMPO-mediated oxidation through partial or complete conversion of the C6 primary hydroxyls to carboxylate groups. The TEMPO-mediated oxidation system was applied to β-glucans in order to prepare water-soluble (1→3)-β-linked polyglucuronic acids. Tamura and co-workers (Tamura et al. 2009; Tang et al. 2018) obtained an oxidized curdlan derivative using the TEMPO/NaBr/NaClO system at room temperature and basic medium (pH = 10). Unfortunately, during the oxidation process there is a significant depolymerization so a high degradation of the polysaccharide chain of the curdlan from 6,790 to 86.

In 2010, Tamura et al reported the regioselective oxidation of primary OH groups of curdlan to carboxylate groups with lower levels of depolymerization using a 4-acetamido-TEMPO/NaClO/NaClO$_2$ system in water at pH 4.7 (Tamura et al. 2010). In Table 10.4 the main properties of oxidized curdlan by the new TEMPO-mediated system are presented, compared with TEMPO/NaBr/NaClO.

TABLE 10.4

Main Characteristics of the Oxidized Curdlan

Sample	pH	M_w (DP_w)	Solubility in Water	DS	Crystallinity Index	References
Curdlan		1,100,000 (6,790)	insoluble	–	0.28	Tamura et al. (2009)
Oxidized Curdlan by:						
TEMPO/NaClO/ NaClO	10	17,000 (86)	soluble	0.28–1.00	0.00	Tamura et al. (2009)
4-acetamide- TEMPO /NaClO/NaClO$_2$	4.7	197,000 (1,020)	soluble	0.95	–	Tamura et al. (2010); Watanabe et al. (2014)

Similar to CMCurd, oxidized curdlan by TEMPO system can be used as both reducing and stabilizing agents for the green synthesis of silver (Yan et al. 2013) and gold (Yan et al. 2015a) nanoparticles or drug retention system (Yan et al. 2015b).

The 4-acetamido-TEMPO-oxidized curdlan was hydrophobically modified by reaction with deoxycholic acid to attain novel amphiphilic curdlan derivatives for the preparation of nano-carriers for antitumour drug like doxorubicin (Yan et al. 2015b).

10.3.5 ACETYLATION AND ACYLATION

Acetylation/acylation of polysaccharides is the introduction of acetyl (-CO-CH$_3$) acyl (-CO-alkyl; -CO-aryl) groups in the polysaccharide molecules. A series of ester derivatives of curdlan with varying alkyl chain lengths (C2–C12) were synthesized by the heterogeneous reaction using trifluoroacetic anhydride (Figure 10.7) (Marubayashi

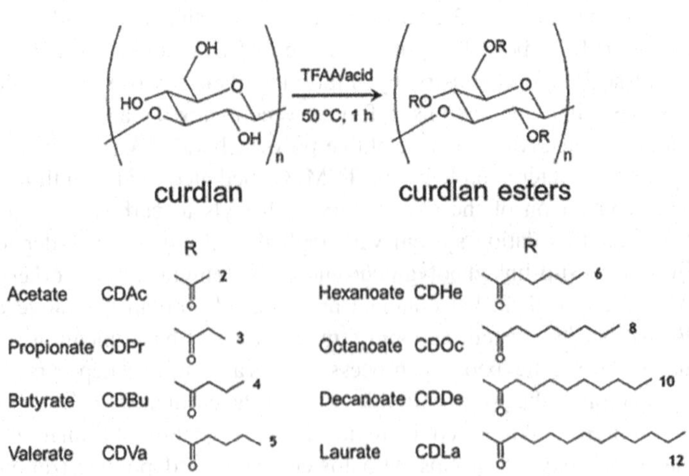

FIGURE 10.7 Syntheses of curdlan esters. Each acyl group is labeled with the corresponding carbon number (Marubayashi et al. 2014).

et al. 2014). The reactions yield was relatively high (>70%). The curdlan esters having two up to six carbon atoms in the alkyl chain are in a crystalline form and those with eight, ten, and twelve atoms are amorphous polymers.

The glass transition temperature (T_g) was evidenced only for derivatives with C2-C4 alkyl radicals (171°C for CDAc (C2), 170°C for CDPr (C3), and 74°C CDBu). Also, CDAc (C2) and CDPr (C3) possessed the melting temperature (T_m) higher than 200°C (287 and 213°C, respectively). The longer ester groups gave the lower T_g (170 → 50°C) and T_m (290 → 170°C) (Marubayashi et al. 2014). These results showed that the properties of curdlan esters can be adjusted by changing the length of the alkyl ester groups. Moreover, these esters with a higher thermal stability compared to the native curdlan can be used as thermoplastic materials.

Acetylated derivatives of curdlan were also synthesized in alkaline medium with different amounts of acetylated reagent (acetic anhydrides), in order to prepare derivatives with variable DS (0.71–1.04) (Chen et al. 2014). These derivatives exhibited a stronger antioxidant abilities on scavenging DPPH (1,1-diphenyl-2-picrylhydrazyl) radical, and inhibitory effects in β-carotene-linoleic acid systems compared with the native polysaccharide although the molecular weight of these derivatives decreased due to the slight degradation of the polysaccharide chain during the acetylation reaction.

Two kinds of regioselective substituted curdlan hetero esters, 2,4-di-O-acetyl-6-O-propionyl-curdlan (CD24Ac6Pr) and 2,4-di-O-propionyl-6-O-acetyl-curdlan (CD24Pr6Ac) were also synthesized (Chien et al. 2017; Chien and Iwata 2018). The reactions occurring by protecting the C6 primary hydroxyl group were followed by the acylation of the secondary hydroxyl groups at C2 and C4. The position of ester groups on secondary hydroxyl group plays a decisive role in the melting behaviour and crystal structure of these curdlan esters. Also, the mechanical properties of curdlan esters are affected by the substitution position of curdlan esters but can be controlled by the adjustment on its molecular structures.

10.3.6 Amination

Over the last two decades, the interest in amino polysaccharides synthesis has grown due to the natural functions of amino polysaccharides and their potential for biomedical applications. The introduction of amino groups on the curdlan chain offers the new derivatives several advantages, including the increase of water solubility, an affinity for nucleic acid, and not least confers antimicrobial activity.

The hydroxyl group at position C6 of the glucan unit of curdlan was substituted with 6-amino-6-deoxy groups in DMSO in two steps (Borjihan et al. 2003; Han et al. 2015). First, the curdlan is azidized by reaction with carbon bromide (CBr_4) in the presence of triphenylphosphine (Ph_3P) and lithium or sodium azide (LiN_3 or NaN_3) under nitrogen protection when a 6-azido-6-deoxycurdlan derivative was obtained. Second, the azido group is reduced to an amino group with sodium borohydride ($NaBH_4$). The 6-amino 6-deoxy curdlan derivative can have a DS between 0.5 and 1.0 by controlling the reaction time and the ratio between the reactants.

This derivative has been also successfully used in various reactions as intermediary product: (a) sulfation when 6-amino-6-deoxy curdlan sulfate was obtained (with a degree of sulfation of 1.44) exhibiting high anti-HIV activity at 0.912 μg mL^{-1} and low cytotoxicity (Borjihan et al. 2003); (b) PEGylation with application in siRNA delivery (Altangerel et al. 2016); (c) acrylation (Zhang and Edgar 2014); (d) acetylation (Zhang et al. 2017); (e) amination when 6-(N,N,N-trialkylammonio)-6-deoxy derivative with DS = 0.89 was obtained in DMSO as optimal solvent (Zhang et al. 2016); (f) TEMPO-oxidation leading to an amphoteric derivative of curdlan with protonatable amino and dissociable carboxyl groups on the C6 position of glucose units (Tamaru et al. 2014).

Polysaccharides with quaternary ammonium groups have some important properties as hydrophilicity, biodegradability, biocompatibility, and bacteriostatic properties, very useful for bioapplications. These cationic polysaccharides were synthesized by the reaction of native polymers with various reagents with quaternary ammonium groups. Glycidyltrimethylammonium chloride and 3-chloro-2-hydroxypropylethyl-ammonium chloride are the most commonly used commercial reagents. Quaternary ammonium groups generally bring antimicrobial effects, so their introduction into the curdlan chain has expanded its use in other fields of medicine. Table 10.5 shows the main quaternization agents and the characteristics of the quaternary ammonium salt of curdlan.

TABLE 10.5
The Quaternization Agents and the Main Characteristics of Quaternary Ammonium Salt of Curdlan

Quaternary Agent	Main Characteristics	References
Glycidyltrimethyl ammonium chloride (GDMAC)	soluble in water; DS = 0.15; $[\eta]_{water}$ = 18.23 - 17.70 L g^{-1}; $[\eta]_{0.1M\,salt}$ = 0.34 L g^{-1}; A good complexation capacity	Suflet 2015; Chen and Liang (2017)
2,3-epoxypropyl triethylammonium chloride	soluble in water; DS = 0.12; $[\eta]_{water}$ = 13.33–11.88 L g^{-1}; $[\eta]_{0.1M\,salt}$ = 0.20 L g^{-1};	Suflet et al. (2015)
Glycidypropyl dimethyloctyl ammonium chloride	soluble in water; DS = 0.2 - 0.4; $[\eta]_{water}$ = 83.75 dL g^{-1}; $[\eta]_{0.05M\,salt}$ = 2.44 L g^{-1}; C* = 0.012 g dL^{-1}; CAC = 0.03 g dL^{-1}; C$_e$ =1.14 g dL^{-1}; Antimicrobial proprieties (*E. coli* and *P. aeruginosa* bacteria, and *C. albicans* fungi)	Popescu et al. (2019)
N-(3-chloro-2-hydroxypropyl) trimethylammonium chloride (CHPTMAC)	soluble in water; DS = 0.034 – 0.43; M$_w$ = 6.07 – 3.67 × 10^5 Da; T$_g$ = 273°C; η_0 = 2.4 - 23.6 (Pa s) Antimicrobial proprieties	Yu et al. (2007); Wang et al. (2012); Suflet et al. (2015)
(3-chloro-2-hydroxypropyl) triethylammonium chloride (CHPTEAC)	soluble in water; DS = 0.06–0.07	Suflet et al. (2015)

C* – the value of overlap concentration; CAC – the first critical aggregation concentration; C$_e$ – entanglement concentration (the limit between semi-diluted unentangled and semi-diluted entangled ranges); η_0 – the zero shear viscosity.

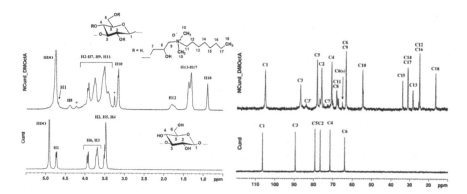

FIGURE 10.8 ¹H NMR (top) and ¹³C NMR (bottom) spectra of 2-hydroxypropyl-3-dimethyloctyl ammonium curdlan derivative and native curdlan in D₂O, with assigned signals (Popescu et al. 2019).

Generally, the quarterization reactions of curdlan were carried at 50–70°C and different molar ratios between AGU and quarterization reagent (1:5–1:10) in order to obtain a high DS. Chemical structure of quaternary curdlan (QCurd) derivatives was extensively studied by FTIR and NMR spectroscopy. In general, FTIR spectroscopy did not show significant bands for these derivatives due to the overlapping bands with the parent polysaccharide. However, specific bands have been identified at 1466–1490 cm⁻¹ for alkyl groups bounded to the quaternary nitrogen, and at 1420 cm⁻¹ for C–N stretching vibration (Pretsch et al. 2010; Suflet et al. 2015; Popescu et al. 2019). In order to have more structural information about the quaternary curdlan derivatives, the NMR spectroscopy (¹H NMR, ¹³C NMR, 2D NMR homo- and heteronuclear correlations) was recorded. Figure 10.8 shows, for exemplification, the NMR spectra of 2-hydroxypropyl-3-dimethyloctyl ammonium curdlan derivative and the specific signals assigned based on 2D NMR homo- and heteronuclear correlations like H,H COSY (Correlation Spectroscopy), H,C HSQC (Heteronuclear Single Quantum Coherence) and H,C HMBC (Heteronuclear Multiple Bond Coherence) (Popescu et al. 2019).

10.3.7 CLICK CHEMISTRY OF CURDLAN

Click chemistry offers many advantages including high yield, regio- and stereo specificity, and rapid reactivity under physiological conditions. Hasegawa and co-workers were the first who investigated the click chemistry of curdlan and developed curdlan derivatives with various functional appendages such as lactoside, ferrocene, pyrene, etc. (Hasegawa et al. 2006). They used the 6-azido-6-deoxycurdlan derivative (described in Section 3.6) for coupling terminal alkynes bearing different functional groups *via* chemo-selective [3+2] cycloaddition. Furthermore, using the 'click chemistry' the curdlan was grafted with cationic polymer by reaction between 6-azido-6-deoxy-curdlan and modified lysine with terminated alkyne. These derivatives can potentially be used as a versatile nano platform for efficient gene delivery in living cells (Han et al. 2017).

10.3.8 GRAFTING

Tukulula et al. (Tukulula et al. 2015) reported a method of conjugating curdlan to PLGA via esterification in order to obtain curdlan-PLGA nanoparticles. The esterification reaction occurs in DMF and in the presence of poly(D,L-lactide-co-glycolide) with terminal carboxylic group. The nanoparticles synthesized from this polymer possess immune-stimulatory activity, as well as sustained drug delivery potential, and were recommended for cellular targeted therapy.

Curdlan was grafted with polyethylene glycol for nanoparticle synthesis to increase the efficiency of in vivo siRNA (small interfering RNA) release. The graft reaction occurred by conjugation of polyethylene glycol monomethyl ether (mPEG) to 6-amino-6-deoxy-curdlan under homogeneous conditions (DMSO/triethylamine), when a water-soluble product (Curd-g-mPEG) was obtained with DS of 0.026 and a molecular weight average (M_w) of 49,399. The reaction takes place with a yield of 58% (Altangerel et al. 2016). The siRNA/Curd-g-PEG nanoparticles were prepared using a solvent displacement method. The average diameter of nanoparticles was 200 nm. These nanoparticles can enter human liver cancer cell line in culture and substantially knock down endogenous genes such as GAPDH (glyceraldehyde 3-phosphate dehydrogenase) and LXRα (Liver X receptor alpha).

The curdlan was grafted to poly(vinyl alcohol) (PVA) in order to prepare porous scaffold with a high water retention capacity and good mechanical properties. PVA and curdlan can be chemically cross-linked in presence of sulfuric acid and formaldehyde by 1,3-dioxane rings, at 80°C (Hsieh et al. 2017). The functional groups of a copolymer were identified by FT-IR analysis when the cross-linking slightly shifted the main peaks of partners. The diagram resulting from the DSC analysis also showed two slightly displaced melting points, indicating that the graft destroyed the crystallization of the original polymer. This type of reaction was performed to obtain porous structures used as scaffolds for medical applications.

Graft copolymerization of methyl methacrylate onto curdlan was first investigated by Yoshida and col. (Yoshida et al. 1996). The grafting reaction can be initiated by different methods: (a) with ammonium persulfate in DMSO, under homogeneous condition, when resulting graft copolymers with low molecular weights (Mn = 0.4–2.2 × 10⁴) and low grafting percentages (93.3%); (b) with ammonium persulfate in water (heterogeneous condition) when a graft copolymer has resulted having relatively higher molecular weight (M_n = 17.7 × 10⁴) and high grafting percentage (548%); (c) with cerium (IV) ammonium nitrate-HNO₃ in water (heterogeneous condition) when the grafted copolymer had the highest grafting percentage of 1620% without degradation of the curdlan chain.

10.4 APPLICATIONS OF CURDLAN DERIVATIVES

Regioselective modified curdlan derivatives find applications in the food industry as food additives and in medical and pharmaceutical field as carriers for the administration of therapeutic agents (drugs, genes,), or as antitumour, anti-infective or anti-inflammatory agents.

10.4.1 Biomedical and Pharmaceutical Applications

Most applications of curdlan and its derivatives are exposed in the field of medicine, due to the special biological properties of these polysaccharides, including antitumour activity, anti-AIDS, anti-viral as well as low toxicity.

10.4.1.1 Inhibitory Activity

Curdlan derivatives have been studied extensively due to their anti-HIV activity (Kaneko et al. 1990; Osawa et al. 1993; Yoshida et al. 1990; Yoshida et al. 1995; Gao et al. 1997; Gao et al. 1998; Gao et al. 1999; Jeon et al. 2000; Borjihan et al. 2003), antiviral (Yamamoto et al. 1990; Evans et al. 1998; Ichiyama et al. 2013), anticoagulant (Yoshida et al. 1995; Alban and Franz 2000), antioxidant (Yang et al. 2005), or antitumour activity (Kakinuma et al. 1981; Zhang et al. 2000a; Chen et al. 2009). Some studies have shown that curdlan derivatives could reduce hyperglycemia and hyperinsulinemia, controlling diabetes (Brenann and Tudorica 2003) and reducing hyperlipidemia and hypercholesterolemia (Maki et al. 2003). There are also studies dedicated to immunostimulatory activity (Sandula et al. 1995; Li et al. 2014; Jin et al. 2020) of curdlan derivatives. Therefore,

- CMCurd with DS of 0.47 and 0.65 showed strongly inhibited growth of *Sarcoma 180* solid-type tumour when injected with daily intraperitoneal (3–40 mg kg^{-1}) or intravenous (5 mg kg^{-1}) for 10 days. The growth of *Sarcoma 180* was strongly inhibited when a single dose intraperitoneal of 20–200 mg kg^{-1} in 7 days was injected after tumour transplantation (Sasaki et al. 1979). The antitumour activity of CMCurd was found to be closely correlated with DS so that a DS = 0.47 resulted in a 100% tumour inhibition ratio, DS = 0.65 inhibited 97.1%, and a DS approximately 0.98 and higher per glucose residue showed lower inhibitory activity.
- CurdS has been studied as an anti-HIV agent since 1990, the studies reaching a clinical stage II, with the permission of the FDA, proving a good efficacy and low toxicity. (Osawa et al. 1993). Curdlan sulfate with high molecular weight and high degree of sulfation exhibited high anti-HIV activity at drug concentration as low as 3.3 µg mL^{-1} in vitro. The half-life of curdlan sulfates in plasma has been found to depend on their molecular masses, being 60 and 180 min for molecular masses of 7×10^4 and 17×10^4, respectively. It was also found that these curdlan sulfates can be transported to tissues such as liver, kidney, lymph node, and bone marrow within 1 h, and remain in the tissues without degradation for 10 days. (Yoshida et al. 1995). The anti-HIV-1 activity of curdlan sulfate is due to both the inhibitory effects on gp120 binding to CD4 and gp120-mediated TNF-α production (Jagodzinski et al. 1994; Takeda-Hirokawa et al. 1997; Jin et al. 2020). The curdlan sulfate exhibited inhibitory effect against Sarcoma 180 tumour cell growth at 0.005–5 mg mL^{-1} concentration. Inhibitory activity is influenced by the degree of substitution of curdlan derivatives (Huang et al. 2006; Sun et al. 2009). CurdS has been demonstrated to have significant activity on macrophages activation and bone marrow derived dendritic cells (BMDCs) maturation through the binding of curdlan sulfate with the primary receptor

dectin-1 on cell surface (Li et al. 2014). Furthermore, this derivative induces the production of cytokines and the expression of molecules associated with immunity without any cytotoxicity and can be used by oral or intravenous injection because it is soluble, supporting the idea that curdlan sulfate is beneficial for immunotherapy. These data highlight the ability of non-TLR receptors to bridge innate and adaptive immunity and suggest that dectin-1 agonists may constitute a useful adjuvant for immunotherapy and vaccination. Despite these advantages, the anticoagulant activity of curdlan sulfate has been shown to be lower (10 units mg^{-1}) than other sulfate polysaccharides such as standard dextran sulfate (20.6 units mg^{-1}) (Yoshida et al. 1995). Preclinical studies have shown that curdlan sulfate inhibits *Plasmodium falciparum* in vitro and down-modulates the immune response. A direct, non-specific effect on cytoadherence and rosetting may be predicted, as has been described with other sulfated polysaccharides, for example, heparin. The anticoagulant effect of CurdS was found to be 10-fold lower than heparin, so that curdlan sulfate has therefore emerged as a candidate for adjuvant medication in the treatment of severe/cerebral malaria (Havlik et al. 2005).

- dibasic curdlan phosphate with only 0.056–0.075 degree of substitution exhibit antitumour activities in vitro and relatively strong inhibition ratios against *Sarcoma 180* (S-180) tumour cells in vivo (Chen et al. 2009).

10.4.1.2 Delivery Systems

Curdlan derivatives were successfully used to prepare the polymeric supports as hydrogels, micro- and nano-particles, semi- and intra-penetrated networks for the controlled release of various therapeutic agents (drugs, proteins, enzyme, vaccine) with medical/pharmaceutical applications. Table 10.6 presented the usually curdlan derivatives used to prepared delivery systems.

10.4.1.3 Scaffold

β-Glucan aerogels can be successfully obtained using supercritical CO_2 drying. Gels dried using this technique have many advantages over those dried at ambient temperature and pressure and by freeze drying, including lower density (0.2 g cm^{-3}) and uniformity of the interior, respectively. As β-glucan aerogels are biodegradable, biocompatible, renewable, and edible, they have great potential for application as delivery vehicles for medicine, nutrition, or pharmaceuticals (Comin et al. 2012). Hsieh et al. prepared 3D curdlan-g-PVA scaffolds for bone tissue engineering with values of compression force of 8–16 × 10^{-3} Pa and a favourable cell proliferation and growth revealed by in vivo tests (Hsieh et al. 2017).

10.4.1.4 Other Medical Applications

The curdlan derivative with monobasic phosphate groups (PCurd) has been studied as a crystallization regulator in the process of calcium orthophosphate crystals formation, in order to be used as filling materials in bone repair applications (Pelin et al. 2017). X-ray diffractometry indicated that a low crystallinity of hydroxyapatite was obtained when PCurd was present in the synthesis process, obtaining a structure

TABLE 10.6

Curdlan Derivatives Used for Delivery Systems

Derivate	Support	Technique	Therapeutic Agent	Loading Capacity	References
CMCurd	Microspheres; 260 μm	w/o/w emulsion	BSA	168 mg g^{-1}	Rafigh et al. (2016)
CMCurd-deoxycholic acid	Nanoparticles; 192–347 nm	sonication	Epirubicin	9.67%	Gao et al. (2008); Gao et al. (2010)
CMCurd	Nanoparticles; 144–233 nm	self-assembled	Epirubicin	13.5–39.6%	Li et al. (2010)
CurdS	Nanoparticle				
SPrCurd	Microspheres	chemical cross-linking	Protein	0.57 meq g^{-1}	Mocanu et al. (2009)
CMCurd/ SPrCurd	Microspheres	chemical cross-linking	Protein	3.24 meq g^{-1}	Mocanu et al. (2009)
PCurd	Microspheres	chemical cross-linking	Diphen-hydramine	4.15 mg g^{-1}	Popescu et al. (2013)
PCurd/Curd	Hydrogel	Chemical cross-linking	Drugs	1.1–3.1 meq g^{-1}	Suflet et al. (2020)
PCurd/PVA	Hydrogel	physical cross-linking	–	–	Suflet et al. (2019)
PCurd/ collagen	Microcapsules;	layer-by-layer	–	–	Suflet et al. (2011)
4-acetamido-TEMPO OxCurd	Nanoparticles; 214–380 nm	sonication	Drug	4.25–14.7%	Yan et al. (2015b)
	Nanoparticles; 330–480 nm	polyelectrolytic complex	Fluorouracil	10.81%	Yan et al. (2018)
Curd-PLGA	Nanoparticles; 301–486 nm	emulsion solvent evaporation	Rifampicin	1.05 μg g^{-1}	Tukulula et al. (2015)
6-amino-6-deoxy-curdlan/PEG	Nanoparticles; 93–200 nm	complexation	siRNA	–	Han et al. (2015); Altangerel et al. (2016)
QCurd/ CMCurd	Microcapsules; 48–145 nm	layer-by-layer	–	–	Zhang et al. (2013)
malylated curdlan-PNIPAM	Microparticles; 60–190 μm	chemical cross-linking	lysozyme	0.69–0.9 meq g^{-1}	Popescu et al. (2018)

BSA – bovine serum albumin

close to the structure of bone apatite. Also, the presence of PCurd positively influenced the morphology of calcium orthophosphate crystals, as well as the rheological properties.

O-Palmitoylcurdlan sulfate (OPCurS) was applied to the surface of the liposomes to improve their stability. OPCurS-coated liposomes were prepared by the solvent evaporation method obtaining spherical forms. The size of OPCurS-coated liposomes increased with the content of OPCurS from 192 nm to 456 nm. The studies have shown that OPCurS-coated liposomes can be used as a drug carrier for oral administration (Lee et al. 2005).

10.4.2 Food Industry

Curdlan is widely used in the food industry due to its unique rheological properties as a bio-thickener. In the year 1966, the FDA acclaimed curdlan for the use in food industries as processing aid, formulation aid, stabilizer, and texturizing agent or thickener (Federal Register 1996). Some of the major food products which use curdlan as their aids include jelly-like foods, edible fibres, and noodles and also as immobilizing supports (Cho 2019). In the meat, dairy, and baking industries it is used as a texturizer because it is tasteless, odourless, and colourless and has a good water retention capacity. The curdlan and their derivatives were also used to prepare blend films with potential applications as edible films and biodegradable food packaging materials (Wu et al. 2012b; Sun et al. 2011).

10.4.3 Other Applications

Carboxymethyl curdlan (CMCurd) has been used in the photoinduced synthesis of Ag nanoparticles with surface enhanced Raman scattering (SERS) applications (Wu et al. 2012). CMCurd had a high chelation capacity and good stabilizing efficiency for Ag nanoparticles. The resulting Ag dispersions with low concentration of CMCurd (0.01–0.05 mg mL^{-1}) can remain stable for two months. The size and size distribution of Ag nanoparticles varied with the concentrations of CMCurd and AgNO$_3$. The 0.05 mg mL^{-1} concentration of CMCurd leads to spheres and uniform size distribution of Ag nanoparticles. When the concentration increased to 0.1 mg mL^{-1} CMCurd, the nanoparticles showed a wide size distribution and a large average diameter of 20 nm. Most particles are spheres or quasi-spheres, but there are also small triangles. Concentrations higher than 0.3 mg mL^{-1} CMCurd lead to large sizes and irregular silver particles (Wu et al. 2012a).

Curdlan derivatives like 6-amino-6-deoxy curdlan were used to prepare luminescent hybrid materials by supplementary modification with succinic acid (SA) or ionic liquid (IL) and complexing with europium (III). These luminescent films are colourless in sunlight but will display intense red-emitting colours when are irradiated with 365 nm ultraviolet. Films can be used in the electronic field as optoelectronic detectors and vapour-sensitive luminescent sensors (Sun et al. 2020).

10.5 CONCLUSIONS

This chapter represents a short overview of the chemical modifications of curdlan and the applications of these derivatives especially in the pharmaceutical and medical field.

The water insolubility of curdlan causes most chemical modification reactions to occur in alkaline or organic solvents.

The introduction of ionic/nonionic groups on the curdlan chain allows for obtaining soluble derivatives, thus expanding the curdlan application fields.

Analysing the methods used for the synthesis of curdlan derivatives, it was concluded that they are generally made without toxic monomer residues. Moreover, curdlan derivatives have shown excellent biocompatibility, which can be considered a decisive advantage for their use in the biomedical or nutritional field.

Chemical modification, for example, carboxymethylation, sulfation, phosphorylation, seems to be a promising approach for curdlan to develop new therapeutic vectors, with the expansion of its practical and pharmaceutical applications to treat modern diseases.

REFERENCES

Alban S. and G. Franz. 2000. Characterization of the anticoagulant actions of a semisynthetic curdlan sulfate. *Thrombosis. Res.* 99(4):377–388.

Altangerel A., Cai J., Liu L., Wu Y., Baigude H., and J. Han. 2016. PEGylation of 6-amino-6-deoxy-curdlan for efficient *in vivo* siRNA delivery. *Carbohydr. Polym.* 141:92–98.

Borjihan G., Zhong G.G., Baigude H., Nakashima H., and T. Uryu. 2003. Synthesis and anti-HIV activity of 6-amino-6-deoxy-(1.3)-ß-D-curdlan sulfate. *Polym. Adv. Technol.* 14(3–5):326–329.

Brenann C.S. and C.M. Tudorica. 2003. The role of carbohydrates and nonstarch polysaccharides in the regulation of postprandial glucose and insulin reponses in cereal foods, *J. Nutraceut. Funct. Med. Foods.* 4:49–55.

Cai Z. and H. Zhang. 2017. Recent progress on curdlan provided by functionalization strategies *Food Hydrocoll.* 68:128–135.

Chan T.-W.D., Chan P.K., and K.Y. Tang. 2006. Determination of molecular weight profile for a bioactive β-(1→3) polysaccharides (Curdlan). *Anal. Chim. Acta* 556:226–236.

Chen M. and P. Liang. 2017. Synthesis and antibacterial activity of quaternized curdlan. *Polym. Bull.* 74:4251–4266.

Chen X.Y., Xu X.J., Zhang L.N., and F.B. Zeng. 2009. Chain conformation and antitumor activities of phosphorylated (1-3)-beta-D-glucan from *Poria cocos. Carbohydr. Polym.* 78(3):581–587.

Chen Y., Zhang H., Wang Y., Nie S., Li C., and M. Xie. 2014. Acetylation and carboxymethylation of the polysaccharide from *Ganoderma atrum* and their antioxidant and immunomodulating activities. *Food Chem.* 156:279–288.

Chien C.-Y. and T. Iwata. 2018. Synthesis and characterization of regioselectively substituted curdlan hetero esters with different ester groups on primary and secondary hydroxyl groups. *Carbohydr. Polym.* 181:300–306.

Chien C.-Y., Enomoto-Rogers Y., Takemura A., and T. Iwata. 2017. Synthesis and characterization of regioselectively substituted curdlan hetero esters via an unexpected acyl migration. *Carbohydr. Polym.* 155:440–447.

Cho S.S. 2019. *Handbook of Dietary Fiber.* 1st Edition, CRC Press/Taylor & Francis eBooks.

Comin L.M., Temelli F., and M.D.A. Saldana. 2012. Barley beta-glucan aerogels via supercritical CO_2 drying. *Food Res. Int.* 48:442–448.

Cremer L., Lupu A-R., Badulescu M.M. et al. 2010. Assessment of two synthesized curdlan derivatives as possible antioxidants and/or modulators of human PMN cells respiratory burst. *Rom. Biotech. Lett.* 15(6):5718–5728.

Deslandes Y., Marchessault R. H., and A. Sarko. 1980. Triple-Helical Structure of (1→3)-β-D-Glucan. *Macromolecules* 3:1466–1471.

Evans S. G., Morrison D., Kaneko Y., and I. Havlik. 1998. The effect of curdlan sulfate on development *in vitro* of *Plasmodium falciparum. Trans. R. Soc. Trop. Med. Hyg.* 92:87–89.

Federal Register/Vol. 61, No. 242/Monday, December 16, 1996/Rules and Regulations, p. 65941.

Futatsuyama H., Yui T., and K. Ogawa. 1999. Viscometry of curdlan, a linear $(1 \rightarrow 3)$-β-D-glucan, in DMSO or alkaline solutions. *Biosci. Biotechnol. Biochem.* 63:1481–1483.

Gao F.-P., Zhang H.-Z., Liu L.-R. et al. 2008. Preparation and physicochemical characteristics of self-assembled nanoparticles of deoxycholic acid modified-carboxymethyl curdlan conjugates. *Carbohydr. Polym.* 71:606–613.

Gao F., Li L., Zang H. et al. 2010. Deoxycholic acid modified-carboxymethyl curdlan conjugate as a novel carrier of epirubicin: *In vitro* and *in vivo* studies. *Int. J. Pharm.* 392:254–260.

Gao Y., Fukuda A., Katsuraya K., Kaneko Y., Mimura T., Nakashima H., and T. Uryu. 1997. Synthesis of regioselective substituted curdlan sulfates with medium molecular weights and their specific anti-HIV-1 activities. *Macromolecules* 30(11):3224–3228.

Gao Y., Katsuraya K., Kaneko Y., Mimura T., Nakashima H., and T. Uryu. 1998. Synthesis of azidothimidine-bound curdlan sulfate with anti-human immunodeficiency virus activity *in Vitro*. *Polym. J.* 30(1):31–36.

Gao Y., Katsuraya K., Kaneko Y., Mimura T., Nakashima H., and T. Uryu. 1999. Synthesis, enzymatic hydrolysis and anti-HIV activity of AZT-spacer-curdlan sulfates, *Macromolecules* 32(25):8319–8324.

Han J., Cai J., Borjihan W., Ganbold T., Rana T.M., and H. Baigude. 2015. Preparation of novel curdlan nanoparticles for intracellular siRNA delivery. *Carbohydr. Polym.* 117:324–330.

Han J., Wang X., Liu L. et al. 2017. 'Click' chemistry mediated construction of cationic curdlan nanocarriers for efficient gene delivery. *Carbohydr. Polym.* 163:191–198.

Harada T., Masada M., Fujimori K., and I. Maeda. 1966. Production of a firm, resilient gel-forming polysaccharide by a mutant of Alcaligenes faecalis var. myxogenes 10C3, *Agric. Biol. Chem.* 30:196–198.

Harada T., Misaki A., and H. Saito. 1968. Curdlan: A bacterial gel-forming β-1,3-glucan, *Arch. Biochem. Biophys.* 124: 292–298.

Hasegawa T., Umeda M., Numata M., et al. 2006. Click chemistry on curdlan: A regioselective and quantitative approach to develop artificial beta-1,3-glucans with various functional appendages. *Chem. Lett.* 35(1):82–83.

Havlik I., Looareesuwan S., Vannaphan S., et al. 2005. Curdlan sulphate in human severe/cerebral *Plasmodium falciparum* malaria. *Trans. R. Soc. Trop. Med. Hyg.* 99:333–340.

Hirashima M., Takaya T., and K. Nishinari. 1997. DSC and rheological studies on aqueous dispersions of curdlan. *Thermochim. Acta* 306:109–114.

Hisamatsu M., Ott I., Amemura A., Harada T., Nakanishi I., and K. Kimura. 1977. Change in ability of Agrobacterium to produce water soluble and insoluble β-glucans. *J. Gen. Microbiol.* 103:375–379.

Hsieh W.-C., Hsu C.-C., Shiu L.-Y., and Y.-J. Zeng. 2017. Biocompatible testing and physical properties of curdlan-grafted poly(vinyl alcohol) scaffold for bone tissue engineering. *Carbohydr. Polym.* 157:1341–1348.

Huang Q., Zhang L., Cheung P.C.K., and X. Tan. 2006. Evaluation of sulfated α-glucans from *Poria cocos* mycelia as potential antitumor agent. *Carbohydr. Polym.* 64:337–344.

Ichiyama K., Reddy S.B.G., Zhang L.F. et al. 2013. Sulfated polysaccharide, curdlan sulfate, efficiently prevents entry/fusion and restricts antibody dependent enhancement of dengue virus infection *in vitro*: A possible candidate for clinical application. *PLOS Negl. Trop. Dis.* 7(4):2188–2205.

Jagodzinski P.P., Wiaderkiewicz R., Kurzawski et al. 1994. Mechanism of the inhibitory effect of curdlan sulfate on HIV-1 infection *in vitro*. *Virology* 202:735–745.

Jeon K.J., Katsuraya K., Inazu T., Kaneko Y., Mimura T., and T. Uryu. 2000. NMR spectroscopic detection of interaction between a HIV protein sequence and a highly anti-HIV active curdlan sulfate. *J. Am Chem. Soc.* 122(50):12536–12541.

Jin Y., Zhang H., Yin Y., and K. Nishinari. 2006. Comparison of curdlan and its carboximethylated derivative by means of rheology, DSC, and AFM. *Carbohydr. Res.* 341:90–99.

Jin Y.M., Mu Y., Zhang S.H., Li, P.L., and F.S. Wang. 2020. Preparation and evaluation of the adjuvant effect of curdlan sulfate in improving the efficacy of dendritic cell-based vaccine for antitumor immunotherapy. *Internat. J. Biol. Macromol.* 146:273–284.

Kakinuma A., Asano T., Torii H., and Y. Sugino. 1981. Gelation of Limulus amoebocyte lysate by an antitumor (1-3)-β-D-Glucan. *Biochem. Biophys. Res. Commun.* 101(2):434–439.

Kalyanasundaram G.T., Doble M., and S.N. Gummadi. 2012. Production and downstream processing of (1-3)-β-D-glucan from mutant strain of *Agrobacterium sp. ATCC 31750. AMB Express* 2:31–41.

Kaneko Y, Yoshida O, Nakagawa R, Yoshida T, Date M, et al. 1990. Inhibition of HIV-1 infectivity with curdlan sulfate *in vitro. Biochem. Pharmacol.* 39:793–797.

Kim Y.-T., Kim E.-H., Cheong C., Williams D.L., Kim C.-W., and S.-T. Lim. 2000. Structural characterization of β-D-(1→3, 1→6)-linked glucans using NMR spectroscopy. *Carbohydr. Res.* 328:331–341.

Lee C.-M., Lee H.-C., and K.-Y. Lee. 2005. *O*-Palmitoylcurdlan sulfate (OPCurS)-coated liposomes for oral drug delivery. *J. Biosci. Bioeng.* 100(3):255–259.

Li L., Gao F.-P., Tang H.-Bo et al. 2010. Self-assembled nanoparticles of cholesterol-conjugated carboxymethyl curdlan as a novel carrier of epirubicin. *Nanotechnology* 21:265–601.

Li P., Zhang X., Cheng Y. et al. 2014. Preparation and in vitro immunomodulatory effect of curdlan sulfate. *Carbohydr. Polym.* 102:852–861.

Maki K.C., Davidson M.H., Ingram K.A., Veith P.E., Bell M., and E. Gugger. 2003. Lipid reponses to consumption of beta-glucan containing ready-to-eat cereal in children and adolescents with mildly-to moderate primary hypercholesterolemia. *Nutr. Res.*, 23:1527–1535.

Marubayashi H., Yukinaka K., Enomoto-Rogers Y., Takemura A., and T. Iwata. 2014. Curdlan ester derivatives: Synthesis, structure, and properties. *Carbohydr. Polym.* 103:427–433.

McIntosh M., Stone B.A., and V.A. Stanisich. 2005. Curdlan and other bacterial (1→3)-β-D-glucans. *Appl. Microbiol. Biotechnol.* 68:163–173.

Miyoshi K., Uezu K., Sakuraia K., and S. Shinkai. 2004. Proposal of a new hydrogen-bonding form to maintain curdlan triple helix. *Chem.Biodivers.* 1:916–924.

Mocanu G., Mihai D., Moscovici M., Picton L., and D. LeCerf. 2009. Curdlan microspheres. Synthesis, characterization and interaction with proteins (enzymes, vaccines). *Int. J. Biol. Macromol.* 44:215–221.

Nakata M., Kawaguchi T., Kodama Y., and A. Konno. 1998. Characterization of curdlan in aqueous sodium hydroxide. *Polymer* 39(6–7):1475–1481.

Ogawa K., Miyagi M., Fukumoto T., and T. Watanabe. 1973. Effect of 2-chloroethanol, dioxane, or water on the conformation of a gel-forming β-1,3-D-glucan in DMSO. *Chem. Lett.* 943–6.

Osawa Z., Morota T., Hatanaka K., et all. 1993. Synthesis of sulfated derivatives of curdlan and their anti-HIV activity. *Carbohydr. Polym.* 21(4):283–288.

Pelin I.M., Maier V., Suflet D.M., Popescu I., Darie-Nita R.N., Aflori M., and M. Butnaru. 2017. Formation and characterization of calcium orthophosphates in the presence of two different acidic macromolecules. *J. Cryst. Growth.* 475:266–273.

Pelosi L., Bulone V., and L. Heux. 2006. Polymorphism of curdlan and (1→3)-β-D-glucans synthesized *in vitro*: A ^{13}C CP-MAS and X-ray diffraction analysis. *Carbohydr. Polym.* 66:199–207.

Popescu I., Pelin I.M., Ailiesei G.L., Ichim D.L., and D.M. Suflet. 2019. Amphiphilic polysaccharide based on curdlan: synthesis and behavior in aqueous solution. *Carbohydr. Polym.* 224, Article Number: UNSP 115157.

Popescu I., Pelin I.M., Butnaru M., Fundueanu G., and D.M. Suflet. 2013. Phosphorylated curdlan microgels. Preparation, characterization, and *in vitro* drug release studies. *Carbohydr. Polym.* 94(2):889–898.

Popescu I., Pelin I.M., and D.M. Suflet. 2018. Dual-responsive hydrogels based on maleilated curdlan-graft-poly(N-isopropylacryl amide). *Int. J. Polym. Mater. Polym. Biomater.* 67(18):1069–1079.

Pretsch E., Buhlmann P., and M. Badertscher. 2010. *Spektroskopische daten zur strukturaufklärung organischer verbindungen.* Springer-Verlag Berlin Heidelberg.

Rafigh S.M., Yazdi A.V., Safekordi A.A. et al. 2016. Protein adsorption using novel carboxymethyl-curdlan microspheres. *Int. J. Biol. Macromol.* 87:603–610.

Saito H., Yokoi, M., and Y. Yoshioka. 1989. Effect of hydration on conformational change or stabilization of $(1{\rightarrow}3)$-β-D-Glucans of various chain lengths in the solid state as studied by high-resolution solid-state ^{13}C NMR spectroscopy. *Macromolecules* 22:3892–3898.

Sandula J., Machoya E., and V. Hribova. 1995. Immunomodulatory activity of particulate yeast β-1,3-D glucan and water-soluble derivatives. *Int. J. Biol. Macromol.* 17:323–326.

Sasaki T., Abiko N., Sugino Y., and K. Nitta. 1978. Dependence on chain length of antitumor activity of $(1{\rightarrow}3)$-β-D-glucan from *Alcaligenes faecalis* var. *myxogenes*, IFO 13140, and its acid degraded products. *Cancer Res.* 38:379–383.

Sasaki T., Abiko N., Nitta K., Takasuka N., and Y. Sugino. 1979. Antitumor activity of carboxymethylglucans obtained by carboxymethylation of (1-3)-β-D-Glucan from *Alcaligenes faecalis* var. *myxogenes* IFO 13140*. *Europ. J. Cancer 15*:211–215.

Shibakami M., Tsubouchi G., Nakamura M., and M. Hayashi. 2013. Preparation of carboxylic acid-bearing polysaccharide nanofiber made from euglenoid β-1,3-glucans. *Carbohydr. Polym.* 98:95–101.

Suflet, D.M., Nicolescu, A., Popescu, I., and G.C. Chitanu. 2011a. Phosphorylated polysaccharides. 3. Synthesis of phosphorylated curdlan and its polyelectrolyte behavior compared with other phosphorylated polysaccharides. *Carbohydr. Polym.* 84:1176–1181.

Suflet D.M., Popescu I., Pelin I.M., Nicolescu A., and G. Hitruc. 2015. Cationic curdlan: Synthesis, characterization and application of quaternary ammonium salts of curdlan. *Carbohydr. Polym.* 123:396–405.

Suflet D.M., Pelin I.M., Dinu V.M., Lupu M., and I. Popescu. 2019. Hydrogels based on monobasic curdlan phosphate for biomedical applications. *Cell. Chem. Technol.* 53(9-10):897-906.

Suflet D.M., Popescu I. Pelin I. et al. 2011b. Preparation and characterization of microcapsules based on phosphorylated curdlan and hydrolyzed collagen. *Dig. J. Nanomater. Biostruct.* 6(2):633–661.

Suflet D.M., Popescu I., Prisacaru A.I., and I.M. Pelin. 2020Synthesis and characterization of curdlan – phosphorylated curdlan based hydrogels for drug release. *Int. J. Polym. Mat. Polym. Biomat.* doi.org/10.1080/00914037.2020.1765360.

Sun Z., He Y., Liang Z., Zhou W., and T. Niu. 2009. Sulfation of $(1{\rightarrow}3)$-β-D-glucan from the fruiting bodies of *Russula virescens* and antitumor activities of the modifiers. *Carbohydr. Polym.* 77:628–633.

Sun Y., Li Q., Wei S., Zhao R., Han J., and G. Ping. 2020. Preparation and luminescence performance of flexible films based on curdlan derivatives and europium (III) complexes as luminescent sensor for base/acid vapor. *J. Lumin.* 225: Article Number 117241.

Sun Y., Liu Y., Li Y., Lv M., Li P., Xu H., and L. Wang. 2011. Preparation and characterization of novel curdlan/chitosan blending membranes for antibacterial applications. *Carbohydr. Polym.* 84:952–959.

Watanabe E., Tamura N., Saito T., Habu N., and A. Isogai. 2014. Preparation of completely C6-carboxylated curdlan by catalytic oxidation with 4-acetamido-TEMPO. *Carbohydr. Polym.* 100:74–79.

Wang J., Guo C., Yue T., Yuan Y., Liu X., and J.F. Kennedy. 2012. Cationization of *Ganoderma lucidum* polysaccharides in concentrated alkaline solutions as gene carriers. *Carbohydr. Polym.* 88:966–972.

Wang Y. and L. Zhang. 2006. Chain conformation of carboxymethylated derivatives of (1→3)-β-D-glucan from *Poria cocos* sclerotium. *Carbohydr. Polym.* 65:504–509.

Wong S.-S., Ngiam Z.R.J., Kasapis S., and D. Huang. 2010. Novel sulfation of curdlan assisted by ultrasonication. *Int. J. Biol. Macromol.* 46:385–388.

Wu J., Zhang F., and H. Zhang. 2012a. Facile synthesis of carboxymethyl curdlan-capped silver nanoparticles and their application in SERS. *Carbohydr. Polym.* 90:261–269.

Wu C., Peng S., Wen C., Wang X., Fan L., Deng R., and J. Pang. 2012b. Structural characterization and properties of konjac glucomannan/curdlan blend films. *Carbohydr. Polym.* 89:497–503.

Takeda-Hirokawa, N., Neohl L.-P., and H. Akimoto. 1997. Role of curdlan sulfate in the binding of HIV-1 gp120 to CD4 molecules and the production of gp120-mediated TNF-α. *Microbiol. Immunol.* 41(9):741–745.

Tamaru, Sichi, Tokunaga D., Hori K., Matsuda S., and S. Shinkai. 2014. Giant amino acids designed on the polysaccharide scaffold and their protein-like structural interconversion. *Org. Biomol. Chem.* 12:815–822.

Tamura N., Wada M., and A. Isogai. 2009. TEMPO-mediated oxidation of (1→3)-β-D-glucans. *Carbohydr. Polym.* 77:300–305.

Tamura N., Hirota M., Saito T., and A. Isogai. 2010. Oxidation of curdlan and other polysaccharides by 4-acetamide-TEMPO/NaClO/NaClO$_2$ under acid conditions. *Carbohydr. Polym.* 81:592–598.

Tang R., Hao J., Zong R., Wu F., Zeng Y., and Z. Zhang. 2018. Oxidation pattern of curdlan with TEMPO-mediated system. *Carbohydr. Polym.* 186: 9–16.

Tao Y., Zhang L., and P.C.K. Cheung. 2006. Physicochemical properties and antitumor activities of water-soluble native and sulfated hyperbranched mushroom polysaccharides. *Carbohydr. Res.* 341:2261–2269.

Tukulula M., Hayeshi R, Fonteh P. et al. 2015. Curdlan-conjugated PLGA nanoparticles possess macrophage stimulant activity and drug delivery capabilities. *Pharm. Res.* 32(8):2713–2726.

Yan J.-K., Cai P.-F., Cao X.-Q., et al. 2013. Green synthesis of silver nanoparticles using 4-acetamido-TEMPO-oxidized curdlan. *Carbohydr. Polym.* 97:391–397.

Yan J.K., Liu J.L., Sun Y.J., Tang S., Mo Z.Y., and Y.S. Liu. 2015a. Green synthesis of biocompatible carboxylic curdlan-capped gold nanoparticles and its interaction with protein. *Carbohydr. Polym.* 117:771–777.

Yan J.K., Ma H.L., Chen X., Pei J.J., Wang Z.B., and J.Y. Wu. 2015b. Self-aggregated nanoparticles of carboxylic curdlan-deoxycholic acid conjugates as a carrier of doxorubicin. *Int. J. Biol. Macromol.*, 72:333–340.

Yan J.-K., Qiu W.-Y., Wang Y.-Y., Wu L.-X., and P.C.K. Cheung. 2018. Formation and characterization of polyelectrolyte complex synthesized by chitosan and carboxylic curdlan for 5-fluorouracil delivery. *Int. J. Biol. Macromol. A* 107:397–405.

Yang Y., Gao X.D., Han T., and R.X. Tan. 2005. Sulfation of a polysaccharide produced by a marine filamentous fungus alters its antioxidant properties in vitro. *Biochim. Biophys. Acta.* 1725:120–127.

Yamamoto I., Takayama K., Gondam T. et al. 1990. Synthesis, structure and antiviral activity of sulfates of curdlan and its branched derivatives. *British Polym. J.* 23(3):245–250.

Yoshida T., Hatanaka K., Uryu T., Kaneko Y., Suzuki E., Miyano H., et al. 1990. Synthesis and structural analysis of curdlan sulfate with a potent inhibitory effect in vitro of AIDS virus infection. *Macromolecules* 23(16):3717–3722.

Yoshida T., Hattori K., Sawada Y., Choi Y., and T. Uryu. 1996. Graft copolymerization of methyl methacrylate onto curdlan. *J. Polym. Sci. Polym. Chem.* 34:3053–3060.

Yoshida T., Yasuda Y., Mirnura T., Kaneko Y., Nakashima H., Yamamoto N., and T. Uryu. 1995. Synthesis of curdlan sulfates having inhibitory effects *in vitro* against AIDS viruses HIV-1 and HIV-2. *Carbohyd. Res.* 276:425–436.

Yu H., Huang Y., Ying H., and C. Xiao. 2007. Preparation and characterization of a quaternary ammonium derivative of konjac glucomannan. *Carbohydr. Polym.* 69:29–40.

Zhang L, Ding Q., Meng D., Ren L., Yang G., and Y. Liu. 1999. Investigation of molecular masses and aggregation of β-D-glucan from *Poria cocos sclerotium* by size-exclusion chromatography. *J. Chromatogr. A* 839:49–55.

Zhang L., Ding Q., Zhang P., Zhu R., and Y. Zhou. 1997. Molecular weight and aggregation behaviour in solution of β-D-glucan from *Poria cocos sclerotium*. *Carbohydr. Res.* 303:193–197.

Zhang L., Zhang M., Zhou Q., Chen J.H., and J.F. Zeng. 2000a. Solution properties of antitumor sulfated derivative of α-(1.3)-D-glucan from *Ganoderma lucidum*. *Biosci. Biotechnol. Biochem.* 64(10):2172–2178.

Zhang R.R., Liu S., and K.J. Edgar. 2016. Regioselective synthesis of cationic 6-deoxy-6-(N,N,N-trialkylammonio) curdlan derivatives. *Carbohydr. Polym.* 136:474–484.

Zhang R. and K.J. Edgar. 2014. Synthesis of curdlan derivatives regioselectively modified at C-6: O-(N)-Acylated 6-amino-6-deoxycurdlan. *Carbohydr. Polym.* 105:161–168.

Zhang R., Liu S., and K.J. Edgar. 2017. Efficient synthesis of secondary amines by reductive amination of curdlan Staudinger ylides. *Carbohydr. Polym.* 171:1–8.

Zhang Y., Chen C., Wang J., and L. Zhang. 2013. Polysaccharide-based polyelectrolytes hollow microcapsules constructed by layer-by-layer technique. *Carbohydr. Polym.* 96:528–535.

Zhang L., Zhang M., Zhou Q., Chen J.H., and J.F. Zeng. 2000b. Solution properties of antitumor sulfated derivative of α-(1.3)-D-glucan from *Ganoderma lucidum*. *Biosci. Biotechnol. Biochem.* 64(10):2172–2178.

Index

Page numbers in *italic* indicate figures. Page numbers in **bold** indicate tables.

Printed in the United States
By Bookmasters